Twenty-One Trends for the 21st Century:

Out of the Trenches and into the Future

Twenty-One Trends for the 21st Century:

Out of the Trenches and into the Future

Their Profound Implications for Students, Education, Communities, and the Whole of Society

Gary Marx

Education Week Press
Editorial Projects in Education

Library of Congress Cataloging-in-Publication Data

Marx, Gary
Twenty-One Trends for the 21st Century: Out of the Trenches and into the Future / Gary Marx
p. cm.
Includes bibliographical references and index.

Paperback ISBN No. 978-1-939864-04-8
E-book ISBN No. 978-1-939864-05-5

1. EDUCATION / General 2. EDUCATION / Leadership 3. EDUCATION / Professional Development 4. EDUCATION / Research 5. BUSINESS & ECONOMICS / Forecasting

Cover design by Laura Baker
Editing and interior design by Jaini Giannovario
Set in Adobe Garamond and Optima LT Std

Education Week Press
6935 Arlington Rd.
Bethesda, MD 20814
www.edweek.org

To order visit: www.edweek.org/go/21trends
or call 1-800-345-6665. Bulk discounts available.

Views expressed in this publication do not necessarily reflect the opinions or positions of Futures Council 21 or Editorial Projects in Education, Inc.

First Edition, April 2014
First Printing

To my family, mentors, friends, and colleagues. Thanks for the inspiration, encouragement, and example.

Among Other Books by Gary Marx
Some with Co-Authors

Future-Focused Leadership

Sixteen Trends…Their Profound Impact on Our Future

Ten Trends…Educating Children for a Profoundly Different Future

Preparing Schools and School Systems for the 21st Century

Excellence in Our Schools…Making It Happen

Public Relations for Administrators

Building Public Confidence in Our Schools

Radio…Your Publics are Listening

Radio…Get the Message

Working with the News Media

FOREWORD

It can be overwhelming to think about the countless forces shaping our lives and work today. We are deluged with media reports, books, blogs, and the pronouncements of experts telling us that the world is changing more rapidly than ever before. That's all true, of course. And it may be especially true, right now, in the world of education. As an industry, schooling looked remarkably similar for hundreds of years. But now we are bombarded by disruptive technologies, drastically shifting student demographics, major budget cuts, new demands for college and career readiness – all while responding to competitive demands on a global playing field.

Certainly, it's important to process what these changes may mean, to react to these shifts, and to look ahead and anticipate what's next. But how do we do all this while keeping the "core" values of who we are, what we aspire to, and how we function?

At Gallup, we are constantly studying. What we study most – and what we find matters most – is how people think and feel. Economic forecasts are one thing. But if you really want to know where the world is headed, you have to tap into the collective voice of the people by understanding their emotions - where they want, or are willing, to go.

George Gallup was famous for saying "if democracy is supposed to be based on the will of the people, then someone should go out and find out what that will is." This is mission-critical advice to leaders in education. Amidst this time of great upheaval, one of the most important things we can do is listen carefully to students, teachers, parents, and other stakeholders in learning. If we don't get the premise right about what's going on, then we can actually do a lot of damage as leaders. Getting the premise right is the first step.

We must also focus carefully on what's not changing. In our education research, we have found that much is not changing about the essential nature of great teaching and learning, what Gallup would describe as the "fundamentals of human development." When Americans describe their best teacher, for example, the number one word they use is "care." That sounds pretty simple, yet is very profound. How are we measuring "caring" in school today? How does technology show a student it cares? Having a teacher who cares about a student's development is crucial to that student's success in schools. And, it turns out the same is true for you – or any adult – in the workplace. If you have someone at work who cares about your development, that is fundamental to your professional success.

We also know that, regardless of how old or young you are, one of the most basic human needs is being able to say that you get to do what you're best at every day. That sounds simple enough too, but we're actually doing a horrible job of this. We waste so much talent in this country because we haven't carefully or intentionally helped enough students figure out what they're good at and to help them do that every day. In fact, you could argue that our educational system is actually better at doing the opposite – figuring out what students don't know!

The rate of knowledge-creation is increasing exponentially while, at the same time, the cost of knowledge is rapidly trending toward free. The number of pages on the Internet is exploding, new scientific discoveries are emerging every day, and vast amounts of information are now accessible for free, through Google, Khan Academy, and the many free MOOC's (massive open online courses). The implication– for ourselves and our students – is that no one will ever again compete on what they know alone. Rote memorization and facts and figures will not be the intellectual currency of tomorrow's world. What will matter most is how we synthesize diverse information and then apply what we know. This changes some of what we do in education, because it means we need to raise the bar on how we teach and learn. If what we do isn't practical, multi- or inter-disciplinary, or applied through experiential and project-based learning, we will be at a disadvantage from the start. Although it won't be easy, this is something we can do in every single subject we teach.

As external forces shape our world, never forget that the most powerful force comes from within. Our educational research continues to reinforce that nothing is more powerful than a talented principal or teacher doing what they do best every day. The great ones build engaged schools and engaged students - whether or not they have ample resources or access to the latest technologies, and despite everything that may be weighing against them. Whatever we do, let's not lose sight of this. It provides great hope to know that a few committed, talented people can still make all the difference.

Looking toward the future, there is one thing that is extremely clear in all of Gallup's research about what people want. What the whole world wants is a good job. This is a really important discovery. If you consider this through the lens of education, it forces us to think very differently about what we mean by "education" and how and when and for whom we deliver it. If our goal is to ensure people have good jobs – meaningful work throughout their lives - we know those jobs will change over time. In fact, many of us may have dozens of different jobs or careers during our lifetime. And, we will need different kinds of education for all of them.

The American Dream may be shifting from the idea of working-to-retire to aspiring to meaningful work for life. As this happens, we will never again think of education as something that only takes place inside a school building or on a college campus during a fixed time in our young lives. Rather, it will be something that continues throughout our lives. Although the need to teach and learn will remain a constant, the result may not always be a diploma, but rather everything from an audited course for fun to a badge, certificate, or a full degree.

Education for life is the new paradigm. This changes a lot of what we do. But I'm not sure it changes how we do it. The importance of caring about students and discovering what they do best will never change.

Brandon H. Busteed
Executive Director
Gallup Education

TABLE OF CONTENTS

Twenty-One Trends for the 21st Century:
Out of the Trenches and into the Future

Education and Learning Sphere

Public and Personal Leadership Sphere

Well-Being Sphere

Introduction

Change is inevitable.
Progress is optional.

Introduction:
Out of the Trenches and into the Future.
Change is inevitable. Progress is optional.

The question is not, "When will things get back to normal?"
The question is, "What will the new normal look like?"[1]

An uncle of mine fought in World War I. He talked about trench warfare. "Soldiers, dig those trenches. They'll protect us from the enemy!" Maybe that's how we coined the term, "entrenched." Some of us keep on digging. We want those trenches to be deep enough to defend the status quo–what we've always done. Our talent and energy is focused like a laser on shielding ourselves from the reality of overwhelming demographic change, a revolution in technology, earth-shaking scientific discoveries, economic and environmental realities, and the constant buzz of new ideas.

Twenty-One Trends for the 21ˢᵗ Century is more than a book. It's a living opportunity to get connected to seismic shifts that will profoundly impact our future. No secret to anyone: There's a big world out there, and we *all* need to be part of it. That means we *all* have to *start building. Digging is not enough.*

"We need to address people in the trenches." That's the hue and cry. While the claim is justified by stark daily reality, we'd better make sure we don't dig those trenches so deep that we lose touch with the outside world. In fact, we need to seize higher ground. Looking at the big picture can open our eyes to strategic perspective and help us better understand the context in which we function.

Visualize this. Deep in our trenches, the clamor from above settles down. One day, a curious scout climbs a tall ladder and peers over the edge. Shocked, the scout rushes back to report, "The war is over. Everyone has moved ahead. They've just gone on and left us behind.

We've won our battle, but we've lost the war." Rather than simply peering from the tops of our trenches, we need to be exploring what lies beyond the horizon. We are, after all, of this world, not separate from it.

Just a Glimpse
Twenty-One Trends for the 21ˢᵗ Century:
Out of the Trenches and into the Future

Here are just a few examples of what's going on outside the trenches. Call it handwriting on the wall. We'll explore these and other world-changing realities in the chapters of this book.

• Following every major economic depression or recession, physical and social infrastructure have been transformed, from transportation and manufacturing to lifestyles and education. No one gets a pass.

• Lifelong education will move forward anywhere, anytime, and any way. Similar expectations will impact many other industries.

• While school curriculum will continue to be aligned with goals, pressure will grow for goals to be aligned with individual students' strengths and the needs of society.

• In the U.S., non-Hispanic Whites are expected to fall below 50 percent of the population by about 2043. For those 18 and under—by 2018.

• Beginning in 2011, Baby Boomers started hitting 65 at the rate of about 10,000 a day. They'll be getting to retirement age at 10,000 a day for about 30 years.

• In 2012, members of the Millennial Generation started turning 30 and will be assuming leadership that will be no less than revolutionary for society and every one of our institutions.

• As growing numbers of Millennials upsize by downsizing, they will insist on quality, style, collaborative leadership, and results.

• Big data and the cloud, coupled with super- and quantum computers will lead to revolutions in everything from education to health care and raise concerns about identity and privacy. Computer speed and capacity will increase exponentially.

• Look for a revolution in energy generation, distribution, storage, and efficiency. *Renewable energy harvesters* will become more commonplace. *Electro-chemists* and *superconducting technologists* will help us increase battery capacity and develop a more efficient and dependable smart grid.

• Scientific instruments have detected that carbon dioxide levels in the atmosphere reached 400 parts per million, a level not seen on earth for three million years, long before the roughly 8,000 years of human civilization.[2] Scientists have already determined what climactic changes we can expect with each degree the average temperature rises.

• Agricultural employment was projected to drop from 69 percent of the workforce in 1840 to 1.2 percent in 2020. Industrial employment rose from 15 percent in 1840 to 35 percent in 1950, and was expected to fall to 11.9 percent by 2020. Service employment is expected to jump from about 17 percent in 1840 to 79.9 percent in 2020.[3][4]

• Economic recovery and sustainability will depend, in part, on *systemic innovation*.

• Anyone who stays in touch with broad societal trends might have known that people would take to the streets in several parts of the world. Converging into a kind of perfect storm are realities such as — a generation of young people who want to solve problems and deal with injustices; soft economies and a lack of jobs; a questioning of authority; and social media that energize and bring people together at a moment's notice.

Where are the talented and resourceful people who will help us deal with these massive changes in society? We all know the answer. They're in our schools and colleges today. Think about the 21 trends revealed in this book. Then, answer the question, "What are their implications for education?"

It's tough enough to be cloistered in the trenches. It's even tougher for educators and students when they are boxed in by high-stakes tests that too often narrow the curriculum in a world that desperately needs depth and breadth. Narrowness and shallowness simply won't do. It's becoming obvious that we can't afford to relegate talented professionals in any field to isolation in the trenches. Sure, we need to be the best specialists possible, but every one of us must also be a generalist.

To pursue the gathering narrowness, in dozens of institutions, we shower inventive people, in essence, fill their trenches, with a virtual rainstorm of nuts and bolts. Occasionally, in one industry or another, we try to do even more perfectly what, for all practical purposes, has become obsolete. Let's face it, the thrill in what we do is in knowing how our genius and hard work fit into the big picture and enrich lives. Dealing with societal trends is an opportunity, as we think about the future, to also demonstrate our *intellectual leadership*, connecting relationships among people, circumstances, and ideas to create new knowledge and inventive solutions to persistent problems.

A prevailing question is *"Why?"* "Why are we doing this?" "Why are we learning this?" "Why do we spend so much of our talent and energy defending the past when we should be creating a future?" A question that comes in a quick second is *"What?"* When we are hit with an avalanche of excuses for not wanting to change, such as: "We'll never get the money;" "We aren't trained to do that;" or "We can't pay attention to the real world because our test scores will suffer;" we need to ask, "What are we going to do about that?" Keep asking and pursuing credible answers to that question until we get to solutions. The future won't wait.

> **What are trends?**
> **The father of issue management, Howard Chase, defined trends as "detectable changes which precede issues."[5]**
> **Webster's Dictionary calls them "a line of general direction or movement, a prevailing tendency or inclination."[6]**

Seizing Higher Ground. Not to overdo the military metaphor, but a rule of the battlefield is often to take the higher ground. That can be a hill, an aircraft, or a spacecraft, sometimes all of the above. From that vantage point, we can get a better perspective. It's a little like seeing the forest for the trees. Such a simple concept…but often so hard to put away our progress-defeating defense against a rising tide of new realities. The world is changing exponentially…and the status quo has become a one-way ticket to obsolescence. In a fast changing world, unless we stay ahead of the curve, we can expect whole companies, industries, and some other institutions to disappear with the lightening speed of a computer's delete button.

If we're an education system, or involved in some part of the process (like a State Legislature, Congress, Governor, President, others), why would we want to freeze the system and disable our students by getting them perfectly ready for yesterday, when we know they will have to move headlong into a very different tomorrow? Every one of those students needs to be ready for life outside the trenches.

A drone of linear goals won't cut it in a world that is changing exponentially. What's happening during our journey into the 21ˢᵗ century is not just a subtle change. It's a reset that will profoundly impact everything from housing patterns and lifestyles to energy sources and economies. We need to repeat what we said at the outset, just to emphasize the sense of urgency, "No one gets a free pass, not even education." In every walk of life, all of us are literally facing a choice, knowing that we can only coast downhill. What will it be? Breakthroughs or breakdowns?

Multitudes of us are ready to get cracking by exchanging our Industrial Age mentalities for Global Knowledge/Information Age realities. This book is, in fact, bursting with positive and inspiring examples and possibilities. Nonetheless, all of us have known a few people who consider nearly any new idea as an attack. A good case of paranoia can drive us to battle stations to fend off change—to dodge the inevitable. That full time, spine-tingling defense can take so much time and energy that there's not much left for shaping a future.

How about celebrating new ideas and societal realities as opportunities? Rather than taking to the trenches and simply hunkering down, we need to ask what the implications of those trends are for us, whatever we do and wherever we are on the planet. Adaptability and resilience are keys to survival. To freeze is to fail.

Twenty-One Trends for the 21ˢᵗ Century. This publication is must reading and an essential guide for anyone involved in education, business, government, nonprofits, community groups, and other types of organizations, industries, or professions. It's an ideal and proven textbook for leadership, education, business, government, planning, citizenship, community and economic development, and futures studies. Each chapter comes complete with suggested readings, enlightening videos, and follow-up activities. Conveniently, the book lays out invigorating processes for putting the information and ideas to work in planning for the future.

Twenty-One Trends for the 21ˢᵗ Century is specifically designed to be dynamic, not static—a beginning, not an end in itself. It's a classic environmental scan, shaped to stimulate thinking and provoke discussion. A lifesaver for everyone who does important work in the trenches, it provides an ongoing way to reconnect with the heartbeat of a world in motion. We discover instantly that what we do reverberates beyond our office, our classroom, and the tops of our desks. We can also sense the vibrations of society to guide us in what we do and how we do it. *Understanding trends is so important to the process of creating a future that we have devoted this entire book to these forces.*

A bonus for educators is that it also provides an array of information and processes ideal for engaging students in active learning, project-based education, real-world education, learning across disciplines, and learning through inquiry. It can hone thinking, reasoning, and problem solving skills, stimulate discussions of ethics, and introduce students to exciting information about everything from the economy and environment to functioning in a global marketplace.

This book, literally years in the making, can help us move toward that higher ground. Think of it as an *intelligence briefing*—as a revealing and invigorating *environmental scan*. It's a *go-to guide for thinking and planning, a virtual catalog of information and ideas* that will stir possibilities. It's *a key for unlocking rigidity and stimulating ingenuity.* No worries: It's *not* a new program. A commitment to staying in touch is mostly a sign of intellectual rigor and enlightened leadership.

Educators are on the front line. They hear the rising chorus of people insisting that our education system get students ready for the future in a world of constant change. Why the pressure? Largely, it's because what educators do is so important to our civil society, our economy, and our ability to play a leadership role within a family of nations. That's why we all need to get and stay in touch with unrelenting political, economic, social, technological, demographic, and environmental forces or trends that are even now impacting the whole of society. They surely have implications for how we run our schools and colleges and for what students will need to know and be able to do if they hope to be primed for the future.

To earn its legitimacy, any organization, including a community, a country, or an education system, needs to be connected with the people it serves…and with the world. If we understand trends and issues, people say we're *in touch*. If we don't understand trends and issues, they say we're *out of touch*. While being firmly connected to people and ideas is crucial for *each of us*, it's just as important for *all of us*. En Español, *nosotros*.

> **"As we scan the environment, nothing stands out more than massive, unrelenting trends. Like the movement of tectonic plates beneath the surface of the earth, they are a signal of seismic shifts."**[7]
> *Sixteen Trends…Their Profound Impact on Our Future*

Massive Trends…
Seismic Shifts that Are Shaping the Future

We're about to bring the curtain up on an array of trends at the very heart of *Twenty-One Trends for the 21ˢᵗ Century*. Consider them just the tip of an iceberg. The list is by no means complete. Chapters are devoted specifically to each of these trends. Our hope is that this leading edge briefing will stimulate an even more expansive discussion and get us on a superhighway toward creating a promising future.

All 21 of these trends reach into every corner of society. They are unrelenting, and won't go away just because we ignore them. In fact, these trends, in one way or another, impact everything we do, wherever we might be on the planet. In each chapter, we've included some possible implications of these trends for schools, school systems, colleges, universities, and other institutions, including communities, nations, and the world. However, it's up to each of us to expand on this list, consider further implications, and bring what we discover home to whatever we do and wherever we are.

Out of the Trenches and into the Future
Twenty-One Trends...that Will Profoundly Impact Education and the Whole of Society

- **Generations**: Millennials will insist on solutions to accumulated problems and injustices and will profoundly impact leadership and lifestyles.
  GIs, Silents, Boomers, Xers → Millennials, Generation E

- **Diversity:** In a series of tipping points, majorities will become minorities, creating ongoing challenges for social cohesion.
  Majority/Minority → Minority/Minority
  Diversity = Division ↔ Diversity = Enrichment   Exclusion ↔ Inclusion
  (Worldwide: Growing numbers of people and nations will discover that if we manage our diversity well, it will enrich us. If we don't manage our diversity well, it will divide us.)

- **Aging:** In developed nations, the old will generally outnumber the young. In developing nations, the young will generally outnumber the old.
  Younger → Older   Older → Younger

- **Technology:** Ubiquitous, interactive technologies will shape how we live, how we learn, how we see ourselves, and how we relate to the world.
  Macro → Micro → Nano → Subatomic   Atoms → Bits
  Megabytes → Gigabytes → Terabytes → Petabytes →
  Exabytes → Zettabytes (ZB)

- **Identity and Privacy:** Identity and privacy issues will lead to an array of new and often urgent concerns and a demand that they be resolved.
  Knowing Who You Are ↔ Discovering Who Someone Thinks You Are
  What's Private? ↔ What's Not?

- **Economy:** An economy for a new era will demand restoration and reinvention of physical, social, technological, educational, and policy infrastructure.
  Industrial Age Mentality → Global Knowledge/Information Age Reality
  Social and Intellectual Capital → 21st Century Products and Services

- **Jobs and Careers:** Pressure will grow for society to prepare people for jobs and careers that may not currently exist.
  Career Preparation ↔ Employability and Career Adaptability

- **Energy:** The need to develop new sources of affordable and accessible energy will lead to intensified scientific invention and political tension.
  Energy Affordability, Accessibility, Efficiency ↔ Invention,
  Investment, and Political Tension

- **Environmental/Planetary Security:** Common opportunities and threats will intensify a worldwide demand for planetary security.
  Personal Security/Self Interest ↔ *Planetary Security*
  Common Threats ↔ *Common Opportunities*
- **Sustainability:** Sustainability will depend on adaptability and resilience in a fast-changing, at-risk world.
  Short-Term Advantage ↔ *Long-Term Survival*
  Wants of the Present ↔ *Needs in the Future*
- **International/Global:** International learning, including relationships, cultural understanding, languages, and diplomatic skills, will become basic.
  Isolationist Independence ↔ *Interdependence*
  (**Sub-trend:** To earn respect in an interdependent world, nations will be expected to demonstrate their reliability and tolerance.)
- **Personalization:** In a world of diverse talents and aspirations, we will increasingly discover and accept that one size does not fit all.
  Standardization → *Personalization*
- **Ingenuity:** Releasing ingenuity and stimulating creativity will become primary responsibilities of education and society.
  Information Acquisition → *Knowledge Creation and Breakthrough Thinking*
- **Depth, Breadth, and Purposes of Education:** The breadth, depth, and purposes of education will constantly be clarified to meet the needs of a fast-changing world.
  Narrowness → *Breadth and Depth*
- **Polarization:** Polarization and narrowness will, of necessity, bend toward reasoned discussion, evidence, and consideration of varying points of view.
  Narrowness ↔ *Open Mindedness*   *Self Interest* ↔ *Common Good*
- **Authority:** A spotlight will fall on how people gain authority and use it.
  Absolute Authority → *Collaboration*   *Vertical* ↔ *Horizontal*
  Power to Impose ↔ *Power to Engage*
- **Ethics:** Scientific discoveries and societal realities will force widespread ethical choices.
  Pragmatic/Expedient → *Ethical*
- **Continuous Improvement:** The status quo will yield to continuous improvement and reasoned progress.
  Quick Fixes/Status Quo → *Continuous Improvement*
- **Poverty:** Understanding will grow that sustained poverty is expensive,

debilitating, and unsettling.
Sustained Poverty ↔ Opportunity and Hope

- **Scarcity vs. Abundance:** Scarcity will help us rethink our view of abundance.
Less ↔ More   What's Missing? ↔ What's Possible?
- **Personal Meaning and Work-Life Balance:** More of us will seek personal meaning in our lives in response to an intense, high tech, always on, fast-moving society.
Personal Accomplishment ↔ Personal Meaning

Twenty-One Trends for the 21ˢᵗ Century: Out of the Trenches and into the Future

How was this book developed?

The second decade of the 21ˢᵗ century was getting under way. The pace of change was becoming increasingly exponential. Author Gary Marx, whose continuous research, writing, and worldwide counsel extend over decades, was primed to develop the third in his series of books about societal trends. Education Week Press, which has earned accolades for its service to education and whose reach has become international, offered to publish this seminal work.

Twenty-One Trends for the 21ˢᵗ Century joins the author's collection of works, including the most recent: *Sixteen Trends…Their Profound Impact on Our Future*;[8] *Ten Trends…Educating Children for a Profoundly Different Future;* [9] and *Future-Focused Leadership…Preparing Schools, Students, and Communities for Tomorrow's Realities.*[10]

Each of these books, including *Twenty-One Trends for the 21ˢᵗ Century*, reflects the author's non-stop observations of massive trends impacting the whole of society. Each is balanced by a constant flow of information, experiences, and insights. Realities and possibilities shared in these publications are often shaped by his ongoing speaking and counsel in a vast number of nations across six continents, as well as his frequent work in local communities of every size, shape, and location. The author's focus is generally on future-focused leadership, including its impact on those of us, in many walks of life, who serve in the trenches. Each one knows that our future depends on staying in touch with a 24/7 world. To be effective, all of us must be leaders, whatever role we play.

A distinguished 26-member international Futures Council 21 enhanced research and writing for *Twenty-One Trends for the 21st Century*. The talented, experienced, and insightful members of that group participated in a modified Delphi process. Each responded to a request to identify three top societal trends and three issues that could be expected to affect education and the whole of society. In a second-round survey, members of the Council speculated on the implications of clusters of no more than two trends each. All were asked to provide insights, based on their observations and experiences, wherever they happened to be in the world. A few combined their comments into one survey and one member, by request, was asked to specifically focus on implications of trends for higher education.

Survey responses were coupled with ideas that emerged from an intensive review of trends, issues, ideas, and processes noted by scholars, thinkers, and commentators across the breadth of society. As part of this sustained process of exploration and information gathering, the author identified and then amplified on the trends included in this book. Obviously, there are many other forces simultaneously affecting society.

Who should be using *Twenty-One Trends for the 21st Century*?

This groundbreaking book emphasizes the benefits of collaboration, of bringing people together in common purpose. In fact, this publication is designed to stimulate discussions about the implications of far-reaching trends for schools and school systems; colleges and universities; accrediting groups and agencies; community, economic, and work force developers; community improvement organizations; professional and trade associations; leadership experts; think tanks; foundations; businesses and industries; journalists; media organizations; demographers; futurists; forecasters; scientists and technologists; social scientists; civic educators; artists; people of various beliefs; politicians; and anyone who enjoys thinking and staying in touch. States and nations, as well as regional, continental, and world organizations, plus governmental, nongovernmental, and civil society organizations, should add what this publication has to say to their list of considerations as they plan for the future.

These trends books have become a staple for professional development and have been used as texts or subtexts for courses across a range of disciplines. What they have to say is regularly featured in articles, keynotes, workshops, conferences, seminars, lectures, professional development programs, online chats, classroom and online courses, book studies, planning sessions, and as a centerpiece of countless discussions.

How is the Book Organized?

Twenty-One Trends immediately cuts to the chase. A full chapter is devoted to each of 21 trends. Specific, fact-filled essays present a rationale, research, and an explanation of possible implications both for society at-large and education in particular.

Each of those chapters concludes with Questions and Activities as well as Readings and Programs that can enhance coursework, enliven a class, pique imagination, and encourage further discussion during professional development sessions or other gatherings. Both readings and videos are included to stimulate further thinking and engagement. In short, the author has tried to breathe even further life into each of the trends, to do some of the groundwork for curriculum developers and professors, and to enliven each topic using a variety of online sources, such as articles, videos, and activities, some interactive. The 900-plus references provide links for those who would like to pursue greater depth and breadth on a host of trends.

A concluding chapter focuses on how educational systems and other organizations and institutions might deal with trends head-on as they think about and plan for the future. Every attempt has been made to make this book as user-friendly, exciting, informative, and motivational as possible.

As you study each of the trends, you might ask, "What's the difference between $\rightarrow$ and $\leftrightarrow$?" The $\rightarrow$ mark indicates a clear, nearly unmitigated trend. A designation of $\leftrightarrow$ indicates a trend that can be expected to develop or continue based on both evidence and the reality that certain conditions are likely unsustainable. For example, we can expect a tug between planetary security and self-interest, and between polarization and open mindedness.

Chapters Grouped into Spheres. Trends chapters are grouped into spheres. For example, aging, diversity, and generations are part of the *demographics sphere*. The *technology sphere* also includes identity and privacy. An *economic sphere* also addresses jobs and careers. The *energy and the environment sphere*, of course, focuses on energy, environmental and planetary security, and sustainability. We've included a chapter devoted to the *international/global sphere*. An *education and learning sphere* explores the depth, breadth, purposes, and personalization of education. A *public and personal leadership sphere* addresses polarization, authority, ethics, and continuous improvement. Finally, a *well-being sphere*, addresses poverty, scarcity vs. abundance, and the ongoing search for personal meaning. Of course, all of these trends intersect, and all have profound implications for education, even for our survival.

Keep in Mind...

A few things to put in your thinking cap as you look toward the future:

Linear Goals in a Non-Linear World. As we develop our plans, linear goals will not be enough. Looking at tomorrow and seeing it only as a little bit more or a little bit less of today just won't cut it. Surrounded by a world filled with discontinuities, we desperately need to stay in touch with societal trends and how they might impact us.

Audacious Goals. Every plan should include at least a few audacious goals, the kind that took us to the moon, reduced computers from a dozen racks of equipment to a single handheld device . . . and envisioned educational opportunity for all.

The Social Environmentalist. Staying in touch means that all of us need to be environmentalists, adapting the organization to the needs of the environment at the same time we're adapting the environment to the needs of the organization.

Out of the Blue. Much of what happens as we break ground on the future will seem to come at us out of the blue.[11] We are in a constant, unrelenting, and exciting race to adapt and lead as we lay the groundwork for a more promising future for our children and ourselves. Staying in touch with trends and issues can put us on higher ground and help us get a glimpse beyond the horizon. That makes it possible for us to anticipate wildcards, game changers, and both intended and unintended consequences of our actions or lack of action. People who move to a higher plane see things in a broader context. Many leaders discover that taking the longer view can even lead to support for ideas they've been putting on the table for years. Instead of only reacting to change, they become effective and thoughtful game changers.

Collaboration. Terrific leaders know they must constantly open their minds to the knowledge, experience, and ideas of diverse groups of people, turning them loose to consider possibilities, to learn from each other, and to help us, across all disciplines, as we think about and plan for the future. Sustainable leadership is connected leadership.

Words of Wisdom from Dick Feynman. Richard Feynman, a fellow educator who won the 1965 Nobel Prize for Physics, said in his own challenging way, "I can live with doubt and uncertainty and not knowing. I think it is much more interesting to live not knowing than to have answers that might be wrong."[12]

How Can I Use *Twenty-One Trends?*
Planning, Engaging, Getting from Today to Tomorrow

> **"An interesting thing happened on the way to**
> **accomplishing our plan.**
> **The world changed."**

Sound familiar? That's why *strategic plans*, no matter how sophisticated, need to become *living strategies* or *strategic visions*. We need to build in flexibility and nimbleness that allow us to deal with problems and opportunities, often at a minute's notice. Constant collaboration and communication are essential. Things move fast.

Your challenge is to turn this book into an action publication. It is meant to stimulate planning, breakthrough thinking, risk taking, and thoughtful action. Put copies in the hands of educators, parents, and students, as well as community, business, government, and other leaders. Use it to stir discussion, even debate, about how schools, school systems, colleges and universities, and other institutions need to be reshaped for the future. You might even want to create a network of *Futures Councils* that are asked to read this and other books and articles suggested throughout the text, hear presentations, and engage in generative thinking about the implications of these and other trends, not necessarily to make decisions, but to provide ongoing counsel.

Build *professional development programs and events* around the need to constantly renew what we do and how we do it. Make the book required reading for staff, board, and key members of the community. Hold *community-wide leadership seminars* or *community conversations* that focus not only on the future of the education system but also on other quality of life issues, as well as economic growth and development. Use what this book has to say and the processes it suggests as a starting point for your *planning.* Make it a key part of your *environmental scan. Devote agenda time* to seriously studying and discussing one or two trends a month, maybe even more. Consider implications, and take action that will keep the organization in tune with society and poised for the future.

As we've mentioned, professors will likely use this book as a *text or sub-text,* perhaps required reading, for both regular classes and as a basis for developing new courses, whether in-classroom, online, or a blend. The courses, for example, might become a staple for preparation of teachers and administrators in any field and at all levels. More specifically, individual courses, courses of study, or units might focus on trends, issues, leadership, planning, communication, or education futures. In some cases, colleges and universities are offering futures studies courses, even degree programs. Some level of futures studies could be required for course completion or graduation.

We expect teachers to draw from the ideas and research included in each chapter to refresh their classes and further excite students about what they are learning. We also look forward to a growth in those futures studies courses or units we mentioned to help students better understand how to deal with issues and consider the implications of trends. *Taking on the future enhances thinking, reasoning, and problem solving skills and contributes to active learning, project-based education, real-world education, learning across disciplines, and learning through inquiry.*

Each of these trends can spark spirited discussions of everything from energy and the environment to poverty, ethics, and sustainability.

The usefulness is found in the creativity of the beholder, but we know that drawing from these trends can make any course more interesting and connected to the larger world. The future, after all, is where those students—all of us for that matter—will live our lives.

You'll find even more suggestions for putting ideas into action in the final chapter of this book. We'll discuss trend analysis, issue analysis, flexibility/innovation analysis, historical/defining moments analysis, and gap analysis, to name a few. However you use *Twenty-One Trends for the 21ˢᵗ Century*, we hope these ideas will ignite interest that challenges the status quo. This forward-looking publication should encourage imagination, creativity, invention, and entrepreneurship that will advance both our economy and civil society. In education, we are often clear about what we're against. Here's a chance to form a clear, inspiring vision of what we're for. We also hope consideration of these trends will make life even more interesting.

Thriving in an Age of Renewal

The world is changing at hypersonic speed. Education systems are expected to prepare their students for the future. They answered the call to get students ready for an agricultural society. Schools and colleges were transformed again as we moved into an Industrial Age. Today, we are entering what seems like the rarified atmosphere of the Global Knowledge/Information Age.

Our education systems, often working against great odds, have traditionally been among the most consistently successful institutions in our society. While schools and colleges continue their heroic efforts, often against a backdrop of higher expectations and limited resources, a sense of urgency is growing. People often hear "change" as a nasty word. Say it, and someone is likely to respond, "Are you telling me I'm not doing a good job? Change makes me uncomfortable."

Rather than simply talking about change, let's focus our energies on developing descriptions of the system we need to help us create an even more promising and sustainable future for everyone, including our students and their education systems. It's one of the most important and uplifting things we could ever do and will become a part of our legacy.

Twenty-One Trends for the 21st Century will help us on our journey toward an increasingly successful and satisfying future. It might even reveal some keys to our very survival.

Driving Questions for Educators and Communities.

After reviewing trends, ask, "What are the implications of these trends: For how we operate our education system (or organization)? For what students need to know and be able to do…their academic knowledge, skills, attitudes, and behaviors? For economic growth and development and quality of life in the community (or country)?" Of course, anyone can substitute other types of implications they'd like to identify.

While we're at it, we might want to consider a few philosophical questions, such as: Do we have a short-range view or a long-range perspective? Are we so intently focused on the bottom line that we've taken our eye off the future? Do we accept the status quo, or do we challenge it?[13] Are we doing things well? Are we doing the right things? What even greater benefit could result from our efforts? Do we have the right answers? Are we asking the right questions?

Introduction
Questions and Activities

1. Use the trends as a first step in creating a future for your organization. To begin, read one or more chapters of this book. We prefer that you study all of them. Then, if you're focusing on an education system, gather a group of colleagues or classmates to brainstorm the implications of the trend or trends for how we run our education institution, what students need to know and be able to do, and possibly for the future of the community, for economic growth and development. Adapt driving questions as you consider the future for any organization, industry, profession, city, county, state, or country.

2. Do a comparison between demographic, economic, and jobs/careers trends in your community and the overall situation described in various chapters of this book.

Readings

1. Marx, G. (2006). *Future-Focused Leadership: Preparing Schools, Students, and Communities for Tomorrow's Realities.* (The companion book to this publication.), ASCD, Alexandria, VA.

2. Marx, G., *Sixteen Trends…Their Profound Impact on our Future*, Educational Research Service, Alexandria, VA, (2006-2008), Editorial Projects in Education, Bethesda, MD, 2012.

3. Florida, R. (2010). *The Great Reset*. Harper, NY.

4. Friedman, T.L. (2005). *The World is Flat*. Farrar, Straus, and Giroux, NY.

5. Gore, A. (2013). *The Future…Six Drivers of Change*. Random House. NY.

6. *The Futurist* magazine. (Various issues.) Bethesda, MD: World Future Society. Available at http://www.wfs.org

---

Note. If you can't connect. Web sites listed in references and in both the *Questions and Activities* as well as the *Readings and Programs* segments of each chapter are likely to change. If you search and draw a blank, first try your favorite search engine, such as Google, Yahoo, or Bing, to find the piece using the title of the work and perhaps the author. In some cases, items may have simply been removed from circulation.

Chapter 1

Generations
Meet the Future

Chapter 1: Generations. *Meet the Future.*

Trend: Millennials will insist on solutions to accumulated problems and injustices and will profoundly impact leadership and lifestyles.
GIs, Silents, Boomers, Xers → Millennials, Generation E

> "The young do not know enough to be prudent, and
> therefore they attempt the impossible...and achieve it...
> generation after generation."
> *Pearl Buck*

Generational Milestones
Count the Candles for Boomers and Millennials

Remember 2011? It was an eventful year, but few things could top this headline: *Boomers Hit 65.*[14] How about 2012? As if born of some imagined intergenerational competition, headlines proclaimed, *Millennials Turn 30.*[15] History was being made before our very eyes.

Both of these generational giants had some things in common as they crossed their individual thresholds. One was the Great Recession. Some senior Boomers had to keep working to support their "retirement." At the same time, legions of Millennials were making their way toward center stage, looking for jobs in a tight economy.

To help them either celebrate, or possibly just deal with dreams yet to be fulfilled, some Boomers dug out a vinyl of Elvis, Bruce Springsteen, or the Rolling Stones. A similar number of Millennials probably put in ear buds and listened to their playlists. A little Beyoncé, Eminem, Lady Gaga, or Tim McGraw might give their minds a break from the promise of their newly minted diplomas balanced against their portfolios of college loans as they groomed their visions of an uncertain future.

Recounting and anticipating the flow of generations is a bit like a real life version of *As the World Turns*, a CBS television soap opera that aired for 54 years, captivating several generations.[16]

> "For the times, they are a-changin' "
> *Bob Dylan*

The Generations
- **G.I. Generation.** Born 1901 and 1924. Often called the "Generation of Heroes."
- **Silent Generation.** Born 1925 and 1945. Smallest generation of the 20th century.
- **Baby Boomers.** Born 1946 and 1964. Largest generation of the 20th century.
- **Generation X.** Born 1965 and 1981. Fewer in number and doubts about the future.
- **Millennial Generation.** Born 1982 and 2003. Moving to center stage.
- **Generation E.** Born 2004-Approx. 2024. May try to cut losses/consolidate gains.

In this chapter, we'll put a spotlight on each of these generations.

History Repeats Itself
The Four Turnings

How would you like to meet some of the people who will be running the country 40 years from now? "They're already here," says Neil Howe, who with the late William Strauss wrote the seminal book, *Generations*. In it, they traced a parade of generations that have populated our planet for all or parts of six centuries. A generation, they suggest, usually covers a period of 17 to 24 years.[17]

Howe and Strauss, based on their observations, wrote a family of books, among them, *The Fourth Turning: An American Prophesy*. Their work has ignited the interest of sociologists, demographers, educators, and planners of every stripe. One of their most compelling discoveries: *Every fourth generation has a tendency to repeat itself.* Consider the four turnings, which reflect the seasonal rhythm of growth, maturation, entropy, and destruction. It's a cycle that will likely continue into the future.[18] Ask yourself, "Where are we and where is our community in this cycle?"

"The *First Turning* is *High*, an upbeat era of strengthening institutions and weakening individualism, when a new civic order implants and the old value regime decays.

"The *Second Turning* is an *Awakening*, a passionate era of spiritual upheaval, when the civic order comes under attack from a new values regime.

"The *Third Turning* is an *Unraveling*, a downcast era of strengthening individualism and weakening institutions, when the old civic order decays and the new values regime implants.

"The *Fourth Turning* is a *Crisis*, a decisive era of secular upheaval, when the values regime propels the replacement of the old civic order with a new one."[19]

Millennials: Taking it to the Streets

It's a worldwide phenomenon. While there are exceptions, Millennials, wherever they are, seem to be finding each other. It's a generation intent on solving problems of the world and dealing with injustices. Expect Millennials to profoundly impact leadership and lifestyles: In fact, they already have. Social media will thunder ahead as a unifying force, something like a generational community, despite concerns that swirl around identity, privacy, and safety. Most in this generation will see these powerful networks as extensions of themselves as they share information, photos, observations, and impulses; stay in touch; meet up; and find common purpose. As for perceived problems and injustices, Millennials have already taken them to the streets. Examples include the *Occupy Movement* and the *Arab Spring*, to name a few. *More on Millennials a bit later.*

Minding the Gap
Generational Co-existence

Let's take a brief tour of our most recent generations. Those generations co-exist, yet each has its uniqueness. Nonetheless, it's a fact that our future depends largely on what we can accomplish together. As we examine the trees (individual generations), let's not forget the forest (intergenerational understanding and collaboration). Each of these generations has been shaped by those who came before them plus the technologies, economies, and lifestyles of their own time.

Yet, some generational changes don't come easy. There are those who stoke fear and use it to raise funds aimed at defending old biases and maintaining both their power and the status quo. Inevitably, life spans being what they are, their power dwindles. However, their unwillingness to open the door to new generations can cause problems that could once have been solved with some ease to explode into crises...even catastrophes. Stephen Aguilar-Millan, director of research for The European Futures Observatory, headquartered in the U.K., adds a word of caution for those who might stoke intergenerational rivalry: "Eventually we may move away from the blame game, where one generation blames the other for the problems of the world, and move to a more constructive phase of building a better position."

Who are we anyway?
GIs, Silents, Boomers, Xers, Millennials, and Es

Generational experts don't always agree on the exact span of years that defines each generation, nor do they fully agree on what to call them. They do, however, agree that there are certain core values that make one generation different from another. The actual characteristics are sharpened by human events or defining moments. In fact, events could even be sparked by generational cycles.

Nearly every organization, if it hopes to survive, has to be adept at communicating across generations. That's especially the case for education systems at all levels. Think of it. Teachers and administrators are multigenerational. So are people in the community. You don't have to tell teachers, who are likely to say, "These kids are different from the ones who came through the door ten years ago."

Consider this a parade of generations. During the next few pages, imagine a drum roll and a bright spotlight that shines on each of these generational groups. When we finish, let's get them all back on stage for a standing ovation. Let our time travel begin!

GI Generation

About 57 million members of the GI Generation were born between 1901 and 1924. At the turn of the 21st century, they were from 76 to 99 years of age. By 2030, when the youngest of this illustrious group will turn 106, we're likely to still have some of them with us. Don't miss the opportunity to get acquainted, listen, and learn.

Widely known as "the generation of heroes," some of them braved two world wars and the Great Depression. Many endured monumental sacrifice and even gave their lives on the battlefield. They produced megatons of food and fiber, manufactured, invented, and helped restore a broken economy. Those who were born early in the 20th century and are still around lived through the Roaring 20s, the consciousness revolution of the 60s and 70s, and saw the turn of a new millennium. They were eyewitnesses to the dawn of radio, television, air travel, and computers. Civic minded, they demonstrated a willingness to make huge sacrifices. Members of this generation were shaped by what they considered the common good, and took loyalty, hard work, patriotism, self-reliance, respect for authority, and a strong sense of civic obligation seriously.[20] These jitterbug specialists, when they weren't on a far-away front line or working their way through the depression, often danced the night away to Glenn Miller's "Juke Box Saturday Night," or sang along with the Modernaires' version of "Don't Sit Under the Apple Tree."

Author and former NBC News Anchor Tom Brokaw, wrote about them in his book, *The Greatest Generation*. "They helped convert a war-time economy into the most powerful peacetime economy in history. They made breakthroughs in medicine and other sciences. They gave the world new art and literature. They came to understand the need for federal civil rights legislation. They gave America Medicare," Brokaw points out. "They became part of the greatest investment in higher education that any society ever made," as the GI Bill made it possible for soldiers to become scholars.[21]

GI Generation contemporaries include: John F. Kennedy, Ronald Reagan, Lyndon Johnson, Thurgood Marshall, George H.W. Bush, Jimmy Carter, Richard Nixon, Gerald Ford, Margaret Thatcher, Louis Armstrong, Bob Hope, Bing Crosby, Lucille Ball, Billie Holiday, Steve Allen, Jack Paar, Ed Sullivan, Jackie Gleason, Milton Berle, Ed McMahon, Thomas P. "Tip" O'Neill, Sidney Poitier, Billy Graham, Pope John Paul II, John Glenn, Charles Lindbergh, Judy Garland, Frank Sinatra, Dean Martin, Hank Williams, Roy Rogers, Roy Acuff, Lawrence Welk, Leonard Bernstein, Ann Landers, Jackie Robinson, Joe DiMaggio, Rachel Carson, J.D. Salinger, John Steinbeck, George Orwell, Arthur Miller, Norman Mailer, Zelda Fitzgerald, Kurt Vonnegut, Henry Kissinger, Robert Oppenheimer, Margaret Mead, Ralph Ellison, Sam Walton, Rosa Parks, Nelson Mandela, Katharine Hepburn, Clark Gable, Lauren Bacall, Eva Marie Saint, Danny Thomas, Ansel Adams,

Isamu Noguchi, William de Kooning, Al Capp, Herblock, Jackson Pollock, Andrew Wyeth, Roy Lichtenstein, and Walt Disney, even Hirohito. *GI and Millennial Generations share similar motivations.*[22]

> **"My father said there were two kinds of people in the world–givers and takers.**
> **The takers may eat better, but the givers sleep better."**
> *Marlo Thomas*

Silent Generation

Born between 1925 and 1945, the 49 million individuals classified as the Silent Generation make up the smallest generation of the 20th century. In 2010, they were between 65 and 85. By 2030, they'll be from 85 to 105. Their parents, largely members of the GI Generation, hardly ever had a dull moment but did have fewer children as they coped with the Great Depression and World War II. Some of the Silents' fresh young faces grace those grainy films of soldiers storming the beaches at Normandy. A vast generational contingent fought in Korea and Vietnam. Silents wanted to their souls to have and to give their children a sound home and the security they had yet to experience.

Silents are generally known for patriotism, hard work, willingness to sacrifice, patience, honor, loyalty, and varying levels of conformity. It's a generation that never had a U.S. president, but its outer conformity camouflaged an intense dedication to purpose, sparked waves of creative genius, and led a civil rights revolution.

While developing new generations of technologies that revolutionized computing and eventually took us to the moon, the so called Silents were watching the Ed Sullivan Show and unveiling new forms of their own music, including a classic array of cultural icons. Voices of the Silents, from "I have a dream today . . ." to "You Ain't Nuthin' But a Hound Dog" continue to electrify the planet. Most took a liking to jazz but grew up to the likes of Patti Page, Ray Charles, Hank Williams, James Brown, Patsy Cline, and Rosemary Clooney. They concocted new moves on the dance floor with guidance from *American Bandstand* and *Soul Train*. While one of the more enduring images of the generation pictured kids all dressed about the same, all wearing Polaroid glasses, watching one of the world's first 3D movies, the Silents proved to be anything but silent.

Silents' contemporaries include or once included people like Martin Luther King, Jr., Neil Armstrong, Elvis Presley, The Beatles, James Brown, Tony Bennett, Bob Dylan, Ray Charles, Willie Nelson, Sammy Davis, Jr., Quincy Jones, Jimi Hendrix, Buddy Holly, Diana Ross, Mick Jagger, Smokey Robinson, Aretha Franklin, Andy Williams, Rosemary Clooney, Patti Page, Joan Baez, Patsy Cline, Loretta Lynn, Glenn Campbell, Kenny Rogers, George Jones, Dick Clark, Johnny Carson, Don Cornelius, Steve Martin, Jackie Kennedy, Muhammed Ali, Marilyn Monroe, Anne Frank, Bob Schieffer, Jim Lehrer, Tom Brokaw, Gloria Steinem, Carl Sagan, Colin Powell, Robert and Ted Kennedy, Kofi Annan, Jonathan Kozol, Marian Wright Edelman, Diane Ravitch, Howard Gardner, Dan Rather, Larry King, Ralph Nader, Bill Moyers, Jerry Falwell, Sonny Bono, Barbara Walters, Julian Bond, Walter Mondale, Jesse Jackson, William F. Buckley, Jr., Malcolm X, Clint Eastwood, Fidel Castro, Che Guevara, Sandra Day O'Connor, Vaclav Havel, Garrison Keillor, Nora Ephron, Judy Blume, Lily Tomlin, Harry Belafonte, Pope Francis, James Dean, Abbie Hoffman, Robert Redford, Audrey Hepburn, Grace Kelly, Frank Stella, Peter Max, Mary Tyler Moore, Elizabeth Taylor, Woody Allen, and Queen Elizabeth.

Eavesdropping

Leonard Bernstein: "Elvis Presley is the greatest cultural force in the twentieth century."

Dick Clurman, an editor at *Time* magazine: "What about Picasso?"

Leonard Bernstein: "No. It's Elvis. He introduced the beat to everything, and he changed everything—music, language, clothes. It's a whole new social revolution—the Sixties comes from it."

A conversation between Time's Dick Clurman and distinguished composer and conductor Leonard Bernstein, recounted by David Halberstam in his book, The Fifties."[23]

Baby Boomers

Born between 1946 and 1964, Boomers, something like 79 million strong, were, up to that time, the largest generation ever produced by the United States...with counterparts in other countries. Theirs was the first generation of the century not to have experienced a world war or the Great Depression. Their parents hailed largely from the Silent and GI Generations. They were confident in "the promise and prosperity of the postwar years," and that "created a sharp rise in births" that continued for 18 years, "when the popularity of birth control pills helped stem the tide," according to Dan Barry, reporting for the *New York Times*.[24]

All generations are young together, and they age together. Starting in 2011 and for the better part of two decades, Boomers will be reaching 65 at about 10,000 a day, according to the Pew Research Center, which adds that "the typical Boomer believes that old age doesn't begin until age 71." By 2030, Boomers will be between 66 and 84, and we'll have only about two people of "working age" for every person drawing benefits from Social Security. Moving ahead to 2050, Boomers will be from 86 to 104.[25]

The generally self-assured, often described as inner-directed, Boomers experienced unprecedented American prosperity, Watergate, the growing impact of television, and the liftoff and splashdown of our first manned space flights and landings on the moon. They were caught in the middle of the civil rights movement and women's liberation and withstood the shocking assassinations of courageous leaders, who were generally members of the GI and Silent Generations and had become symbols of their time. Early Boomers fought in Vietnam, a war that shaped a counterculture, what some called an awakening, a call for peace, and a declaration that we should never trust anyone over 30. Woodstock, a countercultural icon, planned in part by music executives in suits, helped shape the marketplace, clothing, hairstyles, and health habits.

Many Boomers were expected to start the greatest wealth transfer in history until the Great Recession put a crimp in some of their plans, stalled their 401(k)s, and sent them back to work to make ends meet. Some joke that Boomers are hard working people who toast their longevity with a bowl of granola and anti-oxidant-rich berries. They have a reputation for being perennially concerned about their health, wellness, and personal growth. Others raise a red flag about the impact retiring Boomers will have on the health care system. Some, unfortunately, are suffering from many years of the high cholesterol and sugary good life.[26]

Contemporaries of Boomers include: Barack and Michelle Obama, George W. and Laura Bush, Bill and Hillary Clinton, John Kerry, Condoleezza Rice, Caroline Kennedy, Princess Diana of Wales, Tony Blair, Bill Gates, Steve Jobs, Jeff Bezos, Katie Couric, John Grisham, David Baldacci, Dan Brown, Steven Spielberg, Oprah Winfrey, Meryl Streep, David Letterman, Jay Leno, Elton John, Cher, Dolly Parton, Bruce Springsteen, Carlos Santana, Billy Joel, George Strait, Michael Jordan, Michael Jackson, Brian Williams, John Stewart, Jerry Seinfeld, Eddie Murphy, Madonna, Lionel Richie, Bono, Tom Hanks, Tom Cruise, Mikhail Baryshnikov, John Belushi, Dan Aykroyd, Denzel Washington, Sigourney Weaver, Robin Williams, Stevie Wonder, George Clooney, Linda Darling Hammond, George Stephanopoulos, Donald Trump, Sarah Palin, Keith Haring, Deepak Chopra, and Prince Charles.[27]

The Case for Generation Jones: It's not necessarily a protest or a movement, but some late Boomers, born between 1954 and 1965, sense that the early Boomers were a post-war phenomenon, while they were the ones who contributed most to a cultural reset. They have claimed "dramatic differences between collective personality traits of Boomers and Jonesers." They have also claimed from Boomers a set of people selected as national Presidents and Prime Ministers among developed countries, such as Barack Obama in the U.S., Stephen Harper in Canada, Nicolas Sarkozy in France, Angela Merkel in Germany, Julia Gillard in Australia, and John Key in New Zealand. Jonesers see themselves as a bridge between the Boomers and Xers.[28]

Generation X

Born during the Baby Bust of 1965 and 1981, the estimated 51 million Xers were sandwiched between the more or less gigantic Boomers and Millennials. Xers have dazzled the world with a kind of low-key genius. In 2010, Xers were between 29 and 45. In 2030, their birthday cakes will blaze with anywhere from 49 to 65 candles and in 2050, from 69 to 85. Only those who live to be 119 will see 2100. It might just happen.

Early on, a lot of Xers were called "latchkey kids," largely because, in many cases, both parents were working or they lived in a single-parent home. Their largely Boomer parents had busy lives and divorce became more common, so Xers developed the traits of free agents, such as being self-reliant, independent, individualistic, resilient, adaptable, and entrepreneurial. As the economy changed, they often saw their parents laid off or face job insecurity. In response to that experience, some Xers gave up on the idea of loyalty to a single organization. If that weren't enough, "they came of age among dire pronouncements that theirs might be the first generation in memory that might not live better than their parents,"[29]

As a generation, they became known for their acceptance of diversity, their peer friendships, individuality, rejection of rules, adept use of technology, desire for some level of flexibility and freedom, and an interest in their own professional development. Some bristled at unearned authority. Diane Thielfoldt and Devon Scheef, co-founders of The Learning Café, wrote in a 2005 issue of *Law Practice Today* that Xers were likely to leave the impression that, "I don't need someone looking over my shoulder." For mentors, they suggested "bringing out strengths, discarding biases and pre-conceived notions, and enjoying generational differences—and similarities."[30]

Historic defining moments for Generation Xers included: the emergence of a new world order when the Berlin Wall came down and the Soviet Union disintegrated; the L.A. riots and an exhortation by Rodney King to "just get along;" and the O.J. Simpson trial. Other defining moments were the Gulf War, Chernobyl disaster, Challenger explosion, and the 9-11 terrorist attacks.[31] Without fanfare, Xers lit up our television screens explaining how we were able to land the earliest rovers on Mars and have made continuing contributions to science and technology. They felt the scars of the post-Vietnam era and were the first recent generation to worry about whether their pension funds, including Social Security, would be there for them.

Xer contemporaries include: David Cameron, Paul Ryan, Ryan Seacrest, Michael Dell, Tiger Woods, Jimmy Fallon, Anderson Cooper, Kevin Smith, Reese Witherspoon, Jennifer Lopez, Jennifer Hudson, Nigel John Taylor of Duran Duran, Brooke Shields, Ben Stiller, Halle Berry, Janet Jackson, Kurt Cobain, Mike Tyson, Peyton and Eli Manning, Michael Buble, Julia Roberts, Kobe Bryant, Sammy Sosa, Cameron Diaz, Macaulay Culkin, Yao Ming, Charlie Sheen, Shania Twain, Shaquille O'Neal, Jewel, Serena and Venus Williams, Cindy Crawford, Kiefer Sutherland, Mary Lou Retton, Jennifer Aniston, Brett Favre, Ice Cube, Marilyn Manson, Alicia Keys, Mariah Carey, Death Cab for Cutie, Lance Armstrong, Julian Castro, Marco Rubio, Tupac Shakur, Eminem, John Mayer, Leonardo DiCaprio, Kanye West, Christina Aguilera, Britney Spears, and Ashton Kutcher.

> **"There are enough Millennials that today's youth could become a political force that could dominate all political factions and institutions if they are united and share beliefs and a long-term planetary orientation."**
> *Eric Greenberg, Author of Generation We*[32]

The Millennials

Born between 1982 and about 2003, the 76 million U.S. Millennials outnumbered most of the generations that came before them. They are the fourth generation out from the GI Generation, and in a 21st century way, share a number of their motivations. People who really listen to them tell us that this is a generation with high aspirations for themselves, particularly focused on dealing with longstanding problems and injustices.

In 2010, members of the Millennial Generation were between 7 and 28. In 2012, the first of them celebrated their 30th. By 2030, they'll be from 27 to 48. By then, they will likely hold down a growing number of key political positions. Move ahead to 2050 and our Millennials will be in prime time, from 47 to 68. The eldest of them will turn 100 in 2082 and the youngest in 2103.

Parents of Millennials, generally Xers or Boomers, drove their kids to every imaginable activity, when they could. Some of their omnipresent parents were intent on steering their progeny away from hazards they had experienced, such as "drugs, alcohol, smoking, profanity, improper TV, un-chaperoned gatherings, aggressive behavior, AIDS, and teen pregnancy."[33] Hyperactivity seemed to increase and Ritalin became a household word. Cases of asthma also took an upward swing, and obesity seemed to increase as daily physical education classes declined.[34]

Perhaps no generation has been more tech-savvy. Smartphones, iPads, and a host of other gear are only gateways to social media that give them second-by-second access to information, ideas, and connections with people sitting next to them or on the other side of the world. With that enthusiastic and unbridled use come the inevitable combination of demands for more bandwidth and concerns about identity theft and personal security. Millennials worldwide are becoming hyper-connected. Ask how to get something done, and they'll very likely tell you, "There's an app for that."

A Millennial himself, Futures Council 21 member Ryan Hunter points out that members of his generation grew up with social networking technologies. This "plugged-in" generation, Hunter says, regularly uses "Facebook pages, re-tweeting of popular ('hit') posts" and other connections. In the process they are able to "facilitate creation and engagement of an online kind of civil society quite literally at their fingertips."

While each member of this ubiquitous generation is unique, a common portrait includes: optimism, focus, high personal expectations, a sense of enormous academic pressure, digital literacy that grew from never knowing a world without computers, unlimited access to information and connections, comfort with teamwork, and seeing the world as a place that's open 24/7. Many have been told that they are special and would prefer to be treated that way, says Jill Novak, writing for *marketingteacher.com*. She adds, "They do not live to work; they prefer a more relaxed work environment with a lot of hand holding and accolades."[35]

"I think people of the future will be more sensitive about problems of the world.
I hope. Otherwise, I'm moving to Mars."
Marina, 14, Brazil[36]

A Pew Research poll confirmed that Millennials' commitment to solving problems seemed to be turning them out at the polls. However, some still needed to be convinced that it makes any difference. That study also revealed *tolerance* on social issues, such as immigration, race, and gay rights. A full 84 percent believed they had better educational opportunities, compared with 20 years ago. They also speculated that they have more casual sex, resort to violence more often, do more binge drinking, and use more illegal drugs. *Nonetheless, some of these indicators seem to be dropping.* In considering top goals for the future, about 80 percent said, "people in their generation think getting rich is either the most important, or second most important, goal in their lives." About half place high value on the possibility of "becoming famous."[37]

A 2010 Pew Research Center study described Millennials as "the American teens and twenty-somethings (also thirty-somethings), making their passage into adulthood."[38] Their predicted commitment to dealing with problems facing the nation and world seemed to be holding true. When their time came, Millennials got involved in political campaigns, voted in significant numbers, and took to the streets to support or protest an array of issues. However, the number of problems they will encounter as they move into the future is growing. That's in part because of a fast-changing world and indecision and gridlock among politicians, who are accused of "kicking the can down the road." That's code for, "We expect the Millennials to take the leadership, muster the courage and fortitude, and endure the risks that will be essential to finding solutions." Those cans that have been kicked over and over again are headed their way.

As we entered the second decade of the 21st century, early Millennials were gainfully employed in their jobs or careers or searching for work in a tight economy. Many had their own children in school. Later Millennials were still working their way through elementary, secondary, and higher education. Some will benefit from the greatest transference of wealth in history as they inherit accumulated resources of their Boomer and Xer parents. Others will be on their own, but like a few generations before them, taking care of their own families while looking after older relatives, many in the range of 85 to 105. As one of the largest generational groups ever to populate the planet, they will also need a social safety net as life expectancies continue to inch upward.

By the way, not everyone calls them Millennials. They are alternately known as: Generation Y, the Echo Boomers, Gen Tech, Generation Text, Dot Com, Nexters, and the Net Generations. Who are some of the better-known Millennials? The list of those who are known on the world stage is growing. Some include: Princes William and Harry, Duchess of Cambridge Kate Middleton, Mark Zuckerberg, Taylor Swift, Adele, Justin Bieber, Lady Gaga, LeBron James, Danica Patrick, Bow Wow, Andy Roddick, Brad Renfro, Daniel Jacob Radcliffe (Harry Potter), Sarah Hughes, Robert Griffin III, Carrie Underwood, and Malala Yousafzai. Feel free to add to this list. As for music, Millennials regularly add tunes from many generations to their playlists.

What Else Do We Need to Know about Millennials?
Leaders in the Making

Typically, some of the motivations of Millennials came in response to the values and behaviors of their parents, their peers, and society-at-large. Add to that the ongoing stream of information and ongoing conversations using social media. What else should we know? The short answer is, "Ask a nearby Millennial." However, for educators, and anyone else concerned about shaping the future, here are a few other things to consider.

• **Defining Moments** have their greatest impact during our formative years. A Millennial born in 1982 was 7 when the Berlin Wall came down; 10 when Bill Clinton was elected President for the first time; 13 when they heard on television about the Oklahoma City bombing, 17 when the Columbine shootings took place; 18 when George W. Bush was elected for the first time; 19 on September 11, 2001; 21 when the U.S. invaded Iraq; 23 when Hurricane Katrina struck the Gulf Coast;

25 when the Great Recession got under way; 26 when Barack Obama was elected to his first term as President, 29 when a massive tsunami hit Japan; and around 30 when Hurricane Sandy flooded the U.S. northeast coast. That was 2012, when students and educators at Sandy Hook Elementary School in Connecticut were gunned down. Tens of thousands of Millennials served in Iraq, Afghanistan, and other parts of the world. Natural disasters can have a profound impact on our lives, but the defining moments that stir us most deeply are those that involve human judgment.

• **Hitting the Streets.** Millennials have taken their values to the streets and to the ballot box as they pursue a belief that the status quo will not cut it—that vertical, top down decision-making needs to become more collaborative. Revolutions that came with the Arab Spring, sparked by Millennials, grew from converging trends. Among those trends were: (1) a generational commitment to justice and a disconnect with often older authoritarian regimes that left them seemingly powerless; (2) limited opportunities for pursuing education, jobs, and an array of personal aspirations; and (3) the convening power of social media, such as Twitter and Facebook. A 2010 study revealed that 75 percent of Millennials had a profile on a social networking site, compared to 50 percent of Gen Xers, 30 percent of Boomers, and 6 percent of Silents.[39]

• **Convergence and Collaboration.** Convergence is, in fact, emerging as the name of the game. Growing polarization is both a turn-off and a turn-on—a turn off for politics as usual and a turn on for finding a better way. The Occupy Wall Street demonstrations that followed a damaging financial meltdown were a symbol of unrest. Even more forceful may be a subtle movement involving bold, energetic, tech-savvy, and sometimes outspoken members of a generation of people who insist that those who are polarized and entrenched move over or move out. Call it a tipping point. Some Millennials, and many others, see the rigidity standing squarely in the way of progress at a time when the world is facing massive issues, such as energy, the environment, the economy, immigration, security, and longer-term solutions for Social Security and Medicare.

• **Lifestyles.** Social observer Richard Florida points out that, for growing numbers of young people, "where they lived was more important than where they worked." What were they looking for in a place to live? Cost of lifestyle, earnings, vitality, after hours life, learning opportunities, ease in getting around town, diversity, low crime rates, opportunities for civic engagement, and answers to long commute times, such as mass transit, or ideally walk- or bike-ability.[40] As Millennials were hit with a perfect storm of yet-to-be-paid college loans, higher prices for a place to live, and long commutes that seemed to eat up time they didn't have, some settled into micro-apartments. These studio apartments, according to *Forbes*, "often run 300 square feet and smaller" in New York and can be as small as a 220 square foot "nest" in San Francisco.[41] Some Millennials witnessed their parents becoming house poor. They saw what they considered excess space, massive loan payments, and, unfortunately, occasional foreclosures. Nonetheless, with jobs scarce, many went back to live with parents or shared rent.

• **Issues.** Fairness is a big concern among Millennials, as are constellations of social issues, such as: education, child abuse, suicide, AIDS, unemployment/the economy, the environment, women's rights, gay rights, and health care, according to Peter Zollo, president of Teenage Research Unlimited (TRU), in a study he published in 2004, titled, *Getting Wiser to Teens*. At the time, their top five favorite things about school were friends, assemblies/special days/field trips, sports, seeing a boyfriend or girlfriend, and extracurricular activities. Their least favorite were getting up early, tests, homework, the length of the school day, and peer pressure.[42]

• **Rights.** Whether in their own communities or through the media, Millennials have experienced the multicultural society. For them, it's not a stretch—it's the way the world ought to be. Ask Millennials how they feel about rights issues and they might consider them a done-deal since they "grew up being told they could do anything they wanted, with laws on the books to support that, especially in corporate America," Hannah Weinberger wrote for CNN. Some making their way through the education system may not realize "that female graduates make $8,000 less than their male counterparts," she reported.[43]

Many Millennials have expressed interest in members of the GI Generation, since their motivations were often similar. However, some were uneasy communicating face-to-face, since so much of their interaction had shifted online.

- **Tough Times.** Approximately one in four members of this diverse generation lived in a single-parent household. About 75 percent have or had working mothers.[44]

- **Identity.** Individual identity has been important to Millennials, who may hope to portray something unique about them as the world population increased by billions. Body piercing, small artistic or message-laden tattoos, clothing and hairstyles, and unique smartphone cases help to set individuals apart.

- **Tall Stories.** Most have never had to "walk across the room to change channels on the television set," a story some parents tell their children, the equivalent of "walking four miles to school, uphill, through snow four feet deep."

- **Compressed Messaging.** Millennials have been leaders in many things, among them developing shortcuts when they're communicating on Twitter. Some of the most widespread include: NBD (No big deal), KIT (Keep in touch), LOL (Lots of love), N2K (Need to know), SUP (What's up), POS (Parents over shoulder), and B4N (Bye for now).

> "All our lives, our parents have protected us, sheltered us.
> Now we must break free of those chains and wander out
> into the world."
>
> *Laura, age 17*[45]

Millennials in the Workplace

The *Globe and Mail* seemed to hit the mark in a March 2013 article, "The Six Ways Generation Y Will Transform the Workplace." Lauren Friese and Cassandra Jowett of TalentEgg (talentegg.ca), a firm that helps college graduates transition from school to work, reminded readers that Canada's "12 million strong (Millennials) represented one-third of the country's population." These "digital natives," they said, are "the most educated and most diverse generation in history, and the first to have more women than men obtain(ing) postsecondary education credentials."

Like it or not, they say, this massive generation will massively transform the workplace. When they move into executive and management positions, they "won't just be rearranging the furniture and hanging their diplomas in the new corner offices—they will be making sweeping changes to the way organizations and their people work." A few of the changes Friese and Jowett suggest include: more women in leadership roles and a blended environment allowing people to work "anywhere, everywhere, and any time of day or night." Punching in 9 to 5, Monday through Friday, "is so 21st century." They foresee more instant messaging, less email, and fewer pointless or redundant meetings. Plus, they tell us to be ready for "immediate, effective feedback" that allows people to produce better results and improve workflow. That means ROWE, "the results-only workplace."[46]

Writing in the *Washington Post*, Emily Matchar concurred that Millennials will shake up the workplace. "Raised by parents who wanted to be friends with their kids, they're used to seeing their elders as peers rather than authority figures. When they want something, they're not afraid to say so," Matchar observed. An MTV survey of Millennials, which they called "No Collar Workers," found that "81 percent of respondents said they should be able to set their own hours." A Deloitte study quoted in the article warned that companies "must foster a culture of respect that extends to all employees, regardless of age or level in the organization."[47] Matchar's parting shot: "Be selfish. Be entitled. Demand what you want. Because we want it, too."[48]

Get ready for the expectations these fresh generations have for quality, style, an instant response to ideas or requests, and an assumption that they are part of the team. Many will be helicopter parents, sort of hovering and staying in touch, connected to whatever happens with their children at school, even from a distance.

Insights for Millennials, Finding a Path Forward. Annmarie Neil, former chief talent officer for Cisco Systems and author of *Leading from the Edge* has sage advice for Millennials. She suggests, "Smaller, incremental goals" can give you the experience you need for that heavy-duty position, when you arrive. Without that experience, you might get to the big goal fast and then simply drop off. "Always have a purpose you're striving toward." While moving too often may not look good on the resume, "demonstrate that you can work in different cultures, with different business models, and solve different business problems." She observes, "The people who are most effective in senior level leadership roles are those who have had multiple life experiences, transitions, failures, and successes." Neil adds, "Build a board of advisors;" and "Be provocative and insightful about who you are, what you want, and how you're going to get it."[49]

Time passes

"The eldest Millennials will reach age 50 in 2032 and 60 in about 2042. By then, their generation will produce a majority of state governors and members of Congress and very likely its first serious presidential candidate", Neil Howe and William Strauss wrote in their 2000 book, *Millennials Rising*.[50] As important as they are, the Millennials will also be faced with adjusting to a new generation, made up primarily of their own children. The cycle continues.

Generation E . . . for Equilibrium

Just as the Silents followed the GI Generation, it's likely something akin to *Generation E* will follow on the heels of the Millennials. *E*, in this case, stands for equilibrium. Members of this new generation will likely be born between 2004 and 2024. That assumes a 20-year generation, which may be a stretch. If that assumption holds, the youngest will celebrate a 76th and the oldest a 96th birthday in 2100.

The Es will grow up experiencing and hearing stories from aging relatives, some older than 100, about the chaotic decades that preceded them and perhaps feel a need to ease worldwide tensions. By then, if world problems such as energy and the environment, poverty, and security are not on their way to solution, they will face a planetary crisis. They'll ruminate about wars, brinkmanship, breaches of ethics, technological developments that outstrip the speed at which they want to adapt, and consider how to stay healthy during their expanded life spans. Genetics, robotics, and exploration of both inner and outer space will increasingly impact their lives.

However, members of Generation E may be feeling a sense of fatigue just keeping up with their Millennial and Xer parents and grandparents. If they can find some type of equilibrium, they will likely focus on discovering even more about human potential. Like the Silents four generations before them, they will probably break through any conformity that may have developed to create a whole new era of arts and culture, which, today, is only at the edges of our imaginations.

As a group, the Es will probably be neither conservative nor liberal. In fact, they might blanch at the very thought of polarization. Civil discourse is so much more productive and collaboration is about the only thing left to sidestep lethal combat and environmental devastation. Stephen Aguilar-Millan of the European Futures Observatory in the U.K. suggests that stability and equilibrium may be hard to come by and foresees a need for "education about the intractability of wicked problems" the world is facing.

We might expect the Es to spawn a new era for politics and world affairs, education, and numerous other fields of human endeavor. For this largely 21st century generation, problem solving skills and futures studies will likely be firmly placed on the list of basics. As for technologies, they will increasingly become nearly invisible as people think of them not necessarily as tools but as extensions of their minds and bodies.

Of course, a great deal depends on whether the planet is rocked by natural or human caused catastrophes. If the Millennials and Es can make it through those challenges, then, high on their generational agendas will be a commitment *to cutting our losses and consolidating our gains* and perhaps settling even for a very short time on what we might call a *21st century version of a new normal*. Part of their duty will be transitional, to explain those of us who are still around to ourselves and launch us on another generational mission of discovery.

Leading across Generations

For leadership, one of the greatest flashes of insight must be that new ideas are not necessarily an attack on the institution. They may, in fact, actually be its lifeblood. We may not be able to pursue every idea, but we need to encourage and reward creativity, innovation, and entrepreneurship. Some think of it as trying to manage a group of cats. Others know that it's more the role of concertmaster and facilitator. It takes encouragement of ideas and giving energy, not draining it, from the team. Defending the status quo, ad-infinitum, is, after all, a sure ticket to obsolescence. Leaders simply must be orchestrators of information and ideas and depend on everyone around them to help an organization, community, or nation stay ahead of the curve.

Most agree that we need to help people learn about a variety of generations and adapt to the reality. The *Wall Street Journal* offered tips on "How to Manage Different Generations." Among them: do cross-generational mentoring; communicate in ways that are comfortable to various groups; make meetings generationally friendly and not just holding meetings for meetings' sake; recognize people across generations and give positive feedback; allow for a variety of work styles and family situations; and give everyone a voice.[51]

"A small body of determined spirits fired by an unquench able faith in their mission can alter the course of history."
Mahatma Gandhi[52]

Convergence in the Classroom

In teaching a hyper-tech-savvy generation of students, we should not forget the importance of content along with interaction. Thinking, reasoning, problem solving, creativity, entrepreneurial, and communication skills, coupled with ethical behavior, will be essential if we hope to even have a future. As we teach everything from STEM to social studies, civic education, and the arts, we generally need to align with revolutionary changes in how new generations of students get and process information.

Writing for ASCD's *Educational Leadership*, Larry Rosen, a professor of psychology at California State University, Dominguez Hills, and author of *Me, Myspace, and I*, proclaims that students "expect technology to be there, and they expect it to do whatever they want it to do. Their WWW doesn't stand for World Wide Web, it stands for Whatever, Whenever, and Wherever." Nearly every waking minute involves "being online, using computers offline, listening to music, playing video games, talking on the telephone, instant messaging, texting, sending and receiving email, and watching television," Rosen observes. They have access to apps galore, and if something is missing, they might just create it. When Rosen wrote his article in 2011, Nielsen had reported that "the typical teenager sends and receives an incredible 3,339 texts a month, which would be an average of about six an hour when he or she is not sleeping, while making and receiving only 191 phone calls during the same period." (Rosen, Larry D., "Teaching the iGeneration," *Educational Leadership Magazine*, Feb. 2011, Vol. 68, No. 5).

Implications for Society
Generations
Boomers have moved into top leadership positions. Coming up to join them are members of Generation X and the Millennials. Communities and countries, including education systems, businesses, and other organizations, will need to stimulate opportunities for people of all ages. What they do or do not do will have a profound impact on their futures.

- Millennials will continue to be a decisive force in political campaigns, if they believe their votes will make a difference.
- Since Millennials will generally be committed to solving problems and dealing with injustices, they will need to understand how to resolve disputes peacefully and democratically, if we hope to avoid unthinkable destruction.
- Members of younger generations, particularly Millennials, may have to compete for resources with seniors, whose numbers will increase dramatically and who will require an array of community services. Since life expectancies continue to rise, growing numbers of Xers and Millennials may be faced with watching over aging parents and grandparents as well as their own children and grandchildren.
- An array of unsolved problems will accumulate as polarized politicians "kick the can down the road." That means younger generations will need to add their unfinished business and the cost of neglect to other urgent issues they will face.

- Look for increasing numbers of people from multiple generations to consider individual and human rights as well as broad-based respect as givens.

- Enlightened countries and communities will need to be organized to tap the strengths of people across generations and all diversities as they develop a common purpose and a vision for the future. Increasingly, members of younger generations will insist on collaboration and having their voices heard as decisions are considered about issues ranging from the economy to the quality of education.

- Communities and countries will be pressed to provide reasonable employment opportunities, support for entrepreneurship, infrastructure, and a compelling quality of life. All are keys to attracting and keeping well-educated young people.

- If consensus on dealing with problems, righting wrongs, and pursuing opportunities becomes elusive, expect well-organized street protests, electronic "mass mobbing," and other forceful means of expression.

- Unless pension programs and social contracts, such as Social Security, are made secure for Millennials, they will be less likely to enthusiastically support them for others or to support politicians who deprived them of their benefits.

"We are now at a point where we must educate our children in what no one knew yesterday, and prepare our schools for what no one knows yet."
Anthropologist Margaret Mead[53]

Implications for Education
Generations
- **Building intergenerational relationships and communication skills.**

Schools and colleges work with multigenerational communities, teams of educators, and other staff. Students of various ages reflect not only their personal strengths but also a variety of values, aspirations, and behaviors common to their generations. Learning about generations can be interesting. Nearly everyone becomes a captive audience as they encounter life stories from people of other generations. Professional development and in-class sessions can stimulate that interest, not to divide people but to help them appreciate each other as they work together in common purpose. Deliberately orchestrating multi-generational teams, planning sessions, cross-generational mentoring, and other activities can build relationships and even add to our wisdom. Advisory councils and targeted online or printed newsletters that feature a broad range of people can help keep everyone engaged and informed.

- **Teaching students how to make change peacefully and democratically.**

As Millennials move into positions of leadership, they will confront longstanding problems and injustices and demand that they be resolved. To deal with issues of this magnitude, they will need a broad understanding of history, culture, and human behavior, as well as thinking, reasoning, and problem solving skills. All of the above and more should be developed with conflict resolution skills and an ethical code as a foundation. In the end, getting rich and famous may not be as important as how you did it.

Civic educators will likely play a role that goes far beyond the classroom as they advise their educational institutions and communities on how to make the democratic process work. Students will need to learn how the political/decision-making system works and have practice in identifying and exploring issues, involving people in developing solutions, and in drafting and promoting changes in public policy. They will also need to know how to find and use information, build a case, communicate effectively, test ideas, and rally support through personal persuasion. Some will practice persuasiveness pursuing a community project. What's at stake? It could be our very survival. Unless current and future generations understand how to deal with differences and resolve conflicts peacefully and democratically (without resorting to an array of lethal weapons), we could end up anywhere from Nirvana to Armageddon.

- **Listening to students . . . giving them a voice.**
  Getting them engaged.

Animated, energetic, committed generations of students will expect their voices to be heard in decisions that affect their education and their individual and collective futures. Listening to students is not just a nice thing to do. They have good ideas. They'll feel a greater sense of ownership. And it is *their* future. Educators need to be sure that generations of people moving into positions of leadership and citizenship understand the importance of voting, getting involved in community projects, helping people across generations, and providing insights for creating an even brighter future.

- **Building media literacy skills.**

Want information about any topic? Ask a student. They'll very likely pull out a smartphone, do a few key strokes, and lay some version of the facts in front of you. That's great, but as generations of all ages attempt to take on community or world problems, they will need to be armed with media literacy skills. How else will they be able to separate truth from fiction or opinion, wheat from chaff, legitimate information from disinformation?

Our cyber-attic is filled with information. Our future depends on the richness and trustworthiness of that information as we use it to make decisions about the future of our community, our country, and the planet. In a fast-moving world, often driven more by technology than values, we need guidelines to help us sort out what deserves consideration. That's true whether we're making personal, family, business, or societal decisions...or just figuring out what should show up in that end-of-semester term paper.

- **Preparing students for realities of the 21st century workplace.**

Whether they aspire to working locally or globally, each needs to be prepared for rigor, self-discipline, languages, teamwork, an eagerness for feedback, a demand for results in the *real world*, and the importance of face-to-face and online communication and relationships across generations. They need to figure out how entrepreneurial they might want to become.

International relations veteran the late Frank Method has observed that Millennials may be the first generation in a century or two to "grow to adulthood in a world environment that is less centered on the United States." In this environment, he suggests "critical thinking skills, problem-based approaches, and collaborative learning modalities may be increasingly important." Students will need to understand that they are also citizens of a planet that extends far beyond their national boundaries.

- **Attracting Millennials into education careers.**

Considering that growing numbers of Millennials want to make positive changes in the world, schools and colleges might want to encourage them to consider careers in education. That's, at least in great part, where tomorrow is being shaped…today.

Generations

Questions and Activities

1. Explore *Capturing Experience*, a series of information and study modules covering generations, The Intergenerational Center, Temple University, http://cil.templecil.org/node/5.[54] Make a short list of 5 to 7 items you found particularly interesting.

2. Study Strauss and Howe's "Four Turnings." Prepare a brief presentation or paper on how this phenomenon has, does, or will affect your organization.

3. Review the general characteristics we've listed for the Millennial Generation. What additional characteristics have you observed?

4. Consider people in your organization who are members of various generations. List 10 things they could and perhaps should learn from each other.

5. Develop a strategy for helping students learn about, appreciate, and enhance communication with generations other than their own.

6. View "Fast Future: The Rise of the Millennial Generation," David Burnstein, founder of Generation 18, *TED Talk*, posted Nov. 30, 2012. (11:34). http://macromon.wordpress.com/2012/11/30/fast-future-the-rise-of-the-millennial-generation-ted-talks/. Share the presentation with a group. Ask them, "Based on this presentation, how would you describe the Millennial Generation?"[55]

7. If Millennials truly are committed to solving the world's problems and dealing with injustices, share up to six ways we can educate them to attempt to make those changes peacefully and democratically.

8. Prepare a five-minute PowerPoint presentation devoted to how schools and colleges might better educate people for building understanding across generations.

9. Consider viewing "The Generation That's Remaking China," featuring Yang Lan, China's equivalent of Oprah. (17:46). What comparisons do you see between what China's younger generation is facing compared to the younger generation in your country? http://www.ted.com/talks/yang_lan.html.[56]

Readings and Programs

1. Strauss, W., and Howe, N. (1998). *The Fourth Turning . . . An American Prophesy.* New York: Bantam Doubleday.[57]

2. Greenberg, Eric, with Weber, Karl. *Generation We.* (2008), Pachatusan, Emeryville, CA.[58]

3. "The Six Ways Generation Y Will Transform the Workplace, Friese, Lauren, and Jowett, Cassandra, *The Globe and Mail*, Canada, March 12, 2013, http://www.theglobeandmail.com/report-on-business/careers/the-future-of-work/the-six-ways-generation-y-will-transform-the-workplace/article9615027/.[59]

4. View *Wall Street Journal* Interview with Facebook founder and Millennial Mark Zuckerberg, posted June 10, 2010, (52:43). http://live.wsj.com/video/d8-facebook-ceo-mark-zuckerberg-full-length-video/29CC1557-56A9-4484-90B4-539E282F6F9A.html#!29CC1557-56A9-4484-90B4-539E282F-6F9A.[60]

Chapter 2

Diversity
E Pluribus Unum

Chapter 2. Diversity. *E Pluribus Unum*

Trend: In a series of tipping points, majorities will become minorities creating ongoing challenges for social cohesion.
Majority/Minority → Minority/Minority

Worldwide: Growing numbers of people and nations will discover that if we manage our diversity well, it will enrich us. If we don't manage our diversity well, it will divide us.
Exclusion ↔ Inclusion
Diversity = Division ↔ Diversity = Enrichment

> "The U.S. will become the first major post-industrial society in the world where minorities will be the majority."
> *Marcello Suarez-Orozco*[61]

E Pluribus Unum

It's the U.S. motto, "Of the Many...One," but it is much more. It is a statement of belief that people of every shape, size, shade, and belief can work together to build a civil society, a sustainable environment, and a viable economy. While we're working together, we can also pursue our own individual dreams.

Skeptics? There are tons of them. Dominance by one group over others has been a way of life for millennia. But things are changing. Today, the streets of free and open societies have become rainbows, a mélange reflecting and celebrating the combined characteristics of "the many" who make up a society.

The *many* include people across ages, genders, races, ethnicities, national origins, languages, social and economic conditions, generations, developmental differences, and political and religious beliefs. *Exclusion* is turning toward *inclusion* and *collaboration*. In the democracy of our dreams, a country doesn't belong to any one of them. It belongs to all of them. The prevailing question is, "How big can we make the tent?"

Now, wherever we live on the planet, we are faced with this stark reality: If we manage our diversity well, it will enrich us. If we don't manage it well, it will divide us. Meeting that challenge is up to each of us and all of us.

Majority Becomes Minority
We're in the midst of a seismic shift!

Get ready. It will likely happen between 2040 and 2050. If projections hold true, every racial or ethnic group will, in essence, become a minority by about 2043. In fact, tipping points for the trend got underway in 2011.

Tipping Points, a Minority/Minority Society...By Age Group.

Brookings Institution Demographer William Frey interpreted December 2012 U.S. Census Bureau projections. He pointed out, by age group, when non-Hispanic Whites are likely to become less than 50 percent of the U.S. population. *Total Population*: 2043. *Age 1*: 2011. *Age 5*: 2013-14. *Under Age 18*: 2018. *Ages 18-29*: 2027. *Ages 30-44*: 2035. *Ages 45-64*: 2051. *Ages 65+*: After 2060.

While there are many types of diversity, let's take a look at relative percentages of several U.S. racial and ethnic groups. Frey summarizes them, comparing 2012 population numbers with projections for 2060. The following table includes those comparisons for the *total population*, those *under 18*, and those *65+* (See figure 2.1).

Who Are These Minorities?
Better Yet...Who Are We?

Thanks to the U.S. Census Bureau, we have even more head-turning news about the richness of our diverse population:

Non-Hispanic White. Expected to rise from 197.8 million in 2012 to a peak of 199.6 million in 2024, then "slowly decrease, falling nearly 20.6 million from 2024 to 2060," when non-Hispanic Whites will likely be 43 percent of the population. For this group, deaths exceeded births in 2012, for the first time.[64]

Hispanic. Slated to "more than double, from 53.3 million in 2012 to 128.8 million in 2060." The Census Bureau points out that this group will become approximately 1 of 3 U.S. residents, up from 1 in 6 in 2012.

Figure 2.1
Projected U.S. General Population by Race/Ethnicity
2012 and 2060[62] [63]

Note: Numbers were extrapolated from a U.S. Census Bureau Report interpreted by William
H. Frey for the Brookings Institution on Dec. 13, 2012. Percentages have been rounded.
AIAN=American Indian and Alaska Native. Two+=Two or More Races.

Race/ Ethnicity	2012 % Total Population	2060 % Total Population	2012 % Under 18	2060 % Under 18	2012 % Age 65+	2060 % Age 65+
Non-Hispanic White	63	**43**	53	**33**	79	**56**
Black	12	13	14	13	9	13
AIAN	1	1	1	1	1	1
Asian	5	8	5	7	4	8
Two+ Races	2	5	4	8	1	2
Hispanic	17	31	24	**38**	7	21

Just for Reference: Total population of the U.S in *2012* was about *314 million*, headed for a projected *420 million in 2060*. "Minorities," which in 2012 made up 37 percent of the population, were expected to move up to 57 percent by 2060, according to the Bureau.

Black. Between 2012 and 2060, "the Black population is expected to increase from 41.2 million to 61.8 million" even though annual percentages moving forward vary only slightly because of overall population increases.

Asian. The U.S. Asian population will likely "more than double, from 15.9 million in 2012 to 34.4 million in 2060."

American Indian and Alaska Native. These native populations are projected to swell from 3.9 million in 2012 to 6.3 million in 2060, slightly increasing their overall percentage from 1.2 to 1.5 percent.

Native Hawaiian and Other Pacific Islander. This group, while not broken out separately on the above illustration, "is expected to nearly double, from 706,000 in 2012 to 1.4 million in 2060."

Two or More Races. This group "is projected to more than triple, from 7.5 million to 26.7 million" during the period of 2012 to 2060.[65]

"Longstanding fights over civil rights and racial equality are going in new directions, promising to reshape race relations and common notions about being a 'minority,'" the *Boston Globe* declared, in response to the Census Bureau's December 2012 report.[66]

Of course, this situation begs the question: What do you call majorities when they become minorities…or minorities when they become majorities? The answers are sure to evolve. In the meantime, let's just call everyone minorities.

> **"I have a dream that one day, on the red hills of Georgia, the sons of former slaves and the sons of former slave owners will sit down at the table of brotherhood. I have a dream today."**
> *Martin Luther King, Jr., 1963*

One-Year-Old "Minorities" Have Company

It's official! All children under 1 year of age in the U.S. have been minorities since July 1, 2011. That's when the "traditional minority" portion of the youngest among us reached 50.4 percent. Close on their heels were 5-year-olds.[67]

"Little children are in the vanguard of all this change coming to America," Kenneth Johnson, a senior demographer and professor of sociology at the University of New Hampshire, told CNN. Reporters Stephanie Siek and Joe Sterling, reflecting on comments from analysts, noted that the changes "portend a future of a more racially diverse America, with new and growing populations playing more important roles politically and economically in years to come."

Brookings Institution demographer William Frey added, "I think we're going to just see that the younger part of the population will have a different vibe. They are sort of needed to help our youthful image and to add to that vitality. They will bring a dimension and element we sorely need."[68]

So What Do We Mean…Diversity?
Coming Together or Falling Apart

Diversity should be an inclusive word, but defining it is not a simple matter. Generally, we think of it as encompassing racial and ethnic groups. However, New Mexico State University and other institutions have developed broader, more inclusive definitions that include but are not limited to: race, ethnicity, national origin, color, gender identity, sexual orientation, age, disabilities, political and religious affiliation, and socioeconomic status. Some add language and linguistic as well as physical and cognitive abilities and qualities, political beliefs, educational background, geographical location, tenure, organizational roles, marital status, parental status, and life experiences.

Kansas State University adds, "We acknowledge that categories of difference are not always fixed but also fluid, respect individual rights to self-identification, and recognize that no one culture is intrinsically superior to another."[69] In the early 2000s, some communities began taking steps to see minorities based more on social and economic differences than on racial and ethnic characteristics.[70] ASCD, in a 2012 book by Thomas Armstrong, identified neurodiversity, such as learning disabilities, ADD/ADHD, autistic spectrum disorders, intellectual disabilities, and emotional and behavioral disorders.[71]

There's even more to our diversity. Add to all of the above the panoply of political parties that pop up in countries all over the world. Some are liberal, some moderate, some conservative, some radical. When it comes to change, we have early adopters and laggards. Then there is coffee. Decisions range from size of cup to whether you prefer a shot of vanilla or a dash of pumpkin spice.

Common Threats and Common Opportunities. The world gets into monumental spats. Yet, we try to come together in common purpose despite longstanding disagreements and out-and-out conflicts. Sometimes, it takes a common threat or common opportunity to unite us.

• *Common threats* might include a pandemic, a terrorist or other lethal attack, a natural disaster, possible environmental devastation, a financial collapse, or even a lack of support for education.

• *Common opportunities* are all around us. Among them: a common commitment to improving education, promoting the well-being of children, building a park, ensuring social justice, launching and maintaining weather satellites, improving the quality of life, or working as a team across national boundaries to help humanity explore the vastness of space and the mysteries of the atom.

With enough commitment to deal with common opportunities and common threats, we might even find ourselves on common ground. That kind of all-hands-on-deck mentality can act as a social adhesive, bringing us together in common purpose and melting what divides us.

What is Social Cohesion?

In essence, it's the glue that holds us together…as families, as nations, as a world. What we have in common helps us form our sense of community.

Exclusion vs. Inclusion, The Danger of Dominance

When domination by one or more groups replaces an acceptance of diversity as the norm, people generally engage in conflicts that involve: tribal, racial, and ethnic divisions; religious differences; and a variety of other social and economic divides. Without some sense of democratic inclusiveness, people who identify with divergent groups will either feel "in power" or "out of power."

How Did We Get So Diverse?

One answer to that question is immigration. The other is the stork—or birth rates. Those are among forces impacting population diversity in many parts of the world. Educators know that the ebb and flow can make a big difference in who shows up for school next year.

Immigration…The Road to Diversity

We live in an age of massive migration, as people cross political boundaries and oceans to pursue what they hope will be an opportunity to pursue their futures. Refinements in immigration policy require constant adjustments to keep them in line with social and economic realities. Millions of people are moving from one place to another around the planet. Of course, the U.S. is a nation of immigrants. Even Native Americans are widely thought to have emigrated across the Bering Strait from Asia. Let's get a glimpse of the trail:

• In the **1920s**, most immigrants to the United States came from Germany, Italy, the Soviet Union, Poland, Canada, Great Britain, Ireland, Sweden, Austria, and Mexico.[72] They joined other Europeans,

Africans, people from across the Americas, Asians, Oceanians, and Native Americans (Canada's First Nations), who were already here.

• According to a 2010 report from the Center for Immigration Studies (CIS), some of the "top immigrant-sending countries" were: Mexico, 11.71 million; China/Hong Kong/Taiwan, 2.16 million; India, 1.78 million; the Philippines, 1.77 million; Vietnam, 1.24 million; El Salvador, 1.21 million; Cuba, 1.10 million; Korea, 1.10 million; Dominican Republic, 879,187; Guatemala, 830,824; Canada, 798,649; and the United Kingdom, 669,794.[73]

• **Foreign-Born Immigrants, Brief Update.** People born outside the U.S. had lived in the country an average of 19 years, according a 2010 *American Community Survey* conducted by the U.S. Census Bureau. California, for example, was home to 10.15 million of those immigrants, 27.2 percent of the state's population.[74] Together, New York, Texas, Florida, and California accounted for more than half of the U.S. foreign-born population.[75] In 2010, according to the Bureau's survey, the U.S. population "included 39.9 million foreign-born residents."[76] Just to compare, in 2000, the total was approximately 31.1 million.[77]

• **Minority/Minority States.** In 2011, traditional minorities had become more than 50 percent of the population in U.S. states or equivalents, such as: Hawaii (77.1 percent), District of Columbia (64.7 percent), California (60.3 percent), New Mexico (59.8 percent) and Texas (55.2 percent).[78]

• **Immigration and the Workforce.** The U.S. Census Bureau has lowered its longer-term immigration projections since the Great Recession, which created a softer job market. "The Bureau assumes that the U.S. will gain a total of 41.2 million net new international migrants from 2012-2050," according to *Pew Research Social & Demographic Trends*. Although Asian Americans have been the largest group of newly arriving immigrants since at least 2009, Hispanics will have the largest net immigration in all future decades through 2060." By 2050, "net immigration by Black immigrants is projected to overtake net immigration by White immigrants, so Black net immigration will rank third by 2060, after Hispanic and Asian net immigration."[79]

While some immigrants move directly into high-paying, traditionally prestigious careers, many bear great hardship as they reach for the first rung on the ladder of opportunity and begin their climb. However, the Center for Immigration Studies (CIS) reported that U.S. natives comprise the majority of workers in virtually every occupational category.

• **Population Ups and Downs.** *More-developed countries* generally have declining or very slowly increasing populations. Between 2000 and 2050, according the U.S. Census Bureau's International Data Base, countries such as Germany, Japan, Poland, Finland, Romania, and Russia were expected to experience a population drop...as people live longer and have fewer children. Among several exceptions are the U.S., Canada, and Mexico, which are expected to show steady gains, at least through 2100.[80] A prime reason for the U.S. population increase is immigration and the children of immigrants.

Birth Rates...Small Differences Make a Big Impact

Want to see the future? Watch the maternity ward. Across the board, birth rates have been going down in the U.S. and other *more developed countries*. Average rates, however, tend to vary among racial and ethnic groups. In 2010, the overall birth rate had reached 1.30 births per 100 population,[81] according to the U.S. National Center for Health Statistics. For U.S. Hispanics, the rate was 2.4; for Blacks, 2.1; and Asians, 1.8.[82] *Again, the replacement rate is 2.1.*

The shift toward a minority/minority population has been in the works for decades. As mentioned earlier, the 2043 tipping point will be a direct result of immigration and birth rates, coupled with the fact that immigrants are often younger and more likely than the general population to be in their child-bearing years. We've said it before: The face of America is changing.

Improving Student Achievement
For Each of Us and All of Us

Perhaps no mantra has echoed so long as, "We need to improve minority student achievement." We sometimes call it "equal educational opportunity for *all*," or the need for "a level playing field for all." Whatever we call it, the sense of urgency keeps growing. When it comes to education and children, no community or country can afford the cost of neglect. Simply defending what may have worked or helped us avoid facing the issue yesterday doesn't wash today. Our future hangs in the balance.

By 2027, the number of 18-29 year-old non-Hispanic Whites will drop below 50 percent of the population. The whole phenomenon was expected to happen among those *younger than 18* even sooner, by 2018. On average, the lion's share of students in our schools and colleges will be minorities. Many already are. It is a fact that we need to fully develop the talents and abilities of *all* students across *all* diversities.

Increasingly, *minority student achievement* will be seen as an essential part of *majority student achievement.* The face of society is changing, and we're all in the same boat. It's a reality that is hard for many to grasp. However, one thing is certain, we need to be sure that *all* students, despite their "diversity group," have equal opportunity to learn, grow, develop, and achieve at the very highest level possible. The Minority Student Achievement Network (MSAN) has concluded that "causes of achievement gaps are complex, including school, community, home, and societal factors. Eliminating the gap is not only the right thing to do, but it is essential to ensure the future of our democracy."[83] Each of us has a stake in the success of everyone else.

Observer James Harvey, executive director of the National Superintendents Roundtable, declares, "Growth in the number of students from minority backgrounds will put enormous pressure on public schools to improve achievement, close the achievement gap, find new pedagogical strategies to respond to diverse learning needs, and greatly expand the proportion of minority and male teachers in the classroom."

The need to get students ready for a diverse world applies in communities that are large, small, rural, urban, or suburban. No one gets a free pass. That's a reality with monumental implications for every education system, community, and nation.

Keeping Score. Test scores of traditional minority groups have shown improvement over time, but the need for new knowledge, creativity, and problem-solving skills may be increasing faster than the progress. Education systems constantly disaggregate testing information. They analyze problems and develop ways to deal with gaps in achievement, knowing full well that solutions sometimes lie outside their school and beyond their grasp. Many educators are increasingly concerned about an over-reliance on testing, which can lead to self-fulfilling prophecies.

Getting past the scoreboard mentality isn't easy. In fact, it's easier and more convenient for most people to focus on the score at the end of the game (summative assessments/report cards) than to pay attention to more complex information that gives us clues about how to improve education for each of our students (formative assessments). Those results are more likely to help us close achievement gaps. According to *Education Week*, those gaps can show up as "disparities in academic performance between groups of students." They can show up "in grades, standardized test scores, course selection, dropout rates, and college completion rates, among other success indicators."

The good news: Between 2009 and 2011, "students across the board greatly increased the average number of course credits they earned." Black and Hispanic students made "great strides in improving performance in 4th and 8th-grade math and reading." The challenge? They "trailed their White peers by an average of more than 20 test-score points." the National Assessment for Education Progress (NAEP) reported.[84]

Respected longtime teacher and member of Futures Council 21 *Milde Waterfall* believes norm-referenced tests often fight against our dealing with diversity. "How do we define, measure, and monitor teacher/ student success in a pluralistic society? Will anyone want to teach weak students if norming continues?"

The Tug Between Advantage and Disadvantage. By 2012, when the Great Recession seemed to be winding down, 22 percent of all children under 18 in the U.S.—approximately 16 million—were living in poverty.[85] Educators and insightful communities know that schools and colleges are faced with students who hope for a better future but bring with them the benefits and challenges posed by their social, economic, and a massive number of other circumstances. Some have more advantage than others. Despite disadvantage, we need to help as many people as possible achieve their personal goals, get constructively involved in civil society, and contribute to a viable economy. We should also keep in mind that dealing with disadvantage or adversity can help us learn our most valuable life lessons.

Graduation and Dropout Rates. The U.S. high school graduation rate for 2009-10 hit a record high of 78.2 percent of students who graduated on time, at grade 12, after beginning at grade 9. That compares to 73.4 percent in 2005-06. A National Center for Education Statistics (NCES) report also estimated that, approximately 500,000 students dropped out in the four years between the Classes of 2006 and 2010.

Across the board, U.S. Asian/Pacific Islanders had the top *graduation rate* of 93.5 percent; followed by White students, 83.0 percent; Hispanic students, 69.1 percent; Native American/Alaska Native students, 69.1 percent; and Black students, 66.1 percent. *Dropout rates* varied substantially across racial and ethnic groups. About 474,000 people in the U.S. passed the General Equivalency Degree (GED) exam in 2010.[86]

Mark Hugo Lopez, associate director of the Pew Center, told the *Washington Post*, "We've seen a surge in the Hispanic high school completion rate." In a January 22, 2013, article written by Lyndsey Layton, Lopez went on to explain that more of those students were likely to have been born in the U.S. NCES Commissioner Jack Buckley added, "When the economy turns down or there are poor economic conditions, there's a lack of available jobs for high school dropouts, fewer jobs that they can actually be qualified for." He concluded that a weaker economy can actually lower the dropout rate.[87]

The White House set a 2020 goal of increasing the nation's number of college graduates by 50 percent. However, the College Board speculated that only about half of the Class of 2012 was "college ready." Noting that assertion, the *Huffington Post* explained that "without significant changes in the rigor of high school, it will be hard for the nation to achieve (the President's) aspirations."[88]

New Students, New Languages, New Cultures. As with earlier waves of immigration from many parts of the world, students come to school speaking an array of languages, but not necessarily the language of the local marketplace. Schools and colleges have not always been ready to deal with the magnitude of differences students bring with them to the classroom, although many have made heroic strides.

In 2010, the Center for Immigration Studies (CIS) reported there were "10.4 million students from immigrant households in public schools, accounting for one in five public school students. Overall, one in four public school students...speaks a language other than English at home." The impact on overall U.S. population? "New immigration (documented and undocumented), plus births to immigrants, added 22.5 million residents to the country (from 2000-2010), *equal to 80 percent of total U.S. population growth*."[89]

It's not just immigration. In 2012 alone, 764,495 international students were studying at U.S. colleges and universities. The number of U.S. students studying abroad in 2010-11 was estimated at 273,996. *See the International/ Global chapter of this book.*

"Places like Canada, the U.S., and many parts of Europe will need more skilled workers. Many will be immigrants," observes *Avis Glaze*, president of Edu-quest International and a counsel on character education. "Countries will have to develop common ground on values if we hope to live peacefully in the future."

A universal challenge is to prepare all students to feel comfortable with many cultures and ready for a global marketplace. Georgia Congressman John Lewis, addressing an Education Writers Association annual conference in 2000, summed up the situation. He intoned that many of "our foremothers and forefathers came to this great country in different ships, but we're all in the same boat now."[90]

Vive la différence.
Long live the difference.

Connecting Educators and Students with Demographic Realities
Educators Raise Their Voices

Numerous members of Futures Council 21, which provided counsel for this book, raised their voices on issues of diversity and inclusion. Here are a few of their comments:

• "Challenges for social cohesion will exist no matter which group is the majority or minority," observes *Laurie Barron*, superintendent of the Evergreen School District in Montana and 2013 MetLife/NASSP National Middle Level Principal of the Year. The focus, she says, "must be on serving all students, no matter the majority or minority group from which they come." Barron encourages educators to use multiple sources of data for each student, do formative assessments, and provide supplemental instructional programs as needed. She warns that some people will "threaten equitable and quality education for all." Barron recommends: "Recognize, appreciate, and value all students. Respect feelings of those who happen to be in different groups."

• *Marcus Newsome*, superintendent of the nearly 60,000-student Chesterfield County Public Schools in Chesterfield, Virginia, provides this counsel. "As majorities become minorities, school districts will need to work closely with colleges and universities to promote teaching as a profession" for those who have traditionally been considered "minorities." Newsome adds, "As schools experience more racial and cultural diversity, the underlying challenge will still be poverty," and "children, regardless of race or ethnicity, may still struggle to adapt to middle class survival."

• *Rebecca Mieliwocki*, an English teacher at Luther Burbank Middle School in Burbank, California, and 2012 National Teacher of the Year, calls for increased ethnic diversity among educators. "Teacher prep colleges and schools will need to become more diverse. We will need to recruit more teachers of color and ethnic diversity who are bilingual." She recommends "more cultural diversity, including European ethnic sensitivity." Mieliwocki wants schools to "bond over a common vision of academic success and character building opportunities for young people." One of this top teacher's bottom lines: "Schools will need to look like society."

• In South Africa, *Jessica Vinod Kumar*, a teacher at M.L. Sultan (Pmb) Secondary School in Pietermaritzburg, points to "the rise of women in society." She foresees "more females who have an opportunity for education and more women in school management positions" as well as in "other positions of authority and as gate-keepers." An issue, Kumar notes, is "limited (availability of) pedagogical expertise in previously disadvantaged schools."

• *Debra Hill*, associate professor at Argosy University in Chicago and 2012-13 president of ASCD, emphasizes that "bilingualism will be seen as an asset rather than a detriment" to students. Hill adds, "We will not be able to successfully function in a global framework until we recognize that multiple language acquisition is occurring everywhere else across the globe with the exception of the U.S. Schools will need to gear up for successfully teaching other languages."

• "If public education is not transformed to guarantee that each and every child will learn what they need to know to lead decent and productive lives, the new majority will not be sufficiently educated to be productive citizens of the country," observes *Joseph Cirasuolo*, executive director of the Connecticut Association of Public School Superintendents (CAPSS).

An Array of Diversity Issues
Broad Concerns...Big Consequences

Kevin Ryan, adjunct professor at Norwich University in Montpelier, Vermont, addressing a conference on ethno-national conflicts in Europe, declared that there are at least two frames for diversity. One is to see diversity as a deviation from the norm that is wrong or not true to form. Another is to understand diversity as natural—and that people can live and work together in common purpose, even though they may have a variety of identities.[91]

Some diversity issues are historic. They have lingered, unresolved over decades and even centuries. They are often the elephant in the room—the issue that isn't addressed or the question that isn't asked. The following are just a few of those issues. All have implications for education and the whole of society.

• **Exploitation of the Vulnerable.** Perhaps no issue is so offensive to the dream of a diverse society as human trafficking. At its very center is the treatment of human beings as property, in essence, exploitation of the vulnerable. It is a form of slavery that has left dark chapters in the history of the U.S. and many other countries. Poor children and adults are forced to fight in clandestine armies, mine *blood diamonds*, or spend their lives in prostitution or other forms of servitude. After promises of a better life, many find themselves captive. Some hierarchies seem to give tacit approval for the powerful to exploit the vulnerable. "Slavery," said Andrew Forrest, founder of the anti-slavery group *Walk Free,* "is the dark side of globalization" and may call for a "new abolitionism."[92]

• **Equal, Separate but Equal.** Whether it's based on law or tradition, justifying the superiority of one group over another simply doesn't work. Apartheid in South Africa legalized discrimination. So did *Plessy v. Ferguson* in the U.S., under the misleading banner, "Separate but Equal." Disregard for women's rights has also proven uncivil and unsustainable. All of the above are anathema to victims. They have deprived communities and countries of their self-respect, and robbed societies of the ingenuity of people who could help create a brighter future for an economy or civil society. Add to this concern: human rights; voting rights; equal pay for equal work; authoritarianism; the undue influence of money on politics; gerrymandering; and corruption.

• **The Exclusive vs. Inclusive Identity.** Battles for dominance, heightened social tensions, economic upset, and loss of life, ring throughout history and continue into the 21st century. Ongoing friction over beliefs between Shia and Sunni; ethnic and racial battles between Tutsi and Hutu, selective memory of events triggering by ongoing mistrust in the Balkans, or the marginalization of Roma in central and eastern Europe and other parts of the world are reminders of the need to pursue a more inclusive future.

Unless we create an *inclusive identity,* we might just end up, even by choice, with *an identity that excludes* "certain others." Often, in the absence of personal life experience, stereotypes take over, when we should be asking, "How can we teach membership and belonging?"

• **Resegregation.** "The U.S. Department of Education has noted that schools are now more segregated than they were when the Civil Rights Act passed," observes Futures Council 21 member Sheldon Berman. He observes, "It's a sad statement both about housing patterns and about community support for desegregation programs." This trend "does not bode well for either education or the economy of the future," Berman says.

• **Preying on the Faithful.** Living with our deepest differences can often be wrapped up in our religious or secular beliefs. Allison LaFave of Legal Outreach in New York observes greater secularization, identification of various religious groups with certain sets of political views, and culture wars that sometimes seem to flare between religious and non-religious Americans. Can we tolerate those who might have other views? Since our beliefs are often quite personal, can we try to mutually understand rather than overtly condemn each other? What can we or should we do when someone or some group perpetrates a horrific act and does it in the name of our (or their) religion? Around the world, wars are regularly being fought in the name of particular tribes or beliefs, when those who are victims would often prefer peace. A cleric, speaking at a Muslim/Christian Summit at the Washington National Cathedral during 2010 observed, "We are particularly good at confessing other people's sins."

Whatever happened to the melting pot?
Reframing Our Identity

When Irish, German, Italian, and other European immigrants came to the United States during a wave of immigration at the turn of the 20th century, the idea of the melting pot had reached mythic proportions. They were captivated by "the promise that all immigrants can be transformed into Americans, a new alloy forged in the crucible of democracy, freedom, and civic responsibility," according to William Booth of *The Washington Post*.[93]

Invitations to the melting pot turned out to be limited. "Separate but equal," a myth enshrined in law, excluded certain groups, such as African Americans. Even schools and universities were segregated. Then, in May of 1954, the U.S. Supreme Court handed down its landmark decision in *Brown v. Board of Education*. Journalist Juan Williams, writing in the *American School Board Journal*, called it the ruling that changed America. The nation aimed to integrate its schools. Busing for racial balance became common and continues in many communities.

Ten years later, the Civil Rights Act became the law of the land. Affirmative action appeared. The struggle for equal rights continues.[94] The Elementary and Secondary Education Act (ESEA), enacted in 1965, and versions of it that followed, such as No Child Left Behind (NCLB), have sought to overcome achievement gaps. The same is true for Race to the Top (RTTT), announced in 2009, and Common Core

State Standards, also initiated in 2009. While the intent has been memorialized, educators have called for reasonable and adequate resources and other support to pursue the dream.

At the turn of the 21st century, observers began questioning whether growing numbers of immigrants were still willing to give up their national and cultural identities. Is "one nation, indivisible" still a logical outcome? If not, is it possible to maintain a polyglot nation that is "fractured into many separate, disconnected communities" or Booth asks, will we "evolve into something in between, a pluralistic society that will hold on to some core ideas about citizenship and capitalism, but with little meaningful interaction among groups?"[95] Will we devolve into a collection of enclaves?

Unless we celebrate and honor differences, we will likely become—or reinforce an already established—cleft community or cleft nation. A shared identity comes only from celebrating and honoring our differences and by constantly looking for what we have in common. Each community and nation, in a highly mobile world, is challenged to constantly reframe its identity or face division, dissension, and lost potential.

Whether the United States or any other nation hopes to be a melting pot or just a magnificent salad or stew, people young and old need to understand the importance of equal opportunity if they hope to maintain some sense of sustainable progress and domestic tranquility. One of the greatest challenges facing the United States and other *democratic* societies is the responsibility to prove to the world that people from all over the globe can live together in harmony.[96]

Enclaves or Inclusive Communities

The 2010 U.S. Census told us that 96 percent of the population of Webb County, Texas, was Hispanic. Blacks made up 43 percent of the population of Wayne County, Detroit, Michigan. Asians were 43 percent of the population of Honolulu County, Hawaii, and 33 percent of the population of San Francisco County, California.[97] The 2010 Census also pointed out that the numbers of people identifying themselves as being of two or more races is expected to increase from 7.5 million to 26.7 million between 2012 and 2060.[98]

On the one hand, we expect melding, melting, and mixing to continue. On the other, we have people living in enclaves, in some cases, borderlands, or comfort zones, on the way to achieving what for them is full membership in mainstream life. We have words of caution from the Civil War treatise, "A nation divided against itself cannot stand."

Former President Jimmy Carter eloquently observed, "We are, of course, a nation of differences. Those differences don't make us weak. They're the source of our strength."

Cosmopolitans and Multi-Country Nations
The Diversity Market

Concentrations of wealth often bring lower birth rates and lower death rates. That means a good share of developed countries are aging, their labor forces dwindling. People of many races, cultures, and nations see these countries as "opportunity magnets."

We are seeing "the emergence of another kind of tribalism, one forged by globally dispersed ethnic groups," says Joel Kotkin, an internationally recognized authority on global, economic, political, and social trends and author of *The Next Hundred Million*. "These global tribes are today's quintessential cosmopolitans, in sharp contrast to narrow provincials," he concludes. Therefore, the United States is becoming the "Greater U.S." India is becoming "Greater India." Mexico is becoming "Greater Mexico."[99]

Of course, the children of these cosmopolitans take their seats each and every day in schools around the world. That raises questions and fires debates in family conversations and international councils, but it's nevertheless an exciting and enriching phenomenon that's a fact of life.

Marketing: What's Your Demographic? We've already been pegged by big data that floats around in the cloud. Marketers differentiate more than ever. They analyze massive amounts of data to sort out how they might appeal to certain individuals or groups—to get them interested in buying a product or service or maybe supporting or opposing an issue. Like it or not, increasing amounts of personal data are available to the highest bidder or the most adept analyst, occasionally a hacker. In an age of increasing diversity, we can expect to be sorted by everything from our shoe size to our zip code.

News Stories . . . Today and Tomorrow

Issues raised by society's ability to refresh itself with first-generation immigrants will likely create lead stories for decades to come. The following are topics of both recent and possible future articles and news reports that probe some aspect of diversity:

Ongoing:

- White population to lose majority status in 2043. *Boston Globe*, December 13, 2012.[100]
- Most children younger than age 1 are minorities, *U.S. Census Bureau Newsroom*, May 17, 2012.[101]
- Push is on for common ways to identify English Language Learners (ELLs), *Education Week*, February 26, 2013.[102]
- Whites are a minority in 1 in 10 counties in the U.S., according to U.S. Census Bureau, *Philadelphia Inquirer*, August 9, 2007.[103]
- Boys Often Show Signs of Academic Disengagement, *Huffington Post*, April 30, 2013.[104]
- Arizona ruling limits ethnic studies. *ABC News* and *Univision*, March 13, 2013. [105]
- Senate blocks bill for young illegal immigrants, *New York Times*, December 18, 2010.
- Denver forum on immigration reform draws packed house, *Denver Post,* February 3, 2013.[106]
- Despite outpacing men in educational attainment, women's pay still lagging, *Diverse Issues in Higher Education*, October 12, 2012.[107]
- Diversity stories during 2010: anti-LGBT bullying and prejudice, anti-immigrant sentiments, increased percentage of children in poverty, references to and treatment of children with disabilities, and discrimination against Muslims. *Teaching Tolerance*, Southern Poverty Law Center, January 6, 2011.[108]

Future:

- More elections determined by voting strength of growing minorities.
- Great American success stories celebrate leading members of society whose parents or grandparents paved the way for them by spreading acres of concrete, laying miles of bricks, hanging a sea

of drywall, clearing bins of asbestos, mowing millions of acres of lawns, harvesting mega-tons of fruits and vegetables, and digging ditches whose combined depth will never be measured or celebrated.

- Inclusiveness replaces exclusiveness on the road to a more sustainable future.

Symbolism of an Election. Whatever our political persuasions, most of us will admit that the U.S. made a bold statement to itself and the world in 2008. Barack Obama was elected President. He was born in Hawaii. His mother was from Kansas. His father from Kenya. He grew up and went to school in Hawaii, Kansas, and Indonesia, then graduated from Harvard with a law degree. And, he became our first African American President. That election symbolized what we have long told ourselves about acceptance of *diversity as a strength* and sent a message that echoed around the world.

Implications for Society
Diversity

We are moving at deliberate speed from a majority/minority to a minority/minority society. That reality has substantial implications for our individual and collective futures.

- In a fast-changing world, communities, nations, and most institutions will need to adjust their identities to become more inclusive.
- Increasingly, communities will be judged based on their demonstrated acceptance of diversity, which will be seen as both a quality of life and an economic issue.
- Those who identify themselves among traditional minorities will have increasing influence in elections and the formulation of laws and policies. Candidates and political parties will need to declare and demonstrate their inclusiveness.
- Political consensus might become more difficult if candidates and office-holders target their appeals for support only to cer-

tain demographic groups and reflect limited commitment to the common good.

- Communities, states/provinces/cantons/oblasts, and nations will focus increased attention on a balance between "what divides us and what unites us."
- Immigration will continue to be a hot-button issue as society ages and people are needed to enhance the workforce in areas ranging from construction to medicine.
- Many from other nations who have built their lives in a country will expect an opportunity to enter a pathway that legitimately leads to citizenship.
- Public-policy issues and concerns such as equal opportunity, equal access, segregation/integration, fair housing, job discrimination, affirmative action, student loans and scholarships, concentrated poverty, and equal treatment and justice under the law, will continue to command significant attention.
- Thoughtful leaders for the future will realize the imperative of emphasizing the concept of inclusion and diversity, versus exclusion and division.
- Social media will play a key role, for better or worse, in bringing people together in common understanding and purpose—or in amplifying the segmentation of society.
- Media and marketers will appeal to language and cultural groups in an ongoing quest to sell their products and services.
- The arts, such as music, dance, drama, writing, musical theater, and the visual arts, will continue to reflect the emergence of an increasingly diverse society.
- Multiple language skills will be expected as they are in many nations.
- Ethnic organizations will continue to bring people with similar backgrounds together to demonstrate and celebrate their heritage for the whole of society.
- Diversity within various demographic groups will grow as people become identified more by their views on issues than by their demographic uniqueness. Those who declare that they "repre-

sent" a certain group may meet resistance from those who have divergent opinions.

- Traditional minorities will increasingly affect the well-being of the traditional majority, as non-Hispanic Whites (as well as others) get older and become more dependent on the productivity of minority workers to support everything from pension plans to geriatric services.

- An increasingly global society will likely lead to greater numbers of expats, cosmopolitans, or multi-country nationals, people who are citizens of one country but who work and live in another. Most will try to be good members of civil society but may not aspire to change their actual citizenships.

- Foreign policy, international economic growth and development, human rights, and other issues will require a substantial understanding of the people, histories, and cultures of the world. Issues such as the environment, energy, global health, and opportunities or threats will bring a diversity of people together in common purpose.

- In many cases, mutual understanding may be the best we can hope for, rather than conversion to another point of view.

Implications for Education
Diversity
- **Getting ready for a minority/minority society.** By 2043, if U.S. Census projections hold, non-Hispanic Whites in the U.S. will drop below 50 percent of the population. We will all be minorities. While the projections might change slightly, that trend is expected to shape the face of society as the U.S. looks to the future. In a variety of ways, other nations may be facing a similar phenomenon. If *exclusion* or *exclusive* has ever been part of the equation, it needs to be adjusted in the direction of *inclusion* and *inclusive*. Students, educators, parents, and communities need to understand and celebrate our combinations of diversities and the fact that we are all in the same boat.

- **Ensuring equal opportunity and improving achievement for** *all* **students. Raising all boats.** Our education systems exist in a complex and exciting time of mass migration and accelerating demographic change. That means increased diversity. Perhaps no other issue raised by this trend is more critical than making sure each and every student has an equal opportunity to receive a sound education. Futures Council 21 member Debra Hill foresees "enormous pressure on public schools to close the achievement gap, find new pedagogical strategies to respond to diverse learning needs, and greatly expand the proportion of minority and male teachers in the classroom." She calls for a commitment to pre-school and full-day kindergarten programs and reminds us that "half of all children in the U.S. by 2050 will be members of today's minorities."

• **Providing inclusive education that reflects a world of diversity.** Consider this comment: "Diversity is important and needs to be addressed. On the other hand, we're a small community with little or no diversity, so we don't have to deal with it." What's wrong with that statement? From the largest metropolitan area to the smallest rural community, all of our students (all of *us*) will have to survive and thrive in a world of multiple cultures, made even smaller by instant communication and rapid transportation. Socially, economically, politically, technically, and personally, there is nowhere to hide. If we cut ourselves off or sell our students short, the world will simply pass us by.

• **Creating postsecondary education opportunities for students across diversities.** *Matthew Moen*, dean of the College of Arts & Sciences at the University of South Dakota, expresses concern about the "lack of affordability for students caused by a decline in state support." That situation "could translate into an unhealthy social division, wherein the children of the more privileged remain able to receive a higher-quality residential education, while children of the less privileged will gravitate to mostly online," and in some cases, "substandard institutions." Moen calls on higher education "to be on the cutting edge of conversations about diversity and inclusive excellence."

• **Renewing our identities, seeking common denominators, and working toward social cohesion.** A 20th century identity may not be sustainable in a 21st century world. That means helping our countries, communities, schools, colleges, staff, and students understand and embrace an identity that reflects evolving demographics and the changing face of the nation and world. It might also mean that we need to consider constantly renewing our curriculum and instruction to make sure that they are inclusive. In short, we need to work together to find common denominators. How would people describe us today? How would we like to have them describe us as a 21st century learning institution committed to getting students ready for a diverse world?

• **Recognizing the high cost of exclusion.** Communities, countries, and the world have too often overtly or just by default marginalized people who were different from themselves. We have all seen the devastating impact when people are excluded because of their tribal, religious, socioeconomic, racial or ethnic group, or their caste. In each case, we lose the ingenuity of people whose voices and ideas aren't heard. Examples of economic loss to society because of de jure (by law), de facto (matter of fact), or self-imposed exclusion are widespread. The overall, society-wide, long-term, unspoken, most expensive item in most budgets could very well be the high cost of neglect.

• **Aligning expectations and testing with the needs of a diverse society.** While most accept the need for standards, concerns will continue to grow about whether those standards are aligned with the needs of a diversity of students. At the same time, we need to be sure we have ongoing, constantly-adjusted alignment with the pressing social and economic demands of society. We will need all hands on deck, across all diversities, to deal with massive issues, such as energy, the environment, our food and water supply, and concerns about security, war, and peace. In light of these pressing and massive needs, we had better get comfortable with questions such as: Do we have the right standards? Do we have the right tests? Do we need more testing that helps improve education rather than focusing so exclusively on an increasing rash of national and international scoreboards (summative assessments)? Are the tests a fair measure of what an individual student, whatever his or her diversity, has to offer society? In considering the direction of education, do we align ourselves with exponentially increasing challenges facing the world outside the education trenches?

• **Aligning laws and policies with new realities.** Law- and policy-makers, from national parliaments and congresses to state and provincial legislatures and school boards, will need to make sure, if they haven't already, that their policies are consistent with new demographic realities and challenges. In their own context, multinational organizations need to address similar issues.

• **Maintaining high expectations for *all* and the resources to pursue them.** Educators and communities need to be talent scouts, searching for talents, interests, abilities, and aspirations among *all* children and adults. Then, they need to focus on helping them grow and develop. While everyone (including parents) needs to be involved, schools and colleges play a central role. Each student deserves the respect that comes with high expectations. Their education institutions also deserve the support they need to produce the best-educated people possible, across all diversities. Supporting education should be seen as a community responsibility—an investment in the future—not as a burden to be placed on graduates who often end up with massive college loans. What's even more tragic? Potential students who never have a chance to pursue their multiple talents because of limited resources.

• **Developing an international focus, cultural sensitivity, language, and other communication skills.** The student who is prepared for the future will need a grounding not just in the pop culture of his or her own country but in the histories, cultures, people, and languages of the world. All students, wherever they live, need a global perspective. What happens thousands of miles away has repercussions for our lives, our security, our economy, and our happiness. Cultural sensitivity, tolerance, and an appreciation for those who may have differences will become more important than ever, if we hope to achieve a peaceful world. Immigrant students, in varying degrees, are often fluent in more than one language, through sheer necessity and will if not through language courses in their native or adopted countries. That is a distinct accomplishment. The ability to speak or interact in two or more languages is like a window on the world. In the U.S., English Language Learning (ELL) programs and international languages should be considered essential. In our book, *Sixteen Trends*, education leader Rosa Smith recommended that all students eventually be required "to be fluent in at least three languages."

• **Investing in all children.** We simply cannot afford to lose the talents, abilities, interests, and energies of any child anywhere in the world because of a lack of opportunity for education. Let's start by seeing education as a worthwhile investment, not as a mere expense.

• **Additional Implications, Diversity Trend.** Schools and colleges will need to attract diverse, talented members of the team as both excellent educators and role models. Preparation and professional development programs at all levels should address diversity and inclusion. Culturally sensitive communication, including language, expressions, and listening, should be treated as essential in building understanding within the school or college and throughout the community. Inclusive leadership should be a heartbeat of the system. Futures Council 21 member Frank Kwan notes that our expanding diversity, "presents schools with immediate challenges for planning and implementing a multicultural society with demographic shifts…and delays among leadership in recognizing the need." Deliberate efforts to get diverse people involved in thinking about the future should become part of normal operation. Social media, Futures Councils, crowdsourcing, and community conversations should contribute to building a spirit of—*we're all in this together.*

Diversity
Questions and Activities
1. Explore *Mapping the U.S. Census,* http://projects.nytimes.com/census/2010/map[109] Click on a variety of counties in various parts of the U.S. Identify five things you found remarkable about the demographics of those counties.
2. Study Table T2.1 in this chapter devoted to "Projected U.S. Population by Race and Ethnicity, 2012 and 2060." Despite the demographic makeup of your own community, what implications do these trends have for education in general, K-12 and postsecondary? For business and industry? For government? For individuals?
3. In your own words, describe social cohesion. Why is it important and how can we enhance it?
4. What is likely to happen when one group dominates all others in a diverse society? Share three brief examples of this situation currently happening in the nation or world.
5. Read President Abraham Lincoln's Emancipation Proclamation, presented January 1, 1863. What power does he mention that the U.S. Consti-

tution gives him to issue this proclamation? http://americanhistory.si.edu/documentsgallery/exhibitions/document_transcripts/transcript_emancipation_proclamation.pdf.[110] If possible, view the movie, *Lincoln*, produced and directed by Steven Spielberg, released November 2012.

6. Identify and briefly discuss factors that affect the diversity of a population.

7. Consider your own community. How would you describe its "identity" compared to 30 years ago? How is that local situation similar to or different from diversity trends noted in this chapter?

8. Suggest six ways you would improve intercultural understanding within a school or college.

9. In a few paragraphs, what criteria would you use to determine whether a school or college is inclusive or exclusive? Can a school or college be both?

Readings and Programs

1. U.S. Census Bureau. (various years). *National population projections, summary files by age, sex, race, and Hispanic origin*. Available at http://www.census.gov/population.

2. Explore, click through, and read, this PBS site devoted to *Human Diversity*, http://www.pbs.org/race/004_HumanDiversity/004_00-home.htm.[111]

3. Also explore, *Precious Children, Diversity in the Classroom*, Teaching Activities and Related Readings, PBS, includes teaching suggestions, materials, and videos, [112]http://www.pbs.org/kcts/preciouschildren/diversity/index.html[113].

4. Kotkin, J. (2010). *The Next Hundred Million: America in 2050*. New York, Penguin Books.[114]

5. "The Foreign-Born Population in the United States: 2010," American Community Survey Reports, U.S. Census Bureau, issued May 2012, http://www.census.gov/prod/2012pubs/acs-19.pdf.[115]

6. Population data from a current world almanac.

Chapter 3

Aging
We're living longer. Ready for 100?

Chapter 3: Aging. *We're living longer. Ready for 100?*

Trend: In developed nations, the old will generally outnumber the young.
In developing nations, the young will generally outnumber the old.
Younger → Older   Older → Younger

> **"You don't stop laughing because you grow older. You grow older because you stop laughing."**
> **"Old age isn't so bad when you consider the alternative."**[116]
> *Maurice Chevalier, French actor, singer, and showman*

Aging: We Can Live with It!

If we're lucky enough, it happens to all of us. We get older. We add to our life experience and, if we have reasonably good health, get bonus time to enjoy our families, nature, and life itself. Some even tell us that, along the way, we become wise.

If demographic projections hold true, we will have multiplying numbers of people whose wisdom will help us shape our future. Life is a continuing story, and all of us stand on the shoulders of those who have come before us. We need to see life in an intergenerational context.

Globally, some parts of the world, on average, are getting considerably older. When that happens, populations tend to drop. At the same time, other parts of the planet are getting younger. Their populations are skyrocketing. That contrast will fuel seismic shifts—socially, economically, and politically.

Make way for a rising tide of issues firmly rooted in aging. On the home front, the solvency of pension funds, the cost and availability of health services, and a host of other concerns have taken center stage as Baby Boomers reach retirement age at the rate of 10,000 a day until about 2030.

|| **The handwriting is on the wall.**

In 2000, when the U.S. population was 282 million—
27 percent were 18 or younger—21 percent were 55 or older.

Moving ahead to 2030, when the U.S. population will be about 358 million
and the Baby Boom Generation will be between 66 and 84 years of age—
22 percent will be 18 or younger—and 31 percent will be 55 or older.[117]

Six Key Factors that Impact Aging
How Can the World Be Getting Older and Younger at the Same Time?

What's the answer to this question of the century? In a word—*demographics.* Putting it simply, we have a shifting balance among six key factors: life expectancy, median ages birth rates, fertility rates, death rates, and immigration. Those shifts have massive implications for all of us and our future. Think of it this way:

- On the one hand, when *life expectancies* and *median ages* go up and *birth rates* and *death rates* go down, a society becomes inevitably older.
- On the other, when *life expectancies* and *median ages* go down and *birth rates* and *death rates* go up, a society becomes inevitably younger.

A major caveat? *Immigration.*

Life Expectancies

In *1789*, when the United States was taking its first steps as an independent nation and George Washington assumed the presidency, life expectancy, at least for white males, was about 35. Among the lucky ones, Washington made it through rampant infant mortality and sparse medical treatment at the time and lived to be 67.[118] By 1930, life expectancy moved up to 59.7.[119] In 2010, the average U.S. life expectancy had climbed to 78.2 years, ranking approximately 51st among nations of the world.[120] Monaco had one of the highest life expectancies on the planet, 89.68 years. In contrast, average life expectancy in Chad was 48.69 years.[121] The U.S. is expected to have a life expectancy of 83 by 2030. Consider this. A female child born in 2010 had a life expectancy more than 19 years longer than if she had been born 100 years earlier, when her longevity would have been 51.8 years.[122]

Median Ages

In *1850*, the *average age* of people in the United States was 19.[123] Remarkably, the average doubled to about 36.5 years by 2000. The *median age*, midpoint between older and younger, increased to 37.2 in 2010. By 2050, it's expected to reach 38.8 and possibly 40.3 in 2100.[124] *Worldwide* in *2011*, the *median age* was 29.5, headed for a projected 38.8 in *2050*.[125] Admittedly, no one knows for sure what breakthroughs in extending life, epidemics, cures for diseases, or natural or human-caused disasters might occur.

Birth Rates

While life expectancies and median ages continue their relentless climb, U.S. birth rates have taken a steady nosedive. In *1910*, the United States registered 3.01 *live births per 100 population*. In *1960*, there were 2.37. By *1990*, the rate had fallen to 1.67[126] and in *2010* dropped even further to 1.30.[127] *Basic replacement rate is 2.1.*

Among nations with *higher birth rates* during *2010* were Niger at 5.0 per 100 population and Uganda at 4.73. Those with *lower birth rates* included Singapore .77 and Japan and Germany, each in the .83 range.[128] After the world population surpassed 7 billion in *2011*, it kept growing to 7.06 billion by the middle of *2012*. Our planetary population is projected to reach 9.28 billion by *2050* before beginning a gradual decline.

To get a glimpse of where primary population growth is taking place and to grasp the challenges that lie ahead, consider this from the Population Reference Bureau (PRB):

- *"Developing countries* accounted for 97 percent of (growth in 2012) because of the dual effects of high birth rates and young populations.

- Conversely, in *developed countries,* the annual number of births barely exceeds deaths, because of the low birth rates and much older populations. By 2025, it is likely that deaths will exceed births in the developed countries, the first time this will have happened in history."[129]

Fertility Rates

While *birth rates* reveal the number of *live births per 100 population, fertility rates* reflect the *average number of live births per 100 women of normal childbearing age, 15 to 44.* Historically, the numbers of children we conceive tends to drop as education levels rise or as economies improve.

Fertility rates in the U.S. have continued on a downward trajectory. In *1910,* the U.S. *fertility rate* was 12.68 per 100 women of childbearing age. That rate dropped like a rock to 8.92 during the Great Depression of the *1930*s. It tumbled again to 6.41 in 2010 and 6.32 in 2011 in the midst of the Great Recession. However, from *1946 to 1964,* when the Baby Boomers were being born, fertility rates were sky high—10.6 in *1950* and 11.8 in *1960,* a stark contrast to times of depression or recession.[130]

Globally, fertility rates in *more developed regions* were substantially lower than those in *less developed regions.* For example, a 2012 estimate placed the average number of children born *per woman of childbearing years* at only .78 in Singapore, 1.39 in Japan, and 1.41 in Germany. By contrast, in less developed countries such as Niger, the average number of births per woman was 7.16 children, in Nigeria 5.38, and in Yemen 4.45.[131]

Death Rates

Of course, death rates play a key role in the shift toward an aging population. In 2010, the United States reported 7.995 deaths per 1,000 population,[132] compared to 14.7 in 1910.[133] Death rates are projected to remain stable at about 8. In 2025, expect a slight uptick to about 9. [134] Because of improvements in medical care and attention to personal health, these rates dropped substantially during the 20th century.

Figure 3.1
Data Bank
A Digest of Demographics

LIFE EXPECTANCIES
Average lifespans in years:
United States[138]

1789	35.0[139]
1930	59.7[140]
2010	78.2[141]
2020 (Proj.)	79.5
2030 (Proj.)	83.0[142]

Internationally
Figures are for 2010-11.
(U.S. ranked 51st at 78.7)

Monaco	89.68
Macau (China)	84.43
Japan	83.91
Singapore	83.75
Canada	81.48
India	67.14
South Africa	49.41
Chad	48.69[143]

MEDIAN AGES
Based on median age of
people living at the time:
United States[144]

1800	16.00
2000	35.30
2010	37.2[145]
2050 (Projection)	38.8
2100 (Projection)	40.3[146]

Worldwide[147]

1950	23.9
2011	29.5
2050 (Projection)	37.9
2100 (Projection)	41.9

BIRTH RATES
Live births per 100 popu-
lation:
United States

1910	3.01[148]
1960	2.37
1990	1.67[149]
2000	1.44[150]
2010	1.30[151]

Internationally[152]
2010 Data for Comparison:

Niger	5.00
Uganda	4.73
Japan	0.83
Germany	0.83

Note: Replacement Rate is
2.1 births per 100 population.

FERTILITY RATES
Live births per 100 women
of normal childbearing age,
15-44.
United States

1910	12.68
1930	08.92[153]
1950	10.60
1960	11.80
1990	07.09
2000	06.59[154]
2010	06.41[155]

DEATH RATES
Deaths per 1,000 popula-
tion.
United States

1910	14.70[156]
2010	07.99[157]

Internationally[158]
2010 Data for Comparison

UAE (Low)	1.00
Kuwait (Low)	3.00
Russia (High)	14.00
South Africa (High)	15.00

**BIRTHS COMPARED WITH
DEATHS**
United States[159]
2010: 3.999 million births
and 2.465 million deaths,
a natural increase of 1.534
million

Worldwide[160]
2010: 132.39 million births
and 56.16 million deaths,
a natural increase of 76.23
million.

**TOTAL U.S. POPULATION
ACTUAL AND PROJECTED:**

2000	281.4 million
2010	308.7 million[161]
2020	333.9 million
2050	399.8 million
2060	420.3 million[162]
2100	570.9 million[163]

PERCENTAGE 65-PLUS:

2000	35.0 m.,	12.4 percent
2010	40.4 m.,	13.1 percent
2020	55.9 m.,	16.7 percent
2050	83.7 m.,	20.9 percent
2060	92.0 m.,	21.8 percent
2100	131.1 m.,	20.5 percent

NUMBERS 85-PLUS:

2000	4.2 m.,	1.5 percent
2010	5.5 m.,	1.8 percent
2020	6.7 m.,	2.0 percent
2050	17.9 m.,	4.4 percent
2060	18.2 m.,	4.4 percent
2100	37.0 m.,	6.4 percent

NUMBERS 100-PLUS[164]

2000	65,000,	.23 percent
2010	129,000,	.43 percent
2020	106,000,	.03 percent
2050	442,000,	.11 percent
2060	690,000,	.16 percent
2100	5,323,000,	.09 percent

(Observe continuing projections.)

WORLD POPULATION[165]
ACTUAL AND PROJECTED:

1804	1.00 billion
1950	2.55 billion
2000	6.08 billion
2010	6.86 billion
2020	7.62 billion
2050	9.28 billion
2100	9.00 billion
2200	8.50 billion

Births Compared with Deaths

According to the U.S. Census Bureau and Centers for Disease Control and Prevention (CDC), the U.S. had 3.999 million births and 2.465 million deaths during 2010, a *natural increase* of 1.534 million.[135] Worldwide, according to the U.S. Census Bureau's International Data Base, 132.39 million live births occurred on the planet during 2010, compared with 56.16 million deaths, a natural increase of 76.23 million.[136] At the beginning of 2013, the U.S. reported one birth in the country every 8 seconds, one death every 12 seconds, one international migrant (net) every 40 seconds, and a net gain of one person every 15 seconds.[137]

Immigration

A population factor that ameliorates the effect of all others is immigration—and it's having a profound impact on the very face of the planet. It is a known fact that people, if they have the fortitude, will use any means available to pursue opportunity—the chance for education, economic advancement, and the freedom to fulfill their dreams.

While U.S. society is feeling its age, it is, on average, a bit younger than some developed countries, largely because of a flow of immigration and the children of immigrants. Think about this. The Center for Immigration Studies (CIS) reported that 80.4 percent of the U.S. population increase between 2000 and 2010 could be attributed to immigration. During that decade, 8.98 million children were born to immigrant (foreign-born) families.[166] An increasing challenge is to provide those immigrants with the education they need to achieve economic success and assume full responsibilities as members of civil society.[167]

Dependency Ratios
What portion of the population is of working age?

Dependency ratios are generally defined as a comparison of people who are of *working age (15 to 65)*...with those who are *under 15 and 65+*. *Pew Research Social & Demographic Trends* reported that, in the U.S., during 2005, for every 100 people of working age, there were 59 potential *dependents*. That number will likely rise to 72 dependents per 100 adults of working age in 2050."[168] That ratio can have enormous implications for programs such as Social Security.

Worldwide, the 2000 total dependency ratio for *more developed regions* was 48.3, which meant that for every 100 people of working age there were 48.3 dependents. That ratio was expected to increase to 68.5 by 2050. In *less developed regions*, the total dependency ratio was 60.3 in 2000 but expected to drop to 52.8 by 2050, according to a 2011 report from the Congressional Research Service.[169]

Social Security
Worker-to-Beneficiary Ratio

In 1950, 16 people were working for every person drawing benefits from the U.S. Social Security Administration (SSA) system. From 1975 to 2000, the worker-to-beneficiary ratio was fairly stable at approximately 3.3. After 2000, however, the ratio dropped to 2.9. SSA is expecting to have only 2.1 people working for every beneficiary by 2030. As Boomers move past 65, the ratio is expected to stabilize at about 2.0, a level that would "support only 60 percent of the benefits that a dependency ratio of 3.3 will support under a pay-go system,"[170] according to a 2011 SSA Technical Panel.

A few more facts for perspective: U.S. life expectancy in 1950 was 68.2 years. In 2010, it had risen to 78.2.[171] That's more years in retirement. While people who make products for mature consumers are salivating, those working to keep Social Security, Medicare, and Medicaid solvent are burning the midnight oil.

Figure 3.2
U.S. Population Projections*
Mid-Year, Middle Series Projections, 2000-2100
U.S. Census Bureau

		←			Percentages of Total Population				→	
Year	Total U.S. Population	Under 5 Years	5-17 Years	18-24 Years	20-54 Years	55-64 Years	65+ Years	85+ Years	Total 0-18	Total 55+
2000	281,421,906	6.8	18.7	9.7	50.2	8.6	12.6	1.5	26.9	21.2
2010[173]	308,745,538	6.7	17.4	9.8	48.4	11.9	13.1	1.8	25.3	25.0
2015	321,363,000	6.5	16.6	9.6	46.6	12.7	14.8	1.9	23.1	27.5
2020	333,896,000	6.5	16.9	8.9	45.1	12.8	16.7	2.0	22.8	29.5
2030	358,471,000	6.2	16.3	8.5	44.0	10.8	20.3	2.5	22.4	31.1
2040	380,016,000	6.0	15.6	8.7	44.1	10.7	20.9	3.7	21.7	31.6
2050	399,803,000	6.0	15.4	8.5	43.5	11.6	20.9	4.4	21.4	32.5
2060[174]	420,268,000	5.9	15.3	8.3	43.3	11.2	21.8	4.3	21.2	33.0
2100[175]	570,954,000	6.6	16.4	8.8	43.6	10.3	22.9	6.4	24.0	33.2

*A variety of age ranges has been developed for the benefit of readers. Some overlap. Therefore, adding the items across the chart should not be expected to provide a total of 100 percent.

U.S. Population Growth
Make Room. There may be twice as many of us by 2100.

The population of the United States is expected to grow from 281.4 million in 2000 to approximately 399.8 million in 2050 and a possible 570.95 million in 2100. During that time, the massive 79 million strong Baby Boomers will move into their advanced years. In 2030, Boomers will be between 65 and 84 years of age. Look for those 65+ to leap from 12.6 percent of total population in 2000 to 20.9 percent in 2050. In 1950, 12.7 million people were 65 or older. In 2010, it was 40.4 million.

Among the fastest-growing segments of the population will be those who are more than *100 years of age,* whose ranks are expected to increase from 65,000 in 2000 to 442,000 by 2050. A 2000 census estimate put their numbers at 5.3 million by 2100, a figure that might be adjusted in future projections. Similarly, those *85+* are expected to grow from 4.25 million in 2000 to 17.9 million by 2050.[172]

U.S. Population Pyramids[176]

These illustrations are traditionally used to graphically show five-year cohorts of the population moving up the age ladder. On this vertical axis the bottom cohort is devoted to children from birth to four years of age. The top cohort represents people who are 100+. On the horizontal axis, men are on the left and women are on the right.

Take a look at the pyramids for the United States in Figure 3.3 (Page 86). You'll be able to see the Baby Boom generation as it begins its climb, creating bulges as it moves along. For example, in 2000, Boomers, born between 1946 and 1964, were between 37 and 54 years of age. In 2025, they will be between 61 and 79, and in 2050, between 86 and 104. These pyramids for the U.S. are included to demonstrate their use. Similar illustrations are available for most states and countries.

World Population Growth, Facing the Facts (See figure 3.4)
Young Outnumbering Old, Old Outnumbering Young

Earth's population was about 2.5 billion in 1950. By 2000, it had reached 6.08 billion. International experts are projecting a world population of 7.62 billion by 2020 and 9.0 billion-plus by 2050. That's a

Figure 3.3
Population Pyramid Summary for United State
Source: U.S. Census Bureau, International Data Base

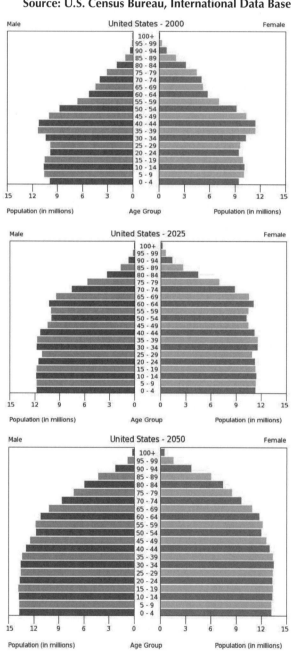

Figure 3.4. Worldwide Population Projections
Older Outnumbering Younger; Younger Outnumbering Older
U.S. Census Bureau, International Data Base, Demographic Overview,
Mid-Year Population Data and Projections by Groups, 2000 and 2050[178]

Country	Population Millions or Billions 2000	Population Millions or Billions 2050	2050 Percent Birth to 19	2050 Percent 55+	2050 Percent 65+	2000/2050 Life Expectancy
Italy	57.78 m.	61.41 m.	18.0	36.8	31.0	80/84
France	61.25 m.	69.48 m.	21.4	37.0	25.8	78/84
U.K	59.14 m.	71.15 m.	21.0	35.9	23.6	78/83
Germany	82.18 m.	71.54 m.	18.3	42.9	30.1	78/83
Rep. of Ireland	3.81 m.	6.33 m.	22.2	34.2	23.3	76/83
Turkey	67.33 m.	100.95 m.	22.8	31.7	19.3	69/80
Russia	147.75 m.	129.90 m.	19.1	29.1	25.7	66/78
Japan	126.77 m.	107.21 m.	14.8	51.6	40.1	81/92
China	1.26 b.	1.30 b.	18.2	42.5	26.8	71/81
India	1.00 b.	1.65 b.	26.6	26.1	14.7	62/77
Indonesia	214.09 m.	300.18 m.	23.4	31.0	19.0	68/80
Australia	19.05 m.	29.01 m.	21.3	34.5	22.5	80/84
United States[179]	281.42 m.	399.80 m.	25.5	32.5	20.5	77/83
Canada	31.10 m.	39.80 m.	20.1	38.7	26.3	80/84
Mexico	99.92 m.	147.90 m.	25.3	30.8	19.0	74/82
Brazil	174.31 m.	232.30 m.	21.6	34.0	21.1	69/80
Saudi Arabia	21.31 m.	40.25 m	25.4	24.0	12.6	72/81
Yemen	17.40 m.	45.78 m.	31.1	17.2	8.0	60/75
Iran	68.63 m.	100.04 m.	21.5	33.7	19.7	66/78
South Africa	45.06 m.	49.40 m.	29.4	21.8	11.4	56/63
Nigeria	124.20 m.	402.42 m.	44.5	10	4.5	45/68
Ethiopia	64.36 m.	228.06 m.	55.2	7.8	5.1	52/73

3.6 times increase from 1950 to 2050.[177]

In the *more developed world*, youth populations tend to be dropping and people are living longer. Where will they get fresh people for the work force? Who will support their pension plans? By contrast, the *developing world* nations are seeing massive increases in their youth populations but have few people older than 55. A persistent shortfall in education, jobs, and other opportunities can lead to anger, unrest, continuous chaos, revolution, civil wars, or even international terrorism. Figure 3.4 is a living drama. We can watch demographic realities unfold before our eyes.

Aging and Education
Growing Number of Students v. Shortage of Educators

Here's the rub. At the same time enrollments in schools and colleges continue to go up, Baby Boom teachers are moving into retirement. Now, consider some of the handwriting on the wall, first for kindergarten through 12th grade (K-12) schools, then for postsecondary education institutions. Will we be able to produce the educators we need and be competitive enough to attract them into schools and colleges?

Enrollment, K-12. In 2010 approximately *54.70 million* students were enrolled in U.S. public and private elementary and secondary schools. By 2020, according to the National Center for Education Statistics (NCES), total enrollment in K-12 schools was expected to reach approximately *57.94 million*, including 52.66 million public and 5.27 million private school students.[180]

Enrollment, Postsecondary. In 2010, approximately 20.58 million students attended postsecondary degree-granting institutions. Those schools were projected to enroll approximately 23.01 million by 2020.[181]

Need for Teachers. The U.S. Bureau of Labor Statistics (BLS) expected an increase of 17 percent or 281,500 in the job outlook for kindergarten and elementary school teachers, not including special education teachers, between 2010 and 2020. Total employment in those positions during 2010 was approximately 1.655 million.[182] Commensurate increases are expected for secondary teachers as demand continues to rise for a workforce requiring high school and college diplomas. The need for postsecondary teachers was expected to rise by 17.4 percent between 2010 and 2020.[183]

Jobs
Need for Jobs and Retraining

Because of tight economic times, many who reach their golden years will want to stay on the job or continue in the workforce well into their "retirements." At the same time, competition for resources between those who are older and those who are younger will shake the political and economic foundations of numerous countries. Traditionally, at a point, people retire, making room for sometimes younger workers.

Many older workers have been laid off. Thousands have taken early retirement. When those seasoned people walk out the door, they take with them more than just a cardboard box. They also walk away with a treasure of talents, skills, connections, and commitment. The cry to "never look back" is anathema even to futurists. History has a way of repeating itself as we cut ourselves off from those who came before us and develop our own self-induced, sometimes terminal case of disconnect called "corporate amnesia." *Global Talent Strategy* reports that many organizations are, in fact, using *retention bonuses* to keep experienced and talented people on board.[184] Being older or younger doesn't automatically mean an unwillingness or willingness to innovate.

Futurist David Pearce Snyder raises a warning flag. "Entry-level labor pools in the developed world will be insufficient to replace retiring Baby Boomers. Looking forward, he sees "a huge opportunity and need for reskilling/upskilling over 40 million adults in the U.S., including, in 2013, 24+ million un- or under-employed workers."

Insightful planners in all institutions and walks of life might be wise to, in some way, engage the talent, wisdom, creativity, and the often pent up bureaucratic frustration of older workers, some of them retired, as they shape their organization's destiny outside the trenches. By the way, people of age represent a growing market. *Also see our chapter devoted to jobs and careers.*

Aging, Health, and Longevity
Are you ready for 100?

After a stress test, the doctor said, "I'll give you another 50 years, but I won't guarantee your knees." One thing we know for sure, people in general are living longer, despite the stress and rat race. For the data driven, consider this: Even the venerable U.S. Census Bureau tells us to expect an increase in the number of people living to be 100 or more—from 65,000 in 2000 to 5.323 million by 2100. Are we ready for that?

Ray Kurzweil, a noted inventor and futurist, has been studying life extension science for decades. He joined Terry Grossman in writing the book, *Fantastic Voyage: Live Long Enough to Live Forever.* In an interview for *Enlightenment Magazine*, Kurzweil concedes that "living long enough to live forever…may require a bit more than simply eating your vegetables and not smoking (although that's definitely a start)." This noted thinker, who looks for age extension to emerge from biotechnology and nanotechnology in the longer run, also suggests attention to diet, consideration of supplements, regular checkups, exercise, and low-stress living. He and a host of others agree on the need to deal with obesity and the diseases it can trigger. [185]

While we think about the prospect of advanced aging personally or get together with a group to discuss it, here are a few things we should keep in mind.

The U.K.'s *Daily Mail* tells us that, as of December 2010, "one in six will live to be 100." Reporter Becky Barrow adds, "More than 10 million of the current population will receive a royal telegram (congratulating them on reaching 100), with many surviving until 110." [186]

In October 2009, *ABC News* quoted an article in the medical journal, *The Lancet,* showing that, "based on current trajectories, more than half of all babies born in industrialized nations since the year 2000 can expect to live into triple digits." Joseph Brownstein of *ABC News'* Medical Unit, added "many western nations will have most people living past 100, with half of all babies born in 2007 in the U.S. likely to live to age 104." The study's lead author, Dr. Kaare Christensen, an epidemiologist with the Danish Aging Research Center at the University of Southern Denmark, added, "I guess it's good news for individuals and a challenge for societies." To further stoke our thinking, Christensen observed, "If you're going to retire when you are 60 or 65, it looks quite different when your life expectancy is 75 or 80 than when it's 100."[187]

Bottom line—Some of us may be around for a while, so we'd better take care of ourselves. Quality of life, for many of us, comes right in there with longevity. Whether you want to guess who among your friends will reach 100 or beyond is up to you.

> **"How dull it is to pause, to make an end, To rust unburnished, not to shine in use! As though to breathe were life!"**
> *Alfred Lord Tennyson, Ulysses*

Implications for Society
Aging

The impact of aging on society will be massive. As the U.S. entered the second decade of the 21st century, nearly 40 million people were 65 or older, about 13 percent of the population. That number was expected to reach between 80 and 90 million by 2050, around 20 percent of the population.[188]

It's a cinch that political, economic, social, and technological concerns will multiply as the nation honors, cares for, continues to educate, technologically connects, and taps the talents of unprecedented numbers of older people. Barring some catastrophe, the aging of the developed world will continue as Baby Boomers spike the number of people reaching retirement age. Consider the following possible implications of aging:

• Pension programs will likely be strained to their limits as the ratio of retirees closes in on the number of actual workers contributing to those funds. Pressure will grow to extend the viability of programs such as Social Security, Medicare, and Medicaid.

• In light of shady business dealings, scams, and questionable oversight that led to losses of retirement savings, demand will grow for assurances of the integrity, safety, and accessibility of those funds. Retirees will be challenged to choose from the most trusted and effective financial service companies.

- Many low-income seniors will struggle to make ends meet and will need high quality public assistance, which will test the values and ethical strength of many communities and nations. Those in *low-income categories* have often seen their home values drop, jobs disappear at the apex of their earning years, retirement and savings accounts largely unproductive, and nest eggs dwindle—as energy, food prices, and the cost of health care and pharmaceuticals, with ups and downs, have steadily risen.

- The need and expectations for medical care will intensify, including both general practice and numerous specialties, to treat a variety of age-related conditions. Research will continue in development of life extension sciences, which is already increasing life expectancies. The use of robotics in providing health care and other services will both offer help and raise

- With massive retirements, several nations will be effectively losing a base of experience and a level of talent that may be difficult to replace. Those retirements will likely lead to opportunities for younger workers who are prepared to be internationally competitive. Some organizations will increase automation. Others will offshore, even though a tendency has been developing toward inshoring.

- Demand will accelerate for products and services directly related to the needs and wants of older citizens. The market for food services, travel, transportation, physical and emotional health, leisure, nostalgia, technology, assistive devices, and education may grow substantially, showing the market clout of seniors. Entrepreneurs will see a "demographic group" with substantial buying power.

- Communities will need to plan ahead for quality geriatric medical and day care, assisted living, nursing homes, and other services for an aging population.

- An aging population may, in many cases, not particularly like to hear the word, "old." Individually, many will constantly seek a fountain of youth and opportunities to improve the quality of their lengthening lives through pharmaceuticals, diet, exercise, and cosmetic surgery. Some, with dated tattoos, will visit tattoo removal parlors.

- As Boomers move even closer toward and into retirement, their political clout will reverberate. Seniors often have the time and the inclination to vote and the life experience to make their case exceedingly convincing to decision makers.

• Older citizens will often want to stay in the work or volunteer force beyond traditional retirement age to continue their contribution to the cause. Many will work full time, part-time, or hourly to produce added income to strengthen their financial viability and retirement funds. More organizations will adjust schedules and working conditions to attract and keep some older citizens on the job.

• Former members of the "regular" work force will give birth to new consultancies that offer their customers a wealth of wisdom and experience not currently on staff.

• Older citizens who have more time they can call their own will look for personalized, lifelong, continuing education opportunities. Whether it is to renew employment skills or pursue intellectual curiosities, we're likely to see a growth in demand for both classroom and online learning opportunities. MOOCs (Massively Open Online Courses) may gain in popularity. School systems, community colleges, and four-year colleges and universities will also be expected to offer appropriate community services and counsel on issues connected to aging.

• Travel and leisure industries will expand. With growing populations of people worldwide who may have the time and resources to see their country and the world, air travel, ocean liners, hotels, timeshares, and entertainment technologies and venues are expected to do well. Family vacations, sometimes involving grandparents and grandchildren, will likely blossom.

• In many developed countries, especially those dealing with massive aging, immigrants will become an even greater portion of the work force, across professional, industrial, and service sectors. Older citizens will see the work and the contributions of these immigrants as vital to maintaining benefit programs such as Social Security, Medicare, and Medicaid.

• A massive transference of wealth will take place as Boomers who have been able to maintain their financial assets pass them along to those who are younger and to fortunate charities or other institutions that serve the common good.

Further Insights: Futures Council 21: *Stephen Aguilar-Millan* suggests we watch Japan as its society ages and younger workers become scarce. He sees a possible wild card if robotics tends to make the young "technologically redundant." *Frank Kwan* anticipates a growing "demand for new teachers to replace the current Baby Boomers" who will be retiring. *Matthew Moen* raises a red flag about the squeeze on residential university students as "taxpayer support to underwrite the cost of higher education continues to erode." Moen forecasts increasing education investments in "developing and rising nations" and believes colleges and universities will be pressed for "more continuing education and lifelong learning."

Implications for Education
Aging

Aging is spawning education issues that demand our attention. In every community, educators and leaders in other industries should be identifying the implications of this immediate trend as they accelerate the process of creating a sustainable future. Rather than thinking just K-12 education, we might be thinking pre-natal through life. Here are just a few additional implications of the aging trend for education. Consider others:

• **Maintaining an education workforce.** As enrollments increase, growing numbers of Boomer educators will retire. Concerns have flourished about the need to make education careers more attractive to younger generations, retain more of the talent that is already there, ensure the quality and depth of preparation for those entering teaching, and improve overall working conditions. Many seasoned retirees will continue to fill needs and gaps. Unless education systems become increasingly competitive for the people they need, they will simply lose them to other employment opportunities.

• **Expanding career education, adult education, community and four-year college and university programs, and other opportunities for lifelong learning.** School systems, colleges, and universities, have been working together in fits and starts to make the education system somewhat more seamless. Some call it the P-16 movement— prekindergarten through the fourth year of college. Those efforts will likely accelerate with increasing demands from an aging population.

The K-12 system may not cover enough ground to meet the expanding needs of older citizens who are looking for new later-in-life careers, trying to satisfy their intellectual curiosities, or upgrading their technology/computer skills. Many schools already offer their own community/adult/continuous education programs. A variety of online courses bring learning to students, including older citizens, anytime and anywhere.

Educators know that demands of the workforce are changing. Shortages develop with every new technology. Futurist David Pearce Snyder suggests development of skills such as, "systemic thinking, problem analysis team work, cybernautics, self-directed learning, and numeracy," to name a few. Snyder observes, "For the foreseeable future, there will be a huge deficit of advanced cognitive skills in the workforce of both the developed and developing worlds that will constrain economic growth." That could mean openings for seasoned workers.

• **Promoting cross-generational communication.** Society is faced with providing services across all or parts of five generations. As life expectancies increase, six might become even more common. At the same time, schools and colleges will be considering how to prepare students for intergenerational understanding—an essential for maintaining a civil society. They need to understand that massive challenges facing the planet can only be solved intergenerationally. In fact, the ability to communicate across generations and cultures is essential to success throughout life. All students need the experience, whether they visit senior centers, interview their elders, or teach members of other generations their technology skills. This commitment should be reflected in policy and included in discussions of budgetary, security, and communication issues.

• **Communicating and collaborating with people who don't have kids in school.** One question that crops up frequently within the education system at all levels is, "How can we inform, engage, and serve people who don't have kids in our schools?" While they may not have regular daily contact with the school, older citizens vote in elections and help determine the policies and resources we will ultimately have to support education. Most educators know that engaged people generally feel a sense of ownership and are among our greatest gifts, while those who are disengaged or uninformed can become our worst nightmares.

That's why targeted communication is essential for empty nesters, who may no longer have children in school. Add to that list those who may never have had children in the first place. Sharing information using newsletters, web sites, targeted email, blogs, social media, radio, newspapers, cable television, podcasting, phone calls, meetings with volunteers, concerts, science fairs, and other special events will be essential.

Of course, listening is basic to effective communication. It's a way to keep an ear to the ground, stay in touch with issues before they become crises, and both identify and close communication or generation gaps. Thoughtful communicators will ask what members of various constituencies would like to know, what they understand, what they don't, and how they prefer to get their information.[189]

• **Maintaining the solvency of pension funds.** The growing phalanx of people poised for either early or on-time retirement has been through booms and busts. Some have taken more economic hits than others. For a variety of reasons, they've seen increased stress on their benefits and retirement investments, if they have them. Couple that stress with the fact that people are living longer. Concerns extend to what some see as threats to Social Security, Medicare, and Medicaid. Education institutions can expect this issue to move to the critical list during the early decades of the 21st century.

• **Balancing political demands of the young and the old.** "Will a population that is both aging and growing press for expensive health and security programs of their own, or will they coalesce with the young and with families to provide adequate resources for education?" asks Arnold Fege, president of Public Advocacy for Kids in Annandale, Virginia. That vexing question will continue to command attention. Our hope would be that each nation will try to deal with both. Keith Marty, now superintendent of the Parkway Schools in Missouri, observed, "School leaders and communities that collaborate and come together for children will be able to garner resources and public support" and better balance the political demands of the young and the old.[190]

> **"I have a wish to die young . . . but as late in life as possible."**
> *Anonymous Greek Sentiment*

Bill Moyers on Retirement

Bill Moyers retired at 70 and started work again a year later. Then, he retired on the eve of his 76th birthday. A year later, he was back at work. His "work" is a demanding television series, which takes time, creativity, and energy. "Retirement, I've heard, can be the enemy of longevity," Moyers remarked, in an *In the Know* opinion piece, "On Not Growing Old" for *aarp.org/bulletin* during April 2012. In it, he reminisced about Walter Cronkite's retiring at 65 and telling him it was a mistake. Cronkite's advice? "Keep going!" "Journalism keeps my curiosity on its toes," Moyers observes. He adds, "We're fascinated by what science and experience are discovering about how to maintain high mental and physical ability as we grow older; how to reduce the risk of disease and disability; why attitude matters; and the importance of wonder, surprise, and joy." His wife Judith, his creative partner, admits that they "have no retirement skills."[191]

Aging
Questions and Activities

1. What are six key factors leading some parts of the world to become, on average, older, and other parts of the world to become, on average, younger?
2. Explain dependency ratios. Gather needed information and compute the dependency ratio for your community, state/province/oblast/canton, or country.
3. Study the figure devoted to "Worldwide Population Projections." Of those listed, which three countries had the lowest life expectancies in 2000? Identify four items in this chart that you believe students should discuss.
4. View "Let My Dataset Change Your Mindset," Hans Rosling," *TED Talk* video, (19:40), addressing an audience at the U.S. State Department, June 2009. Rosling is a Swedish public health professional. http://www.ted.com/talks/hans_rosling_at_state.html.[192] What are four things you learned from his explanation of demographic change in the world? Any disagreements?
5. Check out the Worldometer at http://www.worldometers.info/world-population/.[193] Share it with others to illustrate how and how quickly the population is changing.
6. Develop a demographic profile of your community, including age, gender, race and ethnicity, social and economic, and any other factors you consider

important.

7. Develop a list of 10 ways schools or colleges in your community could better reach out to and serve older people.

8. If you were developing a legitimate new product or service that would be helpful to older people, what would it be? In what way would it be helpful?

Readings and Programs

1. Review the U.S. Census Bureau's International Data Base. Follow directions and click on the selection of information you'd like to review, year or years you want to cover, and the country you want to explore. http://www.census.gov/population/international/data/idb/informationGateway.php.

2. Read Christensen, Clayton M.; Allworth, James; and Dillon, Karen. (2012). *How Will You Measure Your Life?* Harper-Collins, NY.[194]

3. Review population data from a recent world almanac.

4. Consider reading "Chasing Immortality, An Interview with Ray Kurzweil," Hamilton, Craig, *Enlightenment Magazine*, http://www.enlightennext.org/magazine/j30/kurzweil.asp?pf=1. You also might want to read the book, Kurzweil, Ray and Grossman Terry. (2010). *Transcend: Nine Steps to Living Well Forever* (Paperback), Rodale, Inc., NY[195]

Chapter 4

Flash Gordon to Flash Drives

Chapter 4: Technology. *Flash Gordon to Flash Drives*

Trend: Ubiquitous, interactive technology will shape how we live, how we learn, how we see ourselves, and how we relate to the world.
Macro → Micro → Nano → Subatomic   Atoms → Bits
Megabytes → Gigabytes → Terabytes → Petabytes →
Exabytes → Zettabytes (ZB)

‖ **"Computer wins on Jeopardy!"**[196]
‖ *New York Times, February 16, 2011*

"Say it ain't so, Joe!"[197] Ken Jennings had won 74 games in a row on the television game show, *Jeopardy*. He returned to the fray only to be knocked off his perch by *Watson*, an IBM computer that was designed to be "a question answering machine." Many of us recoiled. Some worried, "What does this say about us mere humans? Are we playing second fiddle to an automated robot?" Others found consolation, "The computer was designed by a human team, and the thing filled a room." The robotic Watson? It simply walked off with $77,147.[198]

The Revolution is On...
And We're All Part of It
Try these headlines:
- Tailored on-demand content revolutionizes "24/7 ubiquitous education."
- Cloud computing enables easy access to a world of information.
- Movement accelerates toward a "fully recorded life."
- Fully 95 percent of 12 to 17 year olds are online.
- Big organizations are losing power because of small interactive technologies.
- Graphene Discovered: Six times lighter, two times harder, and ten times stronger than steel. [199]

Ever wonder how it feels to live during a revolution? Well, look around and take a deep breath. It's happening again. It's a technological revolution that is pushing us forward at a speed that tests our adaptability, ingenuity, patience, and even our budgets as we try to keep up with the Joneses and stay ahead of the curve.

Changes have become exponential. Convergence, miniaturization, speed, memory, power, connectivity, and price are the name of the game. Emerging technologies take us beyond what most of us would even have imagined only a few decades ago. We carry a world of information and worldwide connections in our pocket or purse. The startling part is that what we see is only the beginning of what's to come. The good news? We can use that technology to help us shape the future rather than just go along for the ride.

Of course, "technology" is not just a device with a keyboard. There are biomedical, environmental, energy, telecommunication, business, transportation, space, robotic, political, and educational technologies, just to name a few. Google, Yahoo, Bing, and other search engines help us find what we want to know. High Definition (HD) television not only fills screens in our homes but has also become interactive. Wikipedia lays out an array of information. Even though further checking might be needed, it's taken a huge bite out of *Encyclopedia Britannica* as we once knew it.

Social networking sites such as Facebook, Twitter, Pinterest, Instagram, Skype, Tumblr, Snapchat, Reddit, and just plain email have put us in instant contact with each other across political, social, economic, and other boundaries. Each of us can originate our own network, share ideas in the blogosphere, get involved in a chat, meet-up, create an Avatar, or conduct and exchange research. We can do *crowdsourcing* to soak up a wealth of ideas. Conversely, we can be instigators of *mass mobbing*—to fight for a cause, sing as a spontaneous choir at a shopping center, or bring down a regime. Online courses or "blended learning" can enhance education. Virtual reality takes us there without actually being there. Telecommuting, telework, and tele-education are a reality. While it remains a topic for heated and esoteric discussion, no one truly knows the limits of technology.

We've seen a virtual explosion in the availability of applications (apps) for personal communication and networking devices—smartphones and iPads. In fact, those hand-held, wireless, even wearable communication devices are, even now, an extension of our bodies and minds. These technologies are the source of what seems like an unlimited array of information and connections. There seems to be an app for nearly everything firmly packed into our Android, Blackberry, iPhone with iOS, or other device. In essence, miraculous gadgets create a ubiquitous worldwide town square.

Futures Council 21 member Ryan Hunter, a social observer and American University's student, sees social media at the center of "what I would identify as an organic, youth-centered online civil society comprising hundreds of thousands of cause-specific action groups set up primarily through these online 'new media' forms."

Futurist David Pearce Snyder foresees information technology (IT) maturing into "cashless commerce and paperless workplaces, including schools." He anticipates the elimination of "millions of clerical and transactional service jobs while increasing the actual information content of every type of employment." Snyder concludes: "Confronted with *both* a growing shortage of entry-level recruits *and* a growing surplus of technologically displaced workers, the U.S. economy will be faced with the existential necessity of up-skilling or reskilling between 40 and 50 million adults so that they are able to make purposeful use of information on the job."

A veteran superintendent of the Howard–Suamico Public Schools in Green Bay, Wisconsin, Damian LaCroix remarks, "We can't educate today's students for tomorrow's challenges with yesterday's schools." He sees "the proliferation of technology flipping the traditional 20th century model of education and transforming it into a new digital high-tech, high-touch landscape." LaCroix forsees greater focus on "21st century skills, knowledge, and literacies" and "digitally rich online media content that captures the imaginations of students and teachers alike." He adds, "The infusion of gaming mechanics into educational software will alter the learning environment and engage more students in taking personal ownership for their education."

Mega, Giga, Tera, Peta, Exa, Zetta, Yotta, Kibi

With dizzying speed, we have raced from megabytes to gigabytes, tera-bytes, and petabytes. What comes next for computer power? By the time you read this, we might be zipping into exabytes and zettabytes. For the record, a *terabyte* is 1,024 gigabytes. A *petabyte* is equal to 1,000 terabytes. An *exabyte* covers a billion gigabytes. A *zettabyte* is equal to one-sextillion bytes. We won't even mention *yottabytes* and *kibibytes*.

A little perspective—the International Data Corporation estimates that the total amount of *global data* was expected to grow to 2.7 zettabytes during 2012. Expect that number to increase exponentially.[200] When it comes to computer speed and capacity, we are, to put it mildly, soaring beyond the known universe.

Flash Gordon to Flash Drives…
In a Flash

Blame the comic books and strips, moviemakers, television produc-ers, and science fiction writers, if you'd like. Flash Gordon, Buck Rog-ers, and Star Trek piqued our imaginations. They left an impression that nearly anything is possible. Some technophiles were even primed by Dick Tracy, the caricature detective who wore a *wrist radio* for in-stant two-way communication. Tracy's idea has morphed into wireless and wearable devices that can connect us with people, information, and ideas almost anywhere and anytime.

In the mid-1940s, ENIAC, the Electronic Numerical Integrator and Computer, emerged as the first operational working computer. It weighed 30 tons and sported 19,000 vacuum tubes, strategically placed across 42 nine-foot tall cabinets.[201] In 1964, IBM System/360 hit the market. Part of its claim to fame was "tiny chip transistors," only a half-inch square—a fraction the size of previous circuits.[202] When the U.S. made its first journey to and from a successful landing on the Moon in 1969, its Apollo Guidance Computer (AGC) operated on 64 Kilobytes (KB) of memory. According to *Computer Weekly*, that was "more basic than the electronics in modern toasters that have computer controlled stop/start/defrost buttons."[203] Steve Jobs' first Apple Macintosh 128K computer was announced in a commercial that aired during Super Bowl XVIII in January 1984.[204] A little more than 30 years later, we were carrying flash drives on our keychains with up to 64 Gigabytes (GB) *or more* of memory. How many Buck Rogers movies could you store on that?

> **"It is difficult to say what is impossible, for the dream of yesterday is the hope of today and the reality of tomorrow."**
> *Robert Goddard*

Big Data and the Cloud

"Big Data and the Cloud" may sound like the name of a 1960s rock group, but, in the technology world, they are a big deal. *Big data*, for example, is basically a term used to describe the massive amounts of information, "growing at 50 percent a year, or more than doubling every two years." That's an estimate from IDC, the International Data Corporation. The *New York Times*, in an article titled, "The Age of Big Data," quotes Rick Smolan, creator of the *Day in the Life* photography series, who calls the phenomenon "humanity's dashboard." Not to exaggerate, but sensors are everywhere, keeping track of nearly everything from the products we buy to the images and video we put on the web, known as "unstructured data." Big data analysts are needed to help companies use that information to develop marketing plans, design cars, stock store shelves, and help politicians shape their messages.[205] Writing for *Foreign Affairs*, Kenneth Cukier and Viktor Mayer-Schoenberger observed, "In the third century BC, the Library of Alexandria was believed to house the sum of human knowledge. Today, there is enough information in the world to give every person alive 320 times as much of it as historians think was stored in Alexandria's entire collection."[206]

Cloud computing is a web-based service offered by companies that have picked it up as a business model. Using clusters of computers, they take care of everything from email and word processing to data analytics. Those companies that offer cloud services must store your data in two different places, just in case of a gigantic glitch. Many of us end up connecting to the cloud, especially if we use certain web-based email services. A central server using "middleware" acts as a sort of traffic cop for both "front end" and "back end" transactions that connect to each other through the Internet, according to Jonathan Strickland on his web site, *How Stuff Works*. The front end is what the user sees. The back end is a complex system we call "the cloud." Advantages range from accessing applications and data from nearly anywhere on the Internet to cutting back on internal hardware, servers, storage, and possibly IT support. Leading concerns include security and privacy.[207] The cloud, by the way, is often the home of big data.

> **"Technology will become the great equalizer across the globe, giving access to education, knowledge, and wealth, to even more people."**
> *Sheldon Berman, Superintendent of Schools, Eugene, Oregon*

It's a Small World After All...
Macro, Micro, Nano, Subatomic

Disney introduced the "It's a Small World After All" ride at the 1964 World's Fair in New York. A couple of years later, in 1966, the ride had its debut at Disneyland.[208] Even with his ingenuity, creativity, and rich imagination, Walt Disney, like the rest of us, very likely had only a vague idea about how tiny and submicroscopic our sense of small would become. We're now working on the *nanoscale*. That is, indeed, small, or short for that matter—a *nanometer* is one-billionth of a meter; a *nanosecond* is one-billionth of a second. Let's take a quick tour from macro to subatomic.

Macro and Micro. Early computers that faintly resemble those at work today, were huge. ENIAC was an example. So were early mainframe and super computers. Transistors and then silicon chips/microprocessors helped us go micro, to converge and miniaturize while increasing computer speed and capacity. Icons Steve Jobs and Bill Gates each played a central role in developing and popularizing technologies that coaxed us toward the micro world.

Nano. What technology will drive the economy of the future? It may be nanotechnology, which gives us the ability to manipulate atoms and molecules, promising stronger, lighter-weight materials than the world has ever known, faster computers, and unprecedented medical breakthroughs. Michael Roco, who has provided leadership for an Interagency Working Group on Nanotechnology at the National Science Foundation (NSF), called nanotechnology "the next revolution."[209]

Eric Drexler, who heads the Foresight Institute, started organizing his work around the nano idea as early as the 1970s. During a 2011 presentation at the University of Oxford, Drexler explained "atomically precise manufacturing to build machines starting at the molecular level." Generally, we refer to that concept as *molecular manufacturing*. Of many possibilities Drexler pointed out "desktop computers with one billion processors...and materials 100 times stronger than steel."[210] Nano will likely bring huge benefits for medicine, economic productivity, and the earth's environment, but "hostile forces may use it to produce new, decisively powerful weapons." Drexler warns.[211] Research continues on any potential risks posed by nanoparticles.[212]

Subatomic. We're not there yet, but we're in hot pursuit and hearing even more intense discussions about subatomic particles, such as neutrinos and quarks. Think about this—If we could build things at the subatomic level, chances are we could drastically reduce or even eliminate mechanical friction. That would speed things up even more, including your computer.

The Hadron Collider, located in Europe and operated by a multi-national team, is on the frontline of particle physics. Its 23 km. tunnel encases a particle accelerator that is testing theories and discovering frontiers of science. By the way, CERN, the European agency for nuclear research, is also home of the collider, which has made the news for discovery of the long suspected Higgs-Boson particle and what might have been a slight variation in the generally accepted speed of light.[213]

Reaching Out or In...A Journey of Discovery

Quantum mechanics and particle physics, coupled with nanotechnology, allow us to move atoms around inside the molecule and probe the subatomic world of inner space. At the same time, powerful telescopes and space vehicles are opening our eyes to the realities of our planet, our solar system, and beyond. Every discovery reminds us of how much we have yet to learn.

- **Intergalactic.** The Hubble Space Telescope, launched in 1990, raised the curtain on an expanding universe. "To me, it's as if the earth were growing a new sensory apparatus," Timothy Ferris, author of *Seeing in the Dark*, told the *Washington Post*. Fraser Cain, publisher of *Universe Today*, estimated, based on scientific research, that our solar system, the Milky Way, has perhaps 200 to 400 billion stars. The number of galaxies in the entire universe? Estimates run from 100 to 200 billion, perhaps even more. Each one, like our own galaxy, has an estimated hundreds of billions of stars.[214]

- **Exoplanets.** We're discovering a cascading number of planets outside our solar system. For example, in 2011, *Kepler-20 e* and *Kepler-20 f* are planets orbiting a star called *Kepler-20*. The big news is that they are nearly the same size as earth, and the discoveries are accelerating. Don't pack your bags yet, since they are hotter than blazes and are about 1,000 light years away.[215] If we devastate our own planet, they would probably not be acceptable alternatives.

- **Martian Rovers.** Among the most inspiring of interplanetary probes have been the robotic *Spirit* and *Opportunity*. These durable robotic landers arrived on Mars in January of 2004. They were followed by *Curiosity*, a larger robotic vehicle with an anticipated longer range for exploration. It arrived in the Mars Gale Crater on August 5, 2012.

- **Probes.** Voyagers 1 and 2, launched in 1977, have explored our solar system's giant outer planets, 48 of their moons, and the rings of Saturn. They have now joined Pioneers 10 and 11 in becoming the most distant human-made objects in space. Voyager 1, for example, during early 2012, was cruising 17.9 billion kilometers from the Sun. Each carries greetings to anyone who might be out there from the people of planet earth.[216]

- **Get Your Tickets.** Space tourism is waiting in the wings. Some dreamers are conceiving of future orbiting space hotels. A number of people, generally wealthy, have bought seats and taken trips to the International Space Station. Expect ticket prices to be pretty steep, and you may not be able to take a lot of luggage.[217]

- **Students Looking Out and Looking In.** Schools and colleges will, we hope, continue to get more powerful telescopes and microscopes that will help students study everything from microbes to spiral nebulae. Expect the number of students interested in bioinformatics, astrophysics, and other related fields to take off like a rocket.[218]

Tackling Transmission

One of the most celebrated technologies of all time was Johannes Gutenberg's movable type printing press. Fighting the darkness of the Middle Ages, he helped democratize knowledge. We have to give some credit to the automobile, airplane, telegraph, telephone, radio, and television. They also played a key role in *transmission*, helping us discover and learn from each other. However, the wedding of computer technology and telecommunication and their child, the Internet, might just top them all.

Of course, earning a patent and the many steps in transmission are essential to the success of inventors and entrepreneurs. Once upon a time, a new technology had eons to determine whether it would be adopted or discarded. Gutenberg's press took about 400 years to reach critical mass, the telephone about 30 years, and Wikis about three to four years. Now, constant arrays of new products and services hit the market every day. Each is followed by instant reviews—suggesting that you "gotta have it" or revealing "a glitch." Of course, those reviews can go universal in the bat of an eye. Even a technological winner can be displaced overnight and made obsolete by something that is more user friendly, cost efficient, or stylish. Think film vs. digital cameras.

The impact of many technologies depends on how broadly accessible they are across society. The "digital divide" which long focused primarily on whether schools, students, and communities had computers, has shifted to an even broader concern—access to high speed Internet.

Universal Service

We are now connected to an information superhighway. However, a clear challenge is *universal service* and a need to ensure that high-speed Internet connections are available to everyone, wherever they are. A September 2011 Pando Networks study ranked countries based on their *Internet download speeds*—the worldwide average at the time was 580KBps. (kilobits per second). South Korea led the pack with 2,202 KBps, followed by Romania, Bulgaria, Lithuania, and Latvia. The U.S. came in 26th with an average download speed of 616 KBps.[219] Of course, download speeds can vary from city to city and neighborhood to neighborhood. By the end of 2013, Hong Kong led the global pack with 62.27

Mbps (megabits per second). The U.S. came in 33rd with 20.5 Mbps, according to Ookla.com's Household Download Index.[220]

In 1996, the U.S. Federal Communications Commission (FCC) set up a Schools and Libraries Universal Services program. A special E-rate was established to get that important project done.[221] Concern has grown that a commitment to high-speed Internet for every community has waned. Progress in the economy and civil society as well as a commitment to a level playing field are all at stake.

Upsides and Downsides

We've learned the hard way that any technology, from gunpowder to computer power, can have both an upside and a downside. In the ancient world, the stone tool that could grind seeds could also do damage to friend and foe. The same airliner that can take us to exotic places can bring down an office tower. The world took a collective gasp when that happened on September 11, 2001.

A smartphone, with its digital camera, can capture memorable family photos. It can also rally people for a noble cause or instruct them in carrying out an act of violence. Technologies that can put us in contact with the world and serve as revolutionary education tools can also become addictive, depriving us of in-person social interaction.

Cyberbullying is a major concern. Not only have young people bullied their fellow students, in some cases leading to suicides, but bullies have also launched online, social media attacks on educators. Steve Eder writes in *The Wall Street Journal* that the State of North Carolina passed a law making it a crime for a student to "build a fake profile or web site with the intent to intimidate or torment a school employee."[222]

Parenting Tweens conducted a study involving 9 to 11-year-olds. The publication found that 11 percent of that age group had been "victimized online or through their cell phones one or two times in the previous year." An additional 10 percent said they had "experienced cyberbullying three or more times" during that period. About 18 percent said it happened through email, 17 percent in a chat room, 13 percent via instant messages, 12 percent through text messages, and 11 percent using a web site.[223]

Of course, *hacking* is on the rise. Concerns about cyber-security are real. AOL warns of online theft of credit card information and personal data, such as Social Security numbers and passwords. *Keyloggers*, for example can record your keystrokes as a way to spot PIN codes and account numbers. *Phishing* involves what can be fraudulent websites that trick you into providing your logins or passwords to personal accounts. *Spyware* can track your Internet habits and also dip into your personal information.[224] A famous hacking event involved the notorious *Stuxnet* malware-delivered computer virus that found its way into the uranium centrifuge system in Iran and led to explosions.[225]

Clandestine release of hacked government information, some classified, by Wikileaks, a government contract employee, an intelligence officer, and others, intensified developing tension between personal privacy vs. personal/national security.

A World of Technology, *A Few More Examples*

Virtual reality, artificial intelligence, holograms and holographics, CT and PET scans, bio bricks and base pairs of DNA, LCD (liquid crystal display), plasma, and High Definition (HD) television, lithium batteries, space-based solar collectors, the cloud, bitcoins (digital currency), smart and driverless cars,[226] drones, hypersonic aircraft,[227] GPS, and edutainment are among the massive number of technology-related terms (and realities) that have become a part of our daily lives. In a world of technology that seems to be developing on steroids, much more is yet to come. Consider the industries, professions, careers, and jobs that flow from each one. In the meantime, here are highlights from a few fields:

• **Medical.** Medical technologies have generally improved the ability of medical doctors to diagnose and treat ailments (or rule them out). Increasingly, they can have ready access to updated vital signs and other medical information, stored in the cloud, to avoid repetition of recent or unneeded tests, monitor trend lines, and avoid mistakes. Robotic and computer-assisted surgery is being used to overcome some limitations of minimally-invasive surgeries. A "digital pill," approved in 2012, includes sensors that will, with consent, wirelessly transmit key information that could be critical to a patient's well-being.[228] Of course, gene therapy is "one of the most rapidly growing fields of medicine.[229] Genomic science is revolutionary.

• **Robotics.** Robotics is expanding exponentially. Robots have been used to automate factories. As just noted, they also assist in medical care and treatment. Among the most dramatic have been Martian rovers. Others, of various sizes, have been used in dangerous search and rescue operations following human caused or natural disasters. The use of drones in conflict zones has expanded to the use of aerial robotics for monitoring traffic, delivering packages, or tracking a suspected criminal. Concerns have been raised about privacy, ethics, and whether robots should be fully automated or always be directed by human judgment.

• **Printing and Imaging.** Printing is experiencing a Gutenberg moment. *Digital printing and imaging* in formats from very large to very small are now at the heart of the industry. Marketers, communicators, and merchants can now target increasingly smaller and more diverse sub-groups with messages and materials customized to reflect geographic, demographic, or other differences.[230] *3D printing,* sometimes called desktop or additive manufacturing, can be pre-programmed to produce three-dimensional items ranging from prototypes to a jawbone, an auto part, a gun, or something you see in an animated movie. Want to make something? Do it on your own 3D printer or email the design and ingredients to a firm that will do it for you. Instead of ink, the printer might use thermoplastics, metal or ceramic powders, alloy metals, photopolymers, paper, foil, liquid resin, almost anything.[231] These desktop manufacturing units are being used in many schools and colleges, where students are discovering their potential.[232]

Technology Natives, Aliens, and Immigrants

We are discovering that the number of *technology natives* is growing with each generation. Yet, we have holdouts, people who even take pride in being *technology aliens*. They have yet to make the transition from atoms to bits that Professor Nicholas Negroponte has so eloquently described. Let's just say that some of us are *technology immigrants*, who are simply trying to catch up. Schools provide a classic example of this tug as they move toward an accelerating array of instructional, administrative, accountability, transportation, energy, and other technologies.

Are we ready for on-demand, constantly-updated curriculum, available 24/7? What are the benefits? What are the challenges and side effects? What are the limitations? Will we see schools and classrooms differently? How are some schools and colleges using these tools to enhance our ongoing need for truly educated people? This chapter further stimulates that discussion.

Emphasis on Intellectual Tasks

In many industries, technology has driven a restructuring that involves automating certain functions, while shifting staff responsibility toward those things that are difficult to computerize—such as expert thinking and complex communication, solutions to new problems, and service jobs. That's the view of Richard Murnane, an economist and professor of education and society at the Harvard Graduate School of Education. He sees parallels for schools. He suggested to *Education Week* that the future of education might shift from traditional routine tasks to include a larger number of higher-paying positions that involve thinking, communicating, and problem solving.[233]

> **"Is chalk and talk the best way of operating in an Internet Age?"**
> *Stephen Aguilar-Millan, The European Futures Observatory*

How Educators are Using Technologies
A Few Examples

New technologies are enlivening schools and colleges. Many are putting them to creative use to enhance student learning. Some are using an array of devices, programs, and apps to do more efficiently what they have always done.

Others are coupling these new technologies with professional development and adjustments in mentalities about the physical set-ups of classrooms long reminiscent of the Industrial Age. Software is out there to flag us if a student is falling behind, to help students capture and play back lecture notes, and to build comprehension skills.[234]

Meera Balachandran, a member of Futures Council 21 from India, sees the need "to integrate technology with teachers as facilitators." She suggests that learning will become "more individualized as students do not need to depend only on the teacher for help and can work at their own pace." That integration is important, says award-winning principal, now superintendent, Laurie Barron, since it is likely that, "students will become more reliant on technology and (perhaps) less reliant on teachers."

If we "hope to compete on a global scale, we will need a technologically adept society," declares veteran school communication executive Frank Kwan. He calls for "funding allocations to accommodate technology" and "more research on how to fully integrate the 'new media' into classroom practice." Kwan believes expectations will grow "for teachers and administrators to be sufficiently skilled to fully exploit what is available through technology."

Here are just a few examples of how a handful of schools and educators are rethinking what they do and how they are connecting, engaging, and motivating students, using a world of learning opportunities.

• **Immersive Technology and MOOCs.** From his post at the Harvard Graduate School of Education, Chris Dede has impacted thinking about education and technology for generations. He draws a contrast between the classroom as "a barren place with flat writing surfaces" to one that is reinforced by one-on-one computing. At a Future of Learning Institute, he described an "immersive learning environment," reinforced with an array of technologies.

ECO MUVE, funded by the U.S. Department of Education, plays out at Black's Nook Pond in Cambridge, Massachusetts. Students learn in the classroom but also go to the pond to study wildlife and track carbon, phosphorous, and oxygen levels. When they see a dead fish, they investigate what might have happened to it. At hot spots around the lake, students can connect to the Internet for vast amounts of information and match images they capture with explanations of what they are seeing.

Dede notes that the experience helps students "develop an understanding of complex causal relationships." The students do science "the way science is done," aided by their smartphones and iPads. It's a part of "jigsaw pedagogy" and provides a high level of stimulation for all students involved. Dede concludes that the classroom should also be seen as an ecosystem where we have a variety of ways for people to learn.[235] He points, for example, to MOOCs (Massively Open Online Courses), with readings, videos, and other possibilities for learning, as a step forward.

• **Student Motivation.** William Ferriter, a National Board Certified Teacher, noted in his blog that just using technology to do "paper-driven work that tends to define traditional classrooms" can limit possibilities for learning. "The novelty of new tools wears off quicker than digital cheerleaders like to admit," he emphasizes. Ferriter says students want to interact in powerful conversations with their peers, motivated by issues such as fairness and justice. "Technology's role in today's classroom…is to give students opportunities to efficiently and effectively participate in motivating activities built around individuals and ideas that matter to them."[236]

• **An Array of Possibilities.** *Brookfield High School in Connecticut* gave iPads to 240 freshmen. Each tablet computer was loaded with "interactive e-textbooks, artwork, and videos to help explain concepts," according to the *Danbury News Times*. Danbury Deputy Superintendent William Glass remarked, "The students are digital natives. They were born after (this) technology existed."[237] The *West Virginia Department of Education* offers a virtual school, which allows students to take online courses that may not be offered in their home schools. For example, the state has vowed to "allow students to use interactive games, virtual field trips, and video lessons" to enhance core subjects.[238] *Huntsville, Alabama's* school system launched a pilot program devoted to putting wireless access on 20 of the district's school buses. A password is posted at the front of the bus."[239]

The Fairfax County Public Schools in Virginia has an FCPS 24-7 Learning (Blackboard) allowing students and parents to stay up to date on schedules, assignments, courses, and other information.[242]

Mary Reiman, director of library media services for the Lincoln Public Schools in Nebraska believes technologies can reinforce thinking, reasoning, and creativity skills and help teachers become facilitators of learning.[240]

Harvard's Justin Reich cautions about the digital divide. He cites research showing "affluent students are more likely to use technology to acquire higher-order thinking skills, to learn to solve problems, and to organize." Reich observes that "students from poorer districts were more likely to use technology to take online courses to recover credits, to work on drills, and to access technology with less adult supervision."[241]

Digital Games Become Learning Tool

The Joan Ganz Cooney Center at Sesame Workshop conducted a survey devoted to "Teacher Attitudes about Digital Games in the Classroom." The study involved 500 K-8 teachers. Here are a few of the results:

- 70 percent of teachers agree that using digital games increases motivation and engagement with content/curriculum.
- 62 percent report that games make it easier for them to level lessons and effectively teach the range of learners in the classroom.
- 60 percent said that using digital games helps personalize instruction, better assess knowledge, and collect helpful data.
- 60 percent of teachers observed that games foster more collaboration among students and enable students to sustain focus on specific tasks.[243]

As for the children's television program, *Sesame Street*, research has consistently shown that it improves cognitive outcomes, learning about the world, and social reasoning and attitudes.[244] We all have a lot to learn from Kermit the Frog.

Technology and the Future of Learning
A View from a Learning Technology Vet

"Human cognition about presence will adapt to dynamic new forms of telecommunication," foresees Gary Rowe, a member of Futures Council 21. For some, the virtual may become the new physical. "This will evolve into the virtual office, the virtual classroom, the virtual enterprise." A former CEO of Turner Education Services/CNN and a respected media producer and consultant, Rowe speculates that "technologies will continue to disrupt traditional school models," based on "doing school." He shares a few of the changes we might see:

- "School systems will become networks that blend online learning with occasional attendance at a physical place.
- "Grade progression from kindergarten through high school will become irrelevant as students may learn their way through the network at their own pace."
- "The act of reading will include images, sounds, dynamic indexing, cross-referencing, and personally customized navigation. Reading with multimedia dramatizes information and assigns relevance at a personal level."

- The evolution of 3D visual media may make virtual travel a possibility. Students might visit Bunker Hill, Bell Labs, the Great Barrier Reef, or any number of places and events while sitting at their own desk."

- "Experience will override resistance as people see greater rates of success, higher cultural literacy, with greater efficiencies, delivered at lower cost. A networked school system will need fewer buses and buildings, and the efficiencies will help pay the best teachers robust salaries for their success."

Connected Teens...Connected World

The Pew Research Center broke the news that, as of July 2011, "95 percent of all teens in the U.S. were online, and 80 percent of those online are users of social media."

There's even more: 77 percent of those teens had cellphones (23 percent had smartphones and 54 percent regular cell phones). "The bulk of teens are 12 to 13 when they get their first cell phone," the study concluded.

As for social media, the report indicated that, among the 80 percent who were social media users, 93 percent had an account on Facebook, 24 percent on MySpace, 12 percent on Twitter, 7 percent on Yahoo, 6 percent on YouTube, 2 percent on Skype, myYearbook, and Tumblr, and 1 percent on Google Buzz.[245]

"Around three-quarters of the world's inhabitants now have access to a mobile phone," according to a World Bank study.[246]

Super and Quantum Computers.
Welcome to the World of Petaflops and Qubits.

Moore's Law. Moore's Law is interpreted to mean that computer power will double every 18 months. That was the gist of a paper written by Intel Co-Founder Gordon Moore back in 1965. Actually, the law was "more of a rule of thumb," according to theoretical physicist Michio Kaku. He points out advances we've seen, such as the doubling of transistor power that can be placed on a computer chip every two years. Yet, he remarks, "There is an ultimate limit set by the laws of thermal dynamics and quantum mechanics as to how much computing power you can do with silicon." Kaku anticipates that Moore's Law could "flatten out completely by 2022." He adds, "Computer power simply cannot maintain its rapid exponential rise using standard silicon technology."[247]

Justin Rattner, Chief Technology Officer at Intel Corporation, the world's largest maker of superconductors,[248] says, "Science and technology have progressed to the point where what we build is only constrained by the limits of our imagination." Brian David Johnson, Intel's Director of Future Casting and Experience Research, asks, "How do we change the future?" His answer, "Change the story people tell themselves about the future they will live in."[249]

Why is this important? Fact is that most of us could care less about terabytes and petabytes. We just want reliability, plenty of speed and memory, and some compelling and useful programs or apps. We also know, deep inside, that the stork will not bring them. They'll have to be developed and perfected.

There are many reasons we need to take those next big steps. One is that the challenges we face are increasing exponentially and we need to be ready to address them thoughtfully and quickly. Another is that the students who are in our classrooms today will be the ones who will invent, develop, operate, maintain, and conceive of new generations of technology. They'll also create a fresh array of industries, professions, and jobs—maybe even continue development of super- or quantum computers. We need to ask, "What are the implications for our schools and colleges and for each of us?"

Supercomputer—Meet the Sequoia. It's "grabbing headlines and making techies drool." That's how *Time Magazine*'s "Techland" described the world's fastest supercomputer during mid-2012. It was called "Sequoia" and located at the Lawrence Livermore National Laboratory. You might be interested in a few specs. This is "a third-generation Blue Gene machine from IBM and runs on 1.6 million processor cores, reaching speeds of up to 20 petaflops." While it can perform 10^{15} operations per second, it "requires 3,000 gallons of water per minute to cool it down." On an average day, it uses about 6 or 7 megawatts of energy. Peak usage? About 9.5 megawatts.[250]

What does it take to be called a "supercomputer?" While definitions might vary, one that seems to have fairly widespread application is—anything that is much more powerful than what we commonly use today. For example, 20 years ago, your present PC, tablet, smartphone, or other device may very well have been considered a supercomputer. Time marches on, right along with gigabytes and terabytes.

A Quantum Computer: Back in 1981, acclaimed and feisty Nobel prize-winning physicist Richard Feynman, who was an acclaimed and now is a posthumous member of quantum mechanics royalty, challenged scientists to come up with a quantum computer.[251] We've all heard the song about "reaching the unreachable dream." Feynman's assignment, if we accepted it, was "actually inventing what then seemed like an unreachable algorithm" capable of transforming computing.

As the "wafers" that hold multiple silicon chips are no longer able to produce enough of an exponential leap forward in computing power all by themselves, the quantum computer will likely be the answer. *"Digital computers* use a binary code and set bits to either '0' or '1.'" However, "in a *quantum computer,* bits can be '0' *and* '1' at the same time, allowing incredibly fast calculations," physicist Michio Kaku explains.[252] He notes that, "quantum computers encode data in the quantum states of subatomic particles known as quantum bits—or qubits."[253] Welcome to the subatomic world.

David DiVincenzo, who heads a group at IBM's Thomas J. Watson Research Center and done theoretical work in the field, declares, "Quantum computing will change the world in unforeseen ways. This is like electricity was in the 1830s."[254] One of the scientists who stepped up to the plate was Geordie Rose, creator of D-Wave One, what he calls the world's first commercial quantum computer. Because of his work, he was declared Canadian Innovator of the Year in 2011.[255]

Implications for Society
Technology

Implications of the technology trend are as ubiquitous as the technology itself. Consider a few we've included here, and let that just be the start of your conversation.

• Career and job opportunities in technology-related for-profit and not-for-profit industries will increase dramatically. The ability of a community to attract and keep the talented people it needs for those industries will directly depend on its education system, coupled with economic opportunity and quality of life.

• Universal Service should mean that high-speed, internationally competitive Internet access with second-to-none download speeds will be distributed to every corner of a community and nation, worldwide. Constant encouragement will be needed to make that necessity a reality and to close the digital divide.

• MOOCs (Massively Open Online Courses) will offer opportunities to take a variety of courses that are free and open to all who connect online. While open source curriculum can provide expanded opportunities for learning, issues of cost, credit vs. non-credit, generation of income when courses are offered free, assessment, and other questions are in need of ongoing attention.[256] Major providers of those online courses include Coursera, Udacity, and edX.

• Our success in a Global Knowledge/Information Age, including economic growth and environmental sustainability, will depend on our ability to develop entirely new (or new generations of existing) technologies.

• Encouragement will be needed to make sure policies are in place to support investments in new technologies coupled with ongoing professional development.

• Organizations will try to show a *return on investment* for their use of social media. Some might attribute it to brand health, marketing optimization, revenue generation, operational efficiency, customer experience, and innovation.[257]

• Countries and communities as well as other public and private organizations and individuals will need to be ready for cyber attacks that could drastically disrupt a cascade of what we consider essential services. Cyber theft is a growing concern.

• Countries and communities will become more sophisticated in high-tech law enforcement as technologies intended to help are used to deceive or inflict harm.

- Because technology can bring us together or further divide us, ongoing techniques or rituals will be needed to help us discover what we all have in common.

- A worldwide knowledge base will be readily available to anyone with the technology to access it, straining our ability to separate truth from deception, information from misinformation, and fact from fiction or opinion.

- Leaders will no longer be able to *rule* by hoarding information, since those they lead might have access to the same information . . . or even more. Of necessity, new forms of increasingly inclusive leadership will blossom.

- Computing power, networks, and portability will continue to make physical place less important as cottage industries develop, consultants and freelancers provide growing numbers of services, and employees of large and small firms spend part of their time working from a variety of locations.

- Online shopping and services are already becoming part of how we get things done. Routine functions, such as having a doctor check vital signs, shopping for groceries, selecting made-to-fit clothing, buying books/e-books, or purchasing tickets to almost anything, anywhere can be, and often is, handled online.

- Working together in common purpose, society will have the potential to develop technologies that will help answer the call for new sources of energy, cures for diseases, and remedies for prevention of environmental degradation. They will be able to develop strategies for avoiding and countering acts of violence and terrorism, lengthening and improving the quality of life, and even developing new forms of entertainment. Along with these developments come jobs and careers.

- Civic and government leaders will develop even more technology-driven approaches for listening to their clients or communities and determining their personal needs. In turn, they will provide information that is of particular interest to various individuals or groups. A challenge will be to create common ground in a diverse society.

- Businesses will grow around the need to help us declutter, eliminate, or archive our e-waste so that we can clearly sort and analyze mountains of data.

• Growing numbers of people, of necessity, will insist on the fine balance between economic gain and social responsibility as new technologies are developed that carry both benefits and consequences.

Observation on Technology from a Leading School Superintendent

Stephen Murley is superintendent of the Iowa City Community School District in Iowa and a member of Futures Council 21. A local leader with a universal interest in ideas and issues, Murley shares observations about the impact of technology on education.

- The increasing ubiquitousness of technology—what today is considered *of interest* will tomorrow be *a given*. Technology immigrants *think about* technology. For digital natives, children and young adults, technology *just is*.
- The expectation that everything has a technology component will become universal.
- Blended learning will likely be a standard form of instructional delivery, K-12.
- Schools will struggle with concern over those who have too little technology at home.
- Personalization will lead to modularization of coursework that students can access as they move through their overall course of study.
- Competency rather than seat time will become the norm. If the university system demonstrates an ability to break away from the Carnegie Unit system, it may have an even larger impact on K-12.
- New formative and summative assessments will reduce seat time needed to demonstrate proficiency and leave more time for the acquisition of knowledge and skills prior to graduation.

Implications for Education
Technology

Perhaps no trend, other than the standards movement, is more pervasive for our education system. Not only are new technologies developing at an exponential pace but they are unrelenting as we reshape our education system for a Global Knowledge/Information Age. Here are a few of those implications:

• **Getting, Employing, Maintaining, and Constantly Updating Technologies.** The tools of technology, such as laptops, tablets, smartphones, and other interactive devices, plus the programs and apps that bring them to life, have become indispensible and pervasive tools for education. From instruction and assessment to administration, budgeting, scheduling, and communication to making strategic decisions, all education systems need to move forward. There is no going back. The sophistication, currency, and use of those technologies might depend on resources, vision, expectations, and leadership. Industrial Age infrastructure and mentalities are due for upgrades or replacement. "Blended learning will result in a need for extensive changes in how teachers teach and how a student wishes to learn," says Marcus Newsome, superintendent of the 58,000-student Chesterfield County Public Schools in Virginia. Curriculum, Newsome suggests, "must be flexible enough to adapt to a fast-changing environment."

• **Ensuring Constantly Renewed Learning Materials in a Fast-Changing World.** While textbooks have significant value, some education systems are supplementing them with *techbooks* that offer connections to constantly updated information as well as videos and a variety of apps to enhance learning. Former Baltimore County, Maryland, Superintendent Joseph Hairston, highlights the "virtual transmission of instructional content materials. It will be seamless, represent a new normal, and customize the learning process." Hairston also points to developments in artificial intelligence (AI), which can interact with humans in a natural language.

• **Becoming Facilitators and Orchestrators of Learning.** Driven by their curiosity and enabled by tech tools, many students come to school with more information on some topics than their teachers have yet discovered. Self-learning should be seen as a bonus—and encouraged. As concertmasters, teachers will increasingly turn the process of teaching and learning into a partnership, with students and teachers constantly learning from each other. Jessica Vinod Kumar, a teacher at Sultan (Pmb) Secondary School in Pietermaritzburg, South Africa, notes that the "rapid escalation of technology and learning tools enhances how teachers demonstrate concepts and assess student progress." Playing the role of facilitator and orchestrator, teachers can use technology to enhance active, project-based, real-world learning; learning across disciplines; and learning through inquiry, opening the door to an array of sources and possibilities.

• **Leveling the Playing Field.** As issues, equal access and opportunity now include concern about a growing number of digital divides. The original divide was primarily focused on whether students did or did not have access to computers. Now, we face divides based on how widespread Wi-Fi hot spots are located, download speeds, and whether certain devices or connections are allowed in the classroom. Another complication is the lack of professional training educators need to effectively use these technologies to enhance learning. Leveling the playing field cries for an insistence that Universal Service overcomes discrepancies in download speeds—nation-to-nation, urban to rural, and neighborhood-to-neighborhood. Futures Council 21 member Michael Usdan notes that technology will have "a transcendent impact on life globally," including the education system. For technology to help level the playing field, he suggests the "entire fragmented education governance structure will have to be reassessed if equality of opportunity is to be more than empty rhetoric for growing numbers of youngsters."

• **Doing Science and Developing New Generations of Technologies.** Schools and colleges need to instill in students the knowledge, skill, and determination to invent new generations of technologies. That challenge includes encouraging creativity, imagination, invention, and entrepreneurship to impact both the public and private sectors. All people should be grounded in a basic understanding of science so that they can make reasoned judgments about the benefits and possible consequences of scientific projects.

• **Considering How Education is Delivered.** Schools and colleges have growing alternatives for delivering education. Choices range from presentations and discussions in the classroom to online, blended, and hybrid courses. Flipping has become either a stated or unstated practice in some schools, with students watching something like a lecture online so that in-class time can be used for various learning activities, questions, and reinforcement. The process, according to Helen Lazarro of the Flipped Learning Network, makes the teacher more of a "guide on the side" rather than just "a sage on the stage."[258] These and other approaches also apply to professional development programs. Futurist John Meagher sees a tendency for "the rise of virtual universities where the greatest minds are on tap, going open-access to students in third world nations, and in poor areas of the U.S."

• **Enhancing Personalization, Reinforcement, and Research.** Like a violin in the hands of a master, various technologies can help us personalize education, deliver instruction, monitor what students have learned and where they need help, provide reinforcement, and serve as tools to conduct research. Personalization should include a personal and human touch that goes beyond mass customization. Technology expert Alan November believes parents must be involved. He asks, "What is your plan for making every home a center of learning?"[259] "Technology is driving change," says Superintendent Keith Marty of the Parkway School District in Chesterfield, Missouri. "Our school district is allowing students to bring their own personal technology to use in schools, and our teachers are using technology for instruction like never before."

• **Making Interactive Assignments.** Education systems are adopting a wealth of online curriculum, some based on a purchase or subscription, sometimes open source. In many cases, students can download materials, select an appropriate app, and use their electronic tool kits to study what's specifically been assigned or do further research. They can prepare papers or presentations using video, audio, Skype, and PowerPoint, do assessments of their progress, and seek help. A talented teacher can suggest or assign a variety of apps to help customize, personalize, and tailor education to more closely match the needs, interests, and strengths of each student. And they are not just memorizing. They are interacting with information and ideas, even discovering or creating their own.

• **Strengthening Media Literacy Skills, Preparing Students for Appropriate Use of Social Media.** Since students, all of us for that matter, are virtually bombarded with information and enticements to grab our attention, media literacy skills have become more important than ever. We need to be quite clear about how we are teaching students to separate wheat from chaff, truth from fiction. Let's include research skills, ethical use of media, and etiquette (netiquette), as well as solid approaches to deal with cyberbullying. When it comes to social media, such as Facebook, Twitter, Skype, Instagram, Tumblr, Second Life, and others, students need an even better understanding of norms. The University of Southern California's Henry Jenkins and colleagues have identified New Media Literacies, which include using technology for: "play, performance, simulation, appropriation, multitasking, distributed cognition, collective intelligence, judgment, transmedia navigation, networking, and negotiation."[260]

• **Reinforcing Face-to-Face Communication, Dealing with Technology Addiction.** Varying clusters of technologies can become so captivating, in some cases so addictive, that people may need special classes or activities devoted to interpersonal, face-to-face communication and breaking their cyber-addiction. While online activities can be enlightening and help build relationships, they can also turn into obsessive Internet disorders. In some cases, these disorders are related to constant gaming that may be designed to be *infinite*. According to "Internet Addiction is a Clinical Disorder," an article in *The Telegraph*, the symptoms might include forgetting to eat and sleep, developing resistance to what others might find pleasurable, exhibiting withdrawal symptoms, becoming fatigued, feeling isolated, and getting lower marks in school.[261]

• **Promoting Two-Way, Interactive Communication.** Schools and colleges may still have printed newsletters and other publications, but they also have web sites. They connect on Facebook, Twitter, My Space, Tumbler, Chatterkick, Reddit, and through other social media, such as email. They post video on YouTube; share something you might like on Pinterest; do blogs; and get involved in chat rooms. Security threats and weather emergencies are delivered online...as well as on local television and radio stations. The pros and cons of online board meetings are getting some airtime. Interactive professional development programs and conferences have become commonplace. Mutual expectations will be needed between schools and parents to ensure the communication system is effective in light of new technologies.

• **Ensuring Constantly Updated Policies That Support and Encourage the Use of Technology in Education.** Like technologies, policies have at least two sides— they can encourage or inhibit. In a fast-changing world, policies adopted for an Industrial Age may no longer work for education in a Global Knowledge/Information Age, even an Age of Knowledge Creation and Breakthrough Thinking.

Of course, there are many other implications of technology for education, such as:

- **Using technology to develop homework and other helplines for students and parents.** Included can be assignments, learning tips, surveys, etc.
- **Motivating learners.** Game-based learning, learning analytics, artificial intelligence, simulations, and mobile learning, are becoming commonplace.
- **Making distinctions about writing.** A challenge is to help students understand that shorthand language and abbreviations, which work for tweets, may need some enhancement in the marketplace. Reaction to a message should be a primary concern.
- **Connecting with the world and our own learning community.** Students and educators can be connected to a world of information and interactions with others.
- **Addressing identity, privacy, and security issues.** Students need to understand the benefits and consequences of how they use various technologies.
- **Emphasizing STEM+.** Science, technology, engineering, and math are vitally important, but a fully educated, employable member of civil society needs more.

The Classroom of Tomorrow, One Teacher's View

Rebecca Mieliwocki, a 7th-grade English teacher at Luther Burbank Middle School in Burbank, California, is the 2012 U.S. National Teacher of the Year. As a member of Futures Council 21, she speculated on how technology will impact her workplace—the classroom:

"The classrooms of tomorrow will be equipped with the tools people need to be successful in today's work spaces. All classrooms must have wireless Internet capabilities, computers for all to use, handheld devices for research and transmitting schoolwork. Teachers will need tools to help collect, read, grade, publicize, and store student work as well as great libraries of lessons, materials, and curriculum supports. Streaming video for learning experiences will be necessary on a daily basis, as will telecommunication tools for interstate/national group work.

The basic work of running a school will need to be streamlined through use of technology that frees people to do the important work of teaching and facilitating learning. So much will change, with technology and innovation leading the way. It's imperative that we be flexible, forward thinking leaders as we move ahead to the future."

Technology
Questions and Activities

1. Identify an education, community, or world challenge and conceive of five ways crowdsourcing could be helpful in dealing with it.

2. What promise do you see for nanotechnology in building the economy of your community? What are the implications of nanotechnology for what we teach in our schools and colleges?

3. Ask three science and/or technology teachers to share three to five ideas for how they would like to even better prepare students for life in a fast-changing world.

4. In computer terms, why should everyone know more about the strengths and vulnerabilities of *the cloud*? Prepare a five-minute presentation explaining your views. Consider hosting a webinar.

5. Identify five ways you believe schools and colleges could make better use of existing and developing technologies to enhance learning.

6. What is Moore's Law? Can the silicon chip keep up with it or do we need a new approach to increasing computer power?

7. View "MindShift: Chris Dede on Cyberlearning and Games" on YouTube. Dede is an international leader in technology at Harvard's Graduate School of Education. Identify three things he mentions that you think should be shared with all educators. http://www.youtube.com/watch?v=nNAZXB0D-nT4&feature=related (5:05)[262]

8. View "Steve Jobs: How to live before you die," TED Talk, Commencement Address at Stanford University, June 2005 (15:04). Ask yourself, "How many young people like Steve Jobs do we have in our school, our community, or even in our family?" "How should we respond to that talent?" http://www.ted.com/talks/steve_jobs_how_to_live_before_you_die.html.[263]

Readings and Programs

1. Collins, Allan and Halverson, Richard. (2009). *Rethinking Education in the Age of Technology*, Teachers College Press, Columbia University, NY and London.

2. Reiser, Robert and Dempsey, John. (2011). *Trends and Issues in Instructional Design and Technology (3rd Edition)*, Pearson, N.J.

3. U.S. Department of Education, Office of Technology, *Transforming American Education, Learning Powered by Technology, National Education Technology Plan 2010*, http://www.ed.gov/sites/default/files/NETP-2010-final-report.pdf.[264]

4. Review Materials and View Video, *New Media Literacies*, University of Southern California, http://www.newmedialiteracies.org/. Identify the literacies needed for students to participate in new media.[265] Also see white paper, *Confronting the Challenges of Participatory Culture: Media Education for the 21st Century* (Jenkins et al., 2006, produced with support from the MacArthur Foundation, http://www.newmedialiteracies.org/wp-content/uploads/pdfs/NMLWhitePaper.pdf.[266]

5. Review Eric Drexler's Blog, *Metamodern*. Drexler is CEO of The Foresight Institute. Pay particular attention to what he has to say about nanotechnology. What are the implications for education, our community, and our country? http://metamodern.com/.[267]

6. View this video of a 30-story Chinese hotel being built in 15 days, from *BSB, Broad Sustainable Building*, https://www.youtube.com/embed/GVUsIlwW-WM8?rel=0 (3:04).[268]

7. Recent works by many authors, as technologies develop rapidly.

Chapter 5

Identity and Privacy
Life is an open book

Chapter 5: Identity and Privacy. *Life is an open book.*

**Trend: Identity and privacy issues will lead to
an array of new and often urgent concerns
and a demand that they be resolved.**

Knowing Who You Are ↔ Discovering Who Someone Thinks You Are
What's Private? ↔ What's Not?

> "Oh I heard it through the grapevine.
> Oh and I'm just about to lose my mind."[269]
> *Motown Classic Recorded by Marvin Gaye, Smokey Robinson, and Gladys Knight*

Reputation Analysis…Who am I?

Part 1: If you ask people you know to describe you, what words would they use?

Part 2: What words would you like them to use?

Part 3: Note the gap and consider what you might do about it, if anything.

Another big question: How would some mega-marketers likely identify you or me? Hypothetically, their description could be based on what's become known as *big data* stored in *the cloud*, some of it captured by *cookies*, perhaps *Trojan horses*, which have been implanted in our computers. Still more information might be collected from a number of other *data mining* sources.

That "big data identity" will likely be based on the demographics of our zip code, our age, our race or ethnicity, whether we have children, our political leanings, our affiliations, where we shop, the types of food products we buy, the age and make of our car, the clothes we wear, the movies we've seen, the music we've downloaded, the pictures we've placed on *Facebook*, our medical records, pharmaceuticals we've been taking, web sites we've visited, our emails and tweets, calls we've made, images from closed-circuit (CCTV) security cameras, and a host of other things.

That dispassionate market-based assessment of our identities could even seem to many of us like an invasion of our privacy. In fact, our first question might be, "How did they get all that stuff in the first place? Am I a person, a target, or just a series of data points?" That crunching sound we hear may be someone analyzing our personal data.

> **"Just because you're paranoid doesn't mean that they aren't after you."**
> *Joseph Heller, Catch-22*[270]

Paranoia? *Maybe Not.* Big Data? *Maybe.*

I grew up in one of the most open, outgoing, engaging families that has ever existed. We shared our joys and sorrows with the community, and they shared theirs with us. However, once in a while, someone would take that openness too far. "Joe asked me how much money we have in the bank. The nerve. It's none of his business."

In the *face-to-face world*, with the exception of gossip and the grapevine, we have more control over our identity and privacy. In the *online world*, everything seems to be fair game. Invasion of privacy and unwanted attention are commonplace. Say or do something outrageous in the town square or even in your back yard and you may end up with umpteen worldwide hits on YouTube.

Media Literacy. Media literacy skills are becoming essential, since we are often faced with finding the truth among concocted identities designed to influence public opinion. Fabrications spread in an instant by all forms of 24/7 omnipresent media. Each of us needs to be able to separate truth from fiction, which occasionally masquerades as news. Without an understanding of trial by evidence, trial by innuendo might easily triumph. Accusations aimed at personal gain can command attention in a flash and inflict untold damage. Those who deal in fabrications or half-truths know that, if an apology is needed, it will get less attention than the original charge. However, too much falsehood, information leaked for personal glory, or nuanced truth eventually leads to a lack of trust. That can be devastating even to the false accuser's identity and reputation.

Losing Control. Growing numbers of people are concerned that our Facebooking, tweeting, and emailing is dashing our ability and willingness to communicate face-to-face. It's a legitimate concern that has to be addressed.

Psychologist Sherry Turkle describes some people who are overly connected as losing control of their own attention, uneasy unless they are constantly manipulating their smartphone, and losing their personal identities. She notes that, too often, when people meet in person, they are simply "alone together," not wanting to risk saying something that they can't edit before they send.[271]

In an article titled, "Silicon Valley Says, 'Step Away from the Device,'" Matt Richtel writes in the *New York Times* that "the lure of constant stimulation—the pervasive demand for pings, rings, and updates—is creating a profound physical craving that can hurt productivity and personal attention."[272]

In 2012, *Bloomberg Businessweek* announced that Facebook had reached one billion users.[273] Looking at the upsides and downsides of social media, Futures Council 21 member Allison LaFave observes that "Facebook offers a great way for people to stay in touch and organize around shared interests, causes, and events. It also gives people freedom to shape their own identities, which are viewed by thousands of others." She cautions, "No longer are we ourselves but the *digital perception* of ourselves. People, young and old, are filtering out their impurities and presenting themselves at their most charming, most witty, most attractive, and most fun." LaFave concludes, "In such a world, it can be difficult to distinguish true reality from perceived reality."

Social media can also be used to villainize, LaFave adds. She points out "villianization of K-12 teachers in popular culture/media," such as movies or programs. Those programs are created to look objective but occasionally just promote a foregone conclusion or point of view. Still another knotty issue: How can we strike a balance between "ensuring safety and civil liberties?" LaFave describes government leaders "stuck in a difficult position: damned if they invade privacy in the name of protection; damned if they appear complacent and allow future gun massacres, terrorist attacks, etc." Sensitive question: "Who should be monitored and what constitutes cause for suspicion?" In the interest of fairness, "should everyone submit to being monitored?"

"Privacy as it has been traditionally understood in the U.S. will be transformed," remarks Futures Council 21 member Matthew Moen. He notes, "The use of mathematical algorithms to track people's personal habits and interests allows for an unprecedented profiling of individuals, incentivized by profits. Most of these incursions into privacy are simply accepted by people. That fosters more of the same."

Even cars get into the act. In the U.S., "proposed new federal highway safety rules would require all new cars by 2014 to come equipped with 'black boxes' to save vehicle information from the final seconds before and after crashes," according to the *Washington Post*. That 'box' would carry a record of information about both the car and driver. "Consumers have little control over who can see this data and how it can be used," writer Craig Timberg observed.[274]

Multiple Identities and the Choices We Make. Most of us have more than one identity, maybe several, and we may not even know what all of those identities are. Your own version might be, "a good person with a heart of gold who would bend over backwards to help anyone in need." We might be seen as heroes by those who support an issue and villains by those who oppose it. Some will make judgments about who they think we are by how courteous we seem to be. Still others will conclude that we are either "with it" or "out of it" based on our personal choice of Mac or a PC.

Speaking of technology—Given the choice of *privacy* vs. *access to emerging interactive technologies*, there's no contest. Few would disagree that technology, and the avalanche of new apps, has generally won hands-down. Many of us have generally proven that we are willing to trade our privacy, safety, and security for greater connectivity. We might, for example, complain about privacy and then upload a bunch of personal photos. A few thousand online *friends* will likely enjoy seeing them. It's a good news/bad news situation.

The miracle of these electronic wonders is more than occasionally tarnished when someone uses them to "borrow" our credit card number and make purchases across town or across an ocean. *Cyberbullying* has led some children and adults to suicide because of *cybertaunting* and assaults on their identity. Nearly anyone can turn a short clip of our lives into a YouTube sensation overnight, for all the world to see. *Cyberbaiting* is sometimes used to provoke a heated response that is then plastered across cyberspace.

"Hacktivist" seems to be a growing profession. So are multiple forms of *eavesdropping*, such as the recording of phone calls. Some *identity thieves* prey especially on the young, the elderly, and the ill, knowing they might be vulnerable. *The Telegraph*, a U.K newspaper, reported that a sample of 2,107 secondary schools and academies in England, Scotland, and Wales had an estimated 47,806 *closed circuit television (CCTV) cameras* inside and outside their buildings, mostly to protect property from vandals and arsonists.[275] *Data mining* can be a double-edged sword. *Wikileaks* and other clandestine hacking and release of information by *insiders* and *outsiders* has put "state secrets" in full view, giving us a window on events and decisions—but sometimes placing people who work in covert activities, perhaps all of us, at risk. You be the judge. *Tip: Some are using peer-to-peer (P2P) networks or meshnets to bypass large telecommunication providers, hoping for greater privacy.*[276]

In some cases, hackers who had gained substantial notoriety revealed intelligence information and prompted an admission that the National Security Agency (NSA) and some counterparts in other countries had been collecting metadata on telephone and online correspondence. All was part of an effort to abort terrorist activities or capture criminals. Much of it intensified after the 9/11 attacks in 2001. Permission to analyze specific communications required permission from the U.S. Foreign Intelligence Surveillance Court (FISA). Concerns grew about surveillance not only by government but also by social networks, cellphone providers, web sites, retailers, and various marketing information aggregators and data miners. A December 22, 2013, *Washington Post* article, "Surveillance: A Fear, A Comfort," reported on a study indicating that, among those who were concerned, few followed through with recommended precautions.[277]

All that said, it is not fair to simply blame technologies for all of our identity and privacy concerns. We need to ask, "How private should private be, especially if it impacts the broader civil society or the public's interest?" Reputations in public and private life are generally based on our behaviors and the choices we make. Perhaps social networking sites and marketing firms just put them on stage for the enjoyment and/or profit of their subscribers or clients. Some might be gripped by paranoia in an emerging world peppered with identity and privacy concerns. Others simply cast it off as just a side effect of big data.

Note: We need to make sure that we're *as connected as prudently possible.* Caution and counsel should be expressed in *policies* and *practices* we adopt to deal with these concerns. However, *not* being connected is *not* a choice. Cutting the connection or putting extreme limitations on access would be a big leap toward being out of touch.

> **"If you are doing something that you would not want a crowd gathered in the town square to know about, consider whether you should be doing it in the first place."**
> *A Rule of Thumb*

A Matter of Propriety and Magnitude. "Every day, 2.5 quintillion bytes of data are being created, resulting in a myriad of data security and cloud computing security concerns." That observation comes from Sreeranga Rajan, software security systems director for Fujitsu Laboratories of America. Rajan shared those insights just as a Cloud Security Alliance (CSA) launched in 2012.[278]

While the cloud is teeming with big data that can impact the identities of people worldwide, many operators know that they, too, must be concerned about the industry's own identity. Sensitivity is growing about information piracy, data hijacking, and corporate intelligence gathering. Open source is good, but potential clients need to know in advance whether their data will be secure or available for analysis by all comers… which could mean just those who are willing and able to pay a price.

The Identity and Privacy Conundrum. How will we deal with this emerging and multiplying challenge? Some are suggesting that we hire *identity consultants* to advise us. Others propose laws, at least regulations, but that very suggestion raises fear about limiting our right to free speech and expression. The ability to deal with identity and privacy may be at least partially connected to life skills we'll need to maintain our balance in a fast-changing world.

Key question: What role should education play in dealing with and helping students cope with this profound trend?

Counsel from Aristotle

Ancient Greek philosopher and educator Aristotle said our ethos, our personality, how people see us, generally depends on three criteria. Those touchstones include: **competence** (How well do we do our jobs?); **good character** (What seems to drive or motivate us—the common good or the pursuit of personal advantage?); and **goodwill** (How do we treat others?).

Those criteria are as fresh today as they were in Artistotle's time. However, the opportunities people have to observe us have increased immensely. In a world of smoke and mirrors, conspiracy theories, and creative video edits, some are becoming skeptical, even wondering, "How authentic is what I'm observing?"

In political campaigns, for example, a common quest is often to define our opponents before they have a chance to define themselves. On top of that, we occasionally see people in public life employ a phalanx of publicists to create a sense of a reality that may not actually exist. What would Aristotle advise if he were around today?

Multi-Layered Identity and Privacy Concerns
The Education System

While concerns about identity and privacy are universal, educators are at the center of a whirlwind as they daily confront and deal with a flurry of knotty concerns. The Internet, for example, can be a conduit for legitimate or wayward sharing of private information, which education systems generally have in abundance.

The challenges don't stop there. Security concerns can lead to the use of metal detectors, surveillance cameras, and biometric devices. Both locker-searches and frisking can be highly controversial, even when those who do the searching feel they have "reasonable suspicion" or "probable cause." The balance between privacy and security is a source of frequent tension, according to the Privacy Rights Clearinghouse.[279]

Few would contest the fact that we have an increasingly litigious society, ready to bring a complaint or lawsuit at the drop of a hat.

Among the many identity and privacy concerns that come up regularly in the education system are surveys or studies either conducted by the school or an outside party, videos and photography, in-school advertising, information shared through involvement in sports and music activities, military recruitment, incidents of violence or abuse, and freedom of information requests. Pressure has grown to release educator evaluations. Like other institutions, schools and colleges are expected to protect what they consider legitimately private information from growing numbers of relentless hackers. Rotating passwords has become as systematic as rotating crops on the farm.

The following is a brief look at only two front-burner laws and policies with direct implications for identity and privacy within and among education systems. Educators must constantly stay on top of these and other existing and emerging laws, make sure that personnel are trained to comply with them, and communicate the intent of those laws or policies to those who are impacted.

• **Family Educational Rights and Privacy Act (FERPA).** This federal law is intended to "protect the privacy of student education records" and "applies to all schools that receive funds under an applicable program of the U.S. Department of Education," according to *ED.gov*.[280]

Under this act, schools can release certain "directory information." That category might include name, address, phone number, date and place of birth, dates of attendance, and honors and awards. Parents or adult students who wish to prevent the release of directory information can complete a non-disclosure form. Personally identifiable information, such as Social Security number, student ID number, grades, GPA, gender, class schedule, race, religion, test scores, or academic standing cannot be disclosed to a third party without appropriate written permission.[281]

Sometimes called the Buckley Amendment, FERPA was implemented in 1974 and is one of the most longstanding of privacy laws. It's not alone. Several other federal laws also impact the privacy of student records.[282] Generally, a review of a school district or college/university policies is the best way to find a more local interpretation of FERPA. Education institutions understand the importance of maintaining the accuracy and security of those records and trying to operate within existing laws and policies.

• **FOIA (Freedom of Information Act).** "The Freedom of Information Act (FOIA) is a law that gives you the right to access information from the federal government. It is often described as the law that keeps citizens in the know about their government." That's the definition of this far-reaching law that is a central building block of an open society. During FY2011, FOIA had received 644,165 requests for information under the act and had a backlog of 83,490, according to FOIA. gov.[283] States also have specific "open meeting" or "sunshine laws."

Is There a Right to Privacy?

Debates about that question stretch back to at least 1877, when late Supreme Court Justices Louis Brandeis and Samuel Warren wrote a seminal piece for the *Harvard Law Review* titled, "The Right to Privacy." [284]

We know that laws and precedents exist for dealing with invasion of privacy, libel, slander, defamation of character or character assassination, false accusation, and intellectual property. In addition to FOIA and FERPA, we have experienced creation of the Heath Insurance Portability and Accountability Act (HIPAA).[285] Laws and regulations protect people from second-hand smoke, excessive noise, and abuse based on a person's race, religion, gender, or sexual orientation. We have attorney-client and doctor-patient privilege. Bottom line: Is there a right to privacy? The jury is still out.

Identity and Privacy…On the Ground
Definitions, Laws, and Challenges

Identity issues have likely been with us since even before an ancient chest-pounder did everything possible to establish dominance within a tribe. Imagine the *privacy concerns* of a cave dweller who found a secret hiding place for a stash of nuts that could be a key to survival during the upcoming winter.

What's changed? Benefits and consequences have been magnified, and news travels faster than ever before. The concerns have become so massive that this brief chapter could be expanded to fill volumes. Here is a sample of the numerous definitions, challenges, reports, and laws that have implications for identity and privacy.

- **Social Media.** Social media, such as Facebook, Twitter, Instagram, Snapchat, texting, Myspace, Pinterest, Ning, YouTube, Second Life, Reddit, and a host of others, have led to multi-millions of online worldwide connections. We've done friending, photo and video sharing, networking, and co-creating. We've created Avatars that sometimes give us an alternative identity in cyberspace. We've also experienced the intended or unintended benefits and consequences of indiscriminately sharing personal information. While social media can help us build connections with classrooms in other parts of the world, lead us to valuable resources that can enhance education, and create virtual book clubs, they should nonetheless be used safely.[286]

Researchers Katie Davis and Carrie James, who have investigated Internet safety, found most kids "are well aware of risks," and "make sophisticated decisions about privacy settings based on advice and information from parents, teachers, and friends." Students "differentiate between people they don't know out in the world (distant strangers) and those they don't know in the community (near strangers)." They add that "while teens share a great deal online, their willingness to share does not mean they care little for privacy." Most said they "withhold some information even from their friends." One said she wanted to protect herself from people "who like to start a drama." Davis and James found that "90 percent of teens said they don't include their full names, addresses, phone numbers, and birth dates (only day and month, not the year)."[28]

Safekids.com has developed "Kids Rules for Online Safety (for preteens). Among them are: not sharing personal information, location, or a picture without parents' permission; not responding and telling parents right away, if they come across any information or receive a message that makes them feel uncomfortable; never agreeing to get together with someone they meet online; never giving out passwords; and never doing anything that would jeopardize family privacy or… hurt people, possibly even be against the law."[288]

In an *Education Week* article, writer Michelle Davis points out that

"Worries about security, advertising, information-sharing, and social interaction…have led some educators to…seek out social networks designed specifically for learning." Mentioned were sites such as Edmodo, Gaggle, ePals, eChalk, and Livemocha. Nancy Willard, director of the Center for Safe and Responsible Internet Use, remarks that some sites are configured for use only by a group of students and their teacher. Their intended purpose is to provide "interactive environments" rather than "social networks,"[289]

In a *New York Times* letter to the editor, headlined "The Antisocial Effects of Social Media," a Massachusetts private school director, Donna Housman, cautioned that social media "enable us to communicate but replace face-to-face time with others and impede real connections." She emphasized, "For our children to truly become successful personally, socially, and academically, we all need to start connecting emotionally. We need to stop looking at our smartphones and smarten up by looking within ourselves and among one another."[290]

A top technology executive who spends "about $2.7 million a year for the security protection of his family," according to *Business Week*, was a bit surprised when one of his children "posted a picture on Instagram and pointed to it via her Twitter account." A "catchy website" included details about an upccoming vacation and even "GPS locations dished out by her cell phone."[291] At a 2013 World Future Society Conference, consultant Jared Weiner suggested "walled social media" that are specifically for friends and family.

• **Identity Theft.** The U.S. Federal Trade Commission (FTC) has estimated that "as many as nine million Americans have their identities stolen each year." That type of theft takes place when someone "uses your personally identifying information, like your name, Social Security number, or credit card number, without your permission, to commit fraud or other crimes."

How do thieves get their hands on our personal information? The FTC points out they might capture it through *dumpster diving* (going through trash), *skimming* (stealing credit card information on a special storage device when a merchant or even your ATM processes your card), *phishing* (pretending to be a legitimate organization with pop-up messages online that ask for your personal information), *changing your address* (diverting billing statements to another location and even filling out a change of address card for you), *old-fashioned stealing* (snatching wallets, purses, mail, credit card statements, passports, bank account numbers, etc.), and *pretexting* (using false pretense to get your personal information from financial institutions, phone companies, or others). *Encryption* too often provides a bump in the road but not a roadblock for some hackers.

If any of this happens, the agency recommends filing an *Identity Theft Report*, in part to put a stop to it and in part to protect your credit and reputation. They also suggest contacting all creditors and keeping records of any conversations or correspondence related to the case.[292] For some of us, our first line of defense is a data management and protection firm, a firewall and virus detection program, and/or a shredder.

• **Consumer Privacy Bill of Rights.** In February 2012, The White House issued a report titled, *Consumer Data Privacy in a Networked World: A Framework for Protecting Privacy and Promoting Innovation in a Global Digital Economy*. In an introduction to that report, U.S. President Barack Obama noted, "Americans have always cherished our privacy. From the birth of our republic, we assured ourselves protection against unlawful intrusion into our homes and our personal papers. At the same time, we set up a postal system to enable citizens all over the new nation to engage in commerce and political discourse. Soon after, Congress made it a crime to invade the privacy of the mails. And later, we extended privacy protections to new modes of communication such as the telephone, the computer, and eventually email." The President added, "Never has privacy been more important than today, in the age of the Internet, the World Wide Web, and smartphones." He called on the nation to "apply our timeless privacy rules to the new technologies and circumstances of our times."[293] Of course, open societies also face striking a balance between privacy and security.

Then Commerce Secretary John Bryson, noting the growing economic impact of the Internet on business, noted that "online retail sales jumped from about $20 billion in the U.S. in 2000 to nearly $200 billion in 2012." He added, "The Internet has become an engine for innovation, business growth, and job creation, so we need a strong foundation of clear protections for consumers, and a set of basic principles to help businesses guide their privacy and policy decisions."[294]

The Many Faces of Identity

We live in a world of infinite identities, and some cluster of them represents our own. We identify with our families, our communities, our nations, and the world. Our identities may be shaped by geography, whether we live or grew up on an island, in a desert, near the mountains, or on the plains.

Nearly all of us have identities that are molded by our beliefs and how we declare or demonstrate them in the marketplace. That includes the concept we have of ourselves and how we relate to others. If we are empathetic, we care about how what we do or don't do will impact our fellow human beings. If we are arrogant, we could care less.

As rich and satisfying as our personal identity might be, we need to be sure that we are accepting of others, whose lives may have taken another path. At the same time, we should do whatever we can to make sure that our identity is not co-opted without our knowledge and approval. In schools, for example, some students may find themselves in an "identity group," sometimes just people with whom they feel they have something in common. More than occasionally, those informal groups can be stereotyped and have their own monikers, such as nerd, geek, bookworm, pot head, loner, class clown, gamer, etc. Another type of identity group is a gang.

> **"I have preserved my identity, put its credibility to the test, and defended my dignity. What good this will bring the world I don't know. But for me it is good."**
> *Vaclav Havel*

Implications for Education and Society
Identity and Privacy

Implications of identity and privacy for education and the community are so deeply interconnected that they are presented here as one set of possibilities for consideration and further discussion.

- **Getting and Staying Connected.** Concerns about identity and privacy should be addressed with policies, procedures, media literacy education, and personal behavior. Those concerns should not be addressed by becoming disconnected from the wealth of online services and connections that are readily available. Becoming disconnected is a ticket to becoming out-of-touch.

- **Building Understanding of Identity and Privacy.** While virtually everyone is impacted by identity and privacy issues, vast numbers of us need to develop a greater understanding of their magnitude, how they impact us today, and how they could affect our future. Teachers will likely want to make these issues a part of classroom discussions. The education system and broader community might convene school assemblies, workshops, and meetings to raise awareness of how our identity as a school or community can become a self-fulfilling prophesy.

- **Stimulating Students to Think about Identity and Legacy.** These issues should always be on the agenda. We might want to offer a seminar or finishing school class to help students think deeply about identity and legacy. *One*, they might consider the identity they would prefer versus the identity they are presenting through social networking sites or their personal behavior. *Two*, they could ask students to consider the personal legacy they would like to leave for future generations.

- **Preventing and Dealing with Online Challenges.** As part of personal and technological literacy, students need to fully understand possible dos and don'ts in maneuvering the online world. They might discuss what constitutes appropriate sharing of personal information. These discussions could stimulate dialog across disciplines, ranging from communication, social studies, law-related education, civics, and ethics to technology and personal judgment. Those who are adept at thinking and reasoning and problem solving skills will understand the concept of benefits and consequences.

- **Expecting the Inevitable.** Realistically, most organizations, despite their protections, can expect to have their data and information systems invaded by clever and inventive hackers. That might include unwanted revelation of personal information traditionally considered private; tampering with essential operational instructions or data ranging from transportation to finance; or altering test results or grades, just for starters. Despite the immediate problems these types of incidents might cause, they will be even worse if redundancy is not a part of the protection system.

A few additional implications of this identity and privacy trend include:

• **Expecting greater interest in identity and privacy fields.** Growing numbers of careers will develop around technology, law, and data analytics. Schools and colleges may want to call attention to these emerging fields. "Data analytics will become a highly valued course of academic study in the sciences and business," observes Futures Council 21 member Matthew Moen.

• **Developing and Renewing Policies While Maintaining Transparency.** Most organizations, including education systems, will shape policies that provide guidance for dealing with potential identity and privacy issues. They should call for the greatest level of openness or transparency possible, while still providing safeguards and procedures for dealing with issues as they arise, such as cyberbullying and invasion of privacy.

Identity and Privacy
Questions and Activities

1. Review "How Companies are 'Defining Your Worth' Online," a report on an interview done by NPR's Terry Gross with Joseph Turow, author of *The Daily You*. If possible, obtain and read a copy of the book. Consider five things you have done or purchased during the past five days that might shape your identity in the marketplace. http://www.npr.org/2012/02/22/147189154/how-companies-are-defining- your-worth-online.[295]

2. View a *TED Talk* titled, "Sherry Turkle: Connected, but Alone?" Identify five major concerns she raises about our identities. Develop a few suggestions for what we might consider to address those concerns. http://www.ted.com/talks/sherry_turkle_alone_together.html, posted April 2012 (19:48).[296]

3. Lead or participate in a discussion devoted to, "Is there a right to privacy?"

4. Develop a one-page clearly written set of instructions for parents on how they can help ensure their child's online safety and security.

5. Compose a brief policy statement that expresses an organization's commitment to identity and privacy.

Readings and Programs

1. Read *Consumer Data Privacy in a Networked World*, The White House, Washington, D.C., Feb. 2012, http://www.whitehouse.gov/sites/default/files/privacy-final.pdf.[297]

2. Read "Silicon Valley Says, 'Step Away From the Device,'" *New York Times*, July 23, 2012, http://www.nytimes.com/2012/07/24/technology/silicon-valley-worries-about-addiction-to- devices.html?pagewanted=all&pagewanted=print.[298]

3. View "Computer Game Addiction," Australian Broadcasting Company (ABC), April 26, 2012, http://www.abc.net.au/catalyst/stories/3488130.htm (10:54).[299]

4. Read "Our Space: Being a Responsible Citizen of the Digital World," a collaboration of The GoodPlay Project at Harvard and Project New Media Literacies at USC, 2011, http://www.goodworkproject.org/ourspacefiles/0__Introductory_Materials_Our_Space.pdf .[300]

Chapter 6

The Economy
Get Ready for a Reset!

Chapter 6: The Economy. *Get Ready for a Reset!*

Trend: An economy for a new era will demand restoration and reinvention of physical, social, technological, educational, and policy infrastructure.

Industrial Age Mentality → Global Knowledge/Information Age Reality
Social and Intellectual Capital → 21ˢᵗ Century Products and Services

> "The economic crisis doesn't represent a cycle.
> It represents a reset.
> It's an emotional, social, and economic reset."
> *Jeff Immelt, CEO, General Electric, 2008*[301]

Building on a Firm Foundation
The Future Isn't What It Used to Be!

We are moving from an Industrial Age into a Global Knowledge/Information Age, accelerating at a rate that could give us whiplash. With barely a stretch, it is also clear that we are headed into an Age of Knowledge Creation and Breakthrough Thinking.

That's the good news. The glitch is that our infrastructure was created to help us build an economy for yesterday, not for today and tomorrow. Much of that infrastructure is aging, breaking down, or isn't compatible with a flood of new economic realities. It's a dilemma—like trying to squeeze a size 12 foot into a size 8 shoe. Unless we get the right fit for a 21ˢᵗ century economy, we'll be missing opportunities and unable to deal with problems that are growing exponentially. Will we agree to invest in the infrastructure we need? Our future may lie in the balance.

Infrastructure in Tough Shape: In 2009, the respected American Society of Civil Engineers (ASCE) released a "Report Card on American Infrastructure." By its nature, ASCE focuses primarily on the *built environment*, its resilience and its capacity to serve the systems that depend on it. This particular study zeroed in primarily on the *condition of physical infrastructure* impacting: aviation, bridges, dams, drinking water, energy, hazardous waste, inland waterways, levees, public parks and recreation, schools, roads, transit, rail, solid waste, and wastewater. The average grade? D.[302]

The study helps us see a stark reality even more clearly—we are trying to build a new economy on an often outdated, sometimes crumbling and dangerous foundation.

Expanding the Concept of Social Infrastructure: Now, let's broaden the ASCE list to encompass a few more essentials in our quest to create a viable infrastructure for the future. Some, but not all, of these items put the pressure on or pave the way for development of our social and intellectual capital. We'll share a few, such as: a sustainable environment; telecommunication; computer speed and capacity, including download speed; education that is both broad and deep; an ability to tap human imagination, creativity, and ingenuity; innovation and entrepreneurship; and economic, financial, and environmental literacy. All will help move us from an Industrial Age mentality toward 21st century realities.

What Comes First?
How Does the Economy Fit into The Puzzle?

Which came first, the chicken or the egg? While that's a question we'll be asking as long as we occupy the planet, it raises a few more profound questions that should be easier to answer: *Should civil society and the environment exist simply to serve needs of the economy?* or *Should socially responsible economic institutions understand that their very existence and success depend on the overall well-being of the world around them?* Most thoughtful observers would agree that a legitimate, sustainable, and just economy is a product of a flourishing civil society and a sound environment.

Council of Economic Advisers. Responding to the 2011 annual report of the Council of Economic Advisers, President Barack Obama declared, "We can build an economy by restoring our greatest strengths: American manufacturing, American energy, skills for American workers, and a renewal of American values—an economy built to last." He added that an "innovative economy" would also require investment "in the technologies of the future, in the companies that create jobs here in America, and in education and training programs that will prepare our workers for the jobs of tomorrow." He challenged the nation to remain "a country in which anyone can do well, regardless of where they start out."[303]

An Investment...in Education. Of course, doing well tomorrow depends on a sound investment in education today. To make sustained progress on all fronts, including the economy, we need education that is broad and deep—available to everyone. If we hope to build a strong middle class and secure our future, all of us (not just a privileged few, but all of us) need to be in the game and have a shot at the golden ring. As we reshape our institutions for a new economy, we need to think of education, front and center, as an investment, not an expense.

Red Alert:

Just because we are moving beyond the Industrial Age doesn't mean we don't have to be industrious. Simply using our smartphones or computers to "find interesting stuff" or "dish gossip" can be enlightening and an interesting pastime. However, a Global Knowledge/Information Age economy will only work if we use this world of information and connections to hatch ideas, negotiate agreements, develop new or improved products and services, conceive of strategies, and pursue possibilities beyond our imaginations.

Big Turnaround and Historic Transitions
We've Been Here Before.

Call it a big déjà vu moment for humanity. We're in the midst of another massive social and economic transition. It's one of those great historic shifts whose magnitude might be compared to the move from a hunter-gatherer to an Agricultural Age...then from an Agricultural to an Industrial Age.

These humungous transitions don't happen in a moment. They emerge gradually over generations. We may have studied the history of these changes. However, that doesn't mean we necessarily have the personal interest, experience, or will to recognize and deal with them, despite their outsized implications for society.

If we don't make constant adjustments to our social and economic infrastructure, we are headed directly toward a brink, a cliff, a crisis, maybe even a catastrophe, whatever term we might choose. One thing we know for sure: even though these transitions have happened throughout history, they can be disorienting, to say the least.

As we move into any new era, every institution has to go through the transition. No one gets a pass. While agriculture and industry are critically important, our economic focus is stretching with development of new technologies and the expanding needs and wants of people.

Nearly every aspect of what makes our society tick will be pressed to adapt as we make the transition. That includes but is not limited to: energy, the environment, transportation, the economy, jobs and careers, communication, computer power, lifestyles, health, education, human ingenuity, agriculture, industries of every shape and size, and the astuteness of our public policy. These massive resets take place when old systems and infrastructure break down because they are mismatched with new realities.

Having it All

A problem, according to Fareed Zakaria, editor-at-large and host of *Fareed Zakaria GPS* for CNN, is that governments "have run deficits during busts and during booms," and "kept rates low in bad times but also in good ones." In an article, "Can America Be Fixed?" in the January-February 2013 issue of *Foreign Affairs*, he adds, "In poll after poll, Americans have voiced their preferences: they want low taxes and lots of government services. Magic is required to satisfy both demands simultaneously."[304]

Challenges and Choices. By 2008, we had moved into a new era, but status quo thinking, a frequent lack of foresight, a shortage of political will, and powerful self-interests whose economic fortunes or personal advantages seemed to be tied to holding up progress, led us to the brink. We were left with at least two choices:

- *One*, we could freeze our social and economic infrastructure, let it decay, squeeze out every last drop of profit before it becomes obsolete or illegal, and fight off change.
- *Two*, we could wake up to the need for a renewed infrastructure, ranging from a smart grid and world-class download speeds to education that taps our greatest resource—human ingenuity.

Turbulent times are not new. They are always disruptive. Our lives may never be quite the same, wherever we live. However, turbulent times can provide an opportunity for each and every one of us to consider where we've been and what we want to strive to become in the future.

Economic Downturns
Triggers for Infrastructure Change

We all know that economic theories can point in all directions. Some point to an *invisible hand* that guides the economy. All that aside, we would like to have you consider a couple of compelling approaches suggested by well-known economists. Their ideas are grounded in hard lessons from history coupled with prospects for the future.

In his book, *The Great Reset,* economist, columnist, educator, and author Richard Florida zeroes in on historic economic crises. Based on his intense study, Florida suggests "lasting recovery hinges on four key factors." Those factors include: "technological innovation," "individual invention and whole new systems of innovations," "substantial upgrades in the education system," and "major changes in the very way we live." Florida looked squarely at "resets" following the *Long Depression* that began in 1873, the *Great Depression* that kicked off with the 1929 stock market crash, and the *Great Recession* that got underway in earnest in 2007-08, and concluded that "millions upon millions of people respond to challenging economic times by changing the way they live." Infrastructure will change. So will our lifestyles.[305]

Economist, author, educator, and social thinker Jeremy Rifkin, in his book, *The Third Industrial Revolution,* addresses "the triple challenge of the global economic crisis, energy security, and climate change." During the early days of the Industrial Revolution, beginning in the 1800s, enlightenment philosophers saw human nature "as acquisitive, utilitarian, and self-interested." Employers were more likely "to measure a man's industriousness in terms of his productivity, and being productive became the defining characteristic of human behavior itself," Rifkin points out. At the time, coal and steam were among the prime sources of energy. Then, in the 1900s, we saw blossoming of "an oil economy and the automobile." Centralized energy companies mined, manufactured, or generated energy products and services and sold them to customers. Now, we are shifting to renewable energy sources, innovations in energy storage, and a more decentralized system of energy generation and harvesting, that will allow all of us to buy and sell electricity on a smart, continental, interactive power grid." [306] Rifkins's *Third Industrial Revolution* vision has been endorsed by and tailored for a number of community, national, and international organizations.

Bubble Bath

When the Housing Bubble Burst, Millions Took a Bath

An overly inflated housing bubble that burst in 2008 was simply the straw that broke the camel's back as we moved into what many have called the *Great Recession*. Similar things had happened before, leading up to the *Long Depression* that began in the 1870s and the *Great Depression* of the 1930s. We should have known that, as new technologies were developed and expectations rose, constant renewal of our infrastructure would be essential. However, massive lobbies, political campaigns, and comfortable habits convinced us that change was too hard, that we could put it off for another day.

The housing bubble was largely created by easy credit and inflated housing prices coupled with home loans that many simply couldn't afford. To make things worse, when some families discovered they were house poor but wanted to maintain their lifestyles anyway, they first consumed their savings, then used inflated home values as an ATM.

A day of reckoning was inevitable. When the housing bubble burst, prices plummeted. Waves of people were caught "underwater," with mortgages far beyond their home's actual value or trapped with an adjustable rate mortgage. Payments began to soar just as their jobs disappeared. Foreclosures escalated. As home values dropped, so did assessed valuations of property and tax revenues for schools, colleges, and other essential public services. The number of homeless students rose dramatically. With money tighter, people scaled down their purchases of everything from cars to groceries. The economy took a massive hit.

Let's give credit where at least some of the credit is due. The financial sector packaged, then bought and sold clusters of home loans as *derivatives* and *credit default swaps*. These presumed *hedges* against bad loans fueled what some have called *a casino mentality*. Bonuses were an incentive to dig the financial hole even deeper. As home prices dropped and the stock market plummeted, many of those financial institutions came close to collapsing. Some did.

Economies worldwide took a severe beating. The economic downturn escalated, hitting communities, companies, and individuals. Unemployment surged. Government stimulus and bailouts tied to demands for re-structuring helped rescue a number of financial institutions, sometimes called "too big to fail," as well as the auto industry. Government even claimed to come close to turning a profit on some of its bailout loans.

A Home Affordability Refinance Program (HARP) presented the opportunity for those who were underwater on their homes to refinance at lower interest rates but few were able to benefit from it, in part because banks had little incentive to make new loans.

Heated arguments continue about what's best—stimulus or austerity. Eventually, the economy started to turn in a more hopeful direction, but lives were forever changed.

If we were aware of history, we might have known. In 1926, another housing bubble burst. Foreclosures multiplied just prior to the Great Depression, according to "The Forgotten Real Estate Boom of the 1920s" from the Harvard Business School. Add to that many family farms, "heavily mortgaged during WWI, in expectation of continued high prices, were overwhelmed by the postwar collapse of the agricultural commodities market."[307]

These nasty things are a living example of why we need to offer economic education and develop financial literacy for students, and very likely for people of all ages.

Lessons from Economic Downturns
Looking Back…Looking Ahead

When society outgrows its physical, social, and political infrastructure, progress stalls. Depressions, recessions, and panics set in. People and their governments search for solutions. Some will fight any changes, while others will admit that the way forward is to fix what is broken and to create an infrastructure that will support us in building a future. Let's trace just a few examples of physical and social infrastructure: transportation, because a strong economy depends on mobility; lifestyles and housing, since they reflect our aspirations; and education, which is the foundation for progress.

• **Transportation.** Interstate Highways; a network of roads, bridges, and tunnels; rail lines; and a massive air transportation system, constructed over decades, are basic parts of an infrastructure that helps us get people and things from here to there. When financial calamity hit with the Great Recession, growing numbers of people realized that much of the transportation infrastructure was in urgent need of repair, getting old, inadequate—built for an earlier time. New generations of transportation infrastructure would be needed to support sustainable economic growth and development. Investments will be essential before the next bridge drops and traffic jams get any longer.

For example, a 2012 report from the U.S. Department of the Treasury and the Council of Economic Advisers highlighted this startling observation from the Texas Transportation Institute: "Americans in 439 urban areas spent some 4.8 billion hours sitting in traffic during 2010."[308] A more time-efficient transportation infrastructure, coupled with more fuel-efficient and hybrid cars and other vehicles, would pay economic, environmental, and quality of life dividends. The people who will solve future generations of these concerns are in our schools today.

• **Lifestyles and Location.** The first U.S. census in 1790 found that only 5 percent of the population lived in cities of at least 2,500 people. Of course, that meant 95 percent lived in rural areas. An overwhelming majority worked in agriculture.[309] By 1820, city dwellers accounted for 7 percent. A century later, in 1920, 51 percent were urban dwellers. The numbers continued to grow, and in 2000, 80.7 percent of the U.S. population lived in urban/suburban areas. New York City, for example, went "from 120,000 people on the lower end of Manhattan in 1820 to 7,910,000 spread across 14 counties just 100 years later, in 1920," according to *answers.com.*

Taking an even longer view, in 2007, for the first time, more than half the world's population lived in cities. [310] In 1972, only three cities had metro populations of ten million or more: Tokyo, New York/Newark, and Mexico City. By 2009, 20 cities had more than ten million inhabitants.[311]

As the Industrial Age got its footing, throngs of rural-dwellers and immigrants moved into cities seeking jobs and the variety of contacts, culture, and other opportunities these urban areas had to offer. With the advent of streetcars, trolleys, subways, and commuter trains,[312] cities expanded, and were often surrounded by suburbs. In the latter part of the 1900s, many city-dwellers flocked to those suburbs, seeking a somewhat more peaceful life. Most of them added the word "commuter" to how they described themselves. The suburbs of one major city intersected with those of another, creating urban corridors. Examples include *Chicago and Milwaukee* and *San Jose and San Francisco*. Social observer Ben Wattenberg noted, "By the end of the 20ᵗʰ century, an urban corridor extended more than 700 miles from Norfolk, Virginia, to Portland, Maine."[313] The education system faced a tug among rural, urban, and suburban communities.

Now, a reverse trend may be in its early stages. Richard Florida has observed that growing numbers of young people who have seen their parents become house and car poor, and who have seen their own college loans add up, are looking for something more compact and affordable. They are demonstrating a preference for a high quality, secure, and lower cost place to live that is within walking or biking distance of work, restaurants, movie theaters, schools and other learning opportunities, cultural amenities, workout facilities, community centers, and places of worship. The preference is often for a diverse and interesting community.

Whether to buy or rent is becoming more of an issue, since some may want or need to move more frequently and not want to deal with selling property. If possible, some would like to jettison the cost of having a car and paying for parking. Speedy, clean, and safe mass transit is growing in importance.[314] In some cases, people who are retired are moving to center cities to be closer to cultural venues. All of this adds up to a challenge for smaller towns, suburban areas, and city centers. These tendencies, directly related to lifestyle and location, should make the agenda as we plan for the future of our schools, colleges, businesses, and communities.

Sprouting from economic need, environmental consciousness, and what many see as common sense, is the *share economy*. Some make extra money by renting out their home; swapping use of a car or bike; sharing garden tools or a ladder; passing along kids' clothes; exchanging sweaters with people the same size; and even sharing jobs. Craigslist and eBay, among others, provide help for this second-hand market.[315]

Who has the wealth?

Around 1776, the South's "top earners" were "large planters, such as George Washington." In the North, merchants came in at the top of the earnings list, followed by lawyers and tavern keepers. "In the northern colonies, the top ten percent of the population owned about 45 percent of the wealth. In some parts of the South, 10 percent (of the people) owned 75 percent of the wealth," according to Thomas Fleming, a former president of the Society of American Historians, in an opinion piece for the *Wall Street Journal*.[316] In contrast, the Congressional Research Service (CRS) reported, "The share of wealth held by the top 10 percent of wealth owners grew from 67.2 percent in 1989 to 74.5 percent in 2010." On the other end of the scale, wealth held by the bottom 50 percent fell from 3 percent in 1989 to 1.1 percent in 2010.[317]

Nobel economist Joseph Stiglitz believes "inequality is squelching our (economic) recovery. He pointed out "the top 1 percent of income earners took home 93 percent of the growth in incomes during 2010." Those in the middle "had lower incomes." Stiglitz warns about "the hollowing out of the middle class since the 1970s," meaning people are spending beyond their means just to maintain and are "unable to invest in their future, by educating themselves and their children and by starting or improving businesses." The broader economy suffers, in part, he says, since "the weakness of the middle class is holding back tax receipts, especially because those at the top are so adroit at avoiding taxes and in getting Washington to give them tax breaks."[318]

- **Education.** Following the *Great Depression of the 1930s* and the upheaval of World War II, a long overdue spotlight fell on inequities and equal opportunity issues. Schools were seen as the crucible for solving multiple problems that had festered in the country since its founding. At the same time, educators had to deal with a mass of new technologies. Yet, public officials seemed intent on codifying a limited menu of Industrial Age skills and imposing narrow standards and high-stakes tests. Valuable areas of study sometimes dropped from the curriculum since they didn't appear on the required exams.

The Great Recession that began in 2007-08 amplified the need to prepare students for a Global Knowledge/Information Age economy and civil society. Interactive technologies and changes in the trajectory of society demanded education that included a range of 21st century knowledge, skills, behaviors, and attitudes. Basics of the Industrial Age were still important, but so were creativity, as well as thinking and reasoning, problem solving, and entrepreneurial skills, plus an understanding of the needs of civil society and ethical behavior. Yet, the idea of renewing education for a whole new era had a hard time getting a foothold in certain communities. Some blamed reform fatigue, an unwillingness to change, narrow self-interests, and a tendency to use education as a political football.

Nonetheless, education is the foundation for progress. Every institution in society was undergoing massive change, and schools and colleges could not be an exception, since they are *of this world, not separate from it.*

History Lessons from the Great Recession
"It's a fine mess you've gotten us into now, Ollie." Laurel and Hardy
The following are a few history lessons we could learn from the Great Recession that began in 2007-08.

1. Don't call this economic problem unprecedented.
2. When *personal consumption* exceeds *productivity* for too long, look for trouble.
3. Getting rich quick is often not sustainable unless we invest in education, imagination, creativity, invention, and systemic innovation.
4. In a fast-changing world, withholding those investments grinds an economy to a halt.
5. The percentages of wealth between the top ten percent and the bottom 50 percent make a difference because of their significance for maintaining a middle class.
6. Everyone needs to become more economically and financially literate and understand that devout political loyalties do not overrule economic realities.
7. Each major depression or recession has led to new lifestyles and changed expectations for the education system.

Social and Intellectual Capital
Develop It and Put It to Work

Economist, MIT Professor, and *New York Times* Columnist Paul Krugman declares, "For all the talk of an information economy, ultimately an economy must serve consumers…"[319] There will be no let-up.

We simply must constantly develop intellectual entrepreneurs and use our expanding intellectual and social capital.

Intellectual Capital. "Intellectual capital is the intellectual material—the knowledge, information, intellectual property, experience—that can be put to use to create wealth." This now-classic definition, penned by author Thomas Stewart, first appeared in a 1997 issue of *Knowledge Inc.*[320] Among our intellectual assets are knowledge, experience, creativity, thinking and reasoning, and problem-solving. They are among products of a liberal education and are becoming the coin of the realm. The ideas, products, and services they generate are part of our intellectual capital.

Richard Florida, also wrote *The Rise of the Creative Class*. In it, he emphasized that land/raw materials, labor, and traditional forms of capital are, in themselves, not enough the keep the economy humming. The missing link is human creativity. Among other things, he calls for "three-Ts"—talent, technology, and tolerance.[321]

Social Capital. True progress is likely only if we cultivate relationships that build our social capital and reinforce social cohesion. Teamwork and communication skills are essential for a dynamic and responsive civil society and economy. So is our ability and willingness to be flexible and stay in touch with a highly diverse, fast-changing world. Expectations will soar for education systems to become centers for development of social and intellectual capital.

Social Architecture. Everything starts with relationships. While we form relationships because it seems like the right thing to do, we seldom consider their value to our bottom-line success. "The problem facing almost all leaders in the future will be how to develop their organization's social architecture so that it actually generates intellectual capital," leadership counsel Warren Bennis points out.[322]

Harvard Professor Robert Putnam placed social capital directly in the spotlight in both his classic works, *Bowling Alone* and *Bowling Together*, as well as in presentations at numerous conferences. At a 2000 White House Conference on the New Economy, Putnam told a gathering of leaders that, "The basic idea of social capital is that networks have value . . . for transmitting information . . . for undergirding cooperation and reciprocity."[323] In a 1995 Organization for Economic Cooperation and Development (OECD) paper devoted to social capital, Putnam noted declining participation in a variety of groups, from parent organizations and service clubs to labor unions, from people who regularly attend religious services to those who show up for bowling leagues or picnics.[324] Synergy is essential.

What often "inhibits people from exploiting economic opportunities that are available to them?" Francis Fukuyama, in his now classic book, *Trust*, explains it as "a deficit of what sociologist James Coleman has called 'social capital,' the ability of people and organizations to work together for common purposes in groups and organizations."[325]

Face-to-Face or *Screen-to-Screen*. Some would argue that social capital has increased geometrically with crowdsourcing and a full array of social media. For many, Skype has become the prime venue for face-to-face communication. Blogs and online surveys lead to give and take. Social media were tools that brought people to the streets during the *Arab Spring* and have led to frequent rallies worldwide, sometimes called mass mobbings, when people come together to dance, sing, or support or oppose an issue.

Now we can convene people quickly to promote a product or start a fight for something we believe in. Crowdsourcing is in. However, big questions stare us directly in the face, such as how we will unleash the power of our social capital to deal with what are often cross-generational issues, like maintaining a sound economy, building sustained support for the best possible education, or securing the integrity of our environment. Limit the circle, and we narrow the range of ideas that will refresh the organization, community, country, or world. Broaden it, and intellectual wealth blossoms, paving the way for the possibility of creating a more promising future.

Creating Intellectual Entrepreneurs and Knowledge Workers
A Key Focus for Education and Society

"Maximize profits! We want an exceptional quarterly statement!" It's a command that's heard in boardrooms across the nation and world. "Our focus is strictly on the bottom line." Sounds tough. Smacks of focused leadership. However, if the team is only focused on what happens in the short-term and not thinking about tomorrow, the whole approach may be unsustainable.

Educators sometimes recoil when they hear, "We need to increase the entrepreneurial skills of our students." They are concerned that they're being instructed to simply turn out workers who have a laser focus on the bottom line. However, entrepreneurial skills are essential in both the for-profit and nonprofit sectors. Sure, it helps to be entrepreneurial if you want to do well economically. On the other hand, if you want to improve education or get support for a park in your neighborhood, you also need to have that entrepreneurial spirit. As educators, if we hope to produce students who will be reasonably well prepared for the rest of their lives, we need to help them become intellectual entrepreneurs who are both curious and persistent as they pursue a world of possibilities.

MIT Economist Lester Thurow makes clear that "without entrepreneurs, economies become poor and weak. The old will not exit; the new cannot enter." The emerging knowledge/information age encourages people to go over, around, or through barriers to get things done.[326]

Cool Ideas and Hot Potatoes

It's all pretty exciting. In this new economy, driven by social and intellectual capital, we'll be able to move in new directions and develop products and services that are beyond our imaginations. In the process, we're likely to raise the dander of some people who "don't want to go there."

New generations of technologies are speeding up everything. What once may have taken decades is now accomplished in minutes. With less time to fully understand and adjust to new and exciting ideas that will open a panoply of economic possibilities, we can expect a range of reactions. Here are a few:

- Early adopters: "This is the coolest thing I've ever seen. Count me in."
- "We did outsourcing, even offshoring. But now, with streamlined and molecular manufacturing and 3D printing, what we've outsourced may come back. We may be headed for more insourcing."
- In an economy driven by intellectual capital, property rights will grow as an issue. A profound concern at the turn of the 21st century was, "Who owns the genome?" "How much control should I have

over what I put on Facebook?" "What happens to what I publish on-line?" Expect disputes over intellectual property.

• The world's greatest researchers are largely online. Broadly sharing information at the time of Copernicus might have taken 80 years. Now it takes milliseconds. Because of nearly instant exchanges of ideas and information—powered by both personal and cyber relationships—medical, transportation, communication, and other technologies are developing in quantum leaps, far outpacing the ability of some people to understand and adjust. In some cases, so many people worldwide are involved in scientific development that it is often difficult to pinpoint who exactly should be credited with an idea or an invention.

• Astute communities and countries will consider how they can attract or build industries that serve as magnets for knowledge workers. In some cases, communities will work with their education systems to grow their own.

• Public and private partnerships may form to encourage targeted investments that lead to improvements in local economies and quality of life. They might also stir concerns about a delicate balance among business, government, and civil society.

• Conflicts will likely develop as new knowledge, which gives birth to innovative products and services, runs headlong into existing values, beliefs, biases, traditions, and lifestyles. Some who are offended will feel the world is moving too fast for them, or even that it's out to get them.

• Corruption may increase as some entrepreneurs who have little commitment to ethics try to make a fast buck at everyone else's expense.[327]

• Concerns about fairness will become a bottom-line issue that will bear on the reputation of a company or a country. For example, our commitment to bridging the gap between the technology-rich and technology-poor in our communities will increasingly stretch worldwide. Legitimacy will depend in large part on our commitment to strengthening civil society and empowering citizens with new economic tools.[328]

Economic Literacy…Our Futures Are At Stake

Some people get into personal financial hot water, and some organizations, including countries, get into a bind, often because they don't know much about economics or personal finance. Along with civic literacy, economic literacy should be considered essential. As important as it might be, requirements for teaching economic and financial literacy are scarce.[329]

The Council for Economic Education, www.councilforeconed.org, explores what some states emphasize. For example, in Oklahoma, House Bill 1476, the "Passport to Financial Literacy Act" calls for financial literacy education to include, but not be limited to: interest, credit card debt, and online commerce; rights and responsibilities of renting or buying a home; saving and investing; planning for retirement; bankruptcy; banking and financial services; balancing a checkbook; loans and borrowing money, including predatory lending and payday loans; insurance; identity fraud and theft; charitable giving; the financial impact and consequences of gambling; earning an income; and understanding state and federal taxes.[330]

Economic literacy is becoming essential as concern grows about people who seem trapped by limited understanding of profit and loss, differences between investments and expenses, needs vs. wants, and a host of other issues. Our understanding of the need to invest in our future is often directly related to the level of our investment in education.

Implications for Education and Society
Economy

Economic trends? Focusing on restoration, reinvention, and expansion of infrastructure. Moving from Industrial Age mentalities to Information Age realities. Developing 21st century products and services. All have profound implications for education and society. The following are just a few more of those implications.

• **Establishing a clear and firm relationship between education and both our economy and quality of life.** Every road toward a sound economy and a more civil society runs directly through our education system. As we move even more deeply into a Global Knowledge/Information Age, driven by knowledge creation and breakthrough thinking, educators need to be part of every discussion of a community's or a country's future. In fact, educators will be expected to be in touch with forces in society that have implications for the regional, national, and even international economy.

Futurist David Pearce Snyder foresees somewhat limited GDP growth coupled with persistent unemployment, which could "shrink the public education revenue base." That situation, Snyder believes, "will require widespread use of information technology to increase productivity through data-enriched decision making."

• **Making education part of our essential infrastructure.** We know that physical infrastructure, such as roads and bridges, must be maintained and repaired. However, we also need to invest in social infrastructure, including education, health, science, research and development, and leadership. Our short- and longer-term economic future depends on producing and encouraging a well-educated workforce. Let's not forget that physical infrastructure includes school and college buildings, technologies, and networks. "Universities will have to rebuild their aging infrastructure, many with state support, to remain appealing to students as an attractive, relevant part of the digital information economy," observes Matthew Moen, dean of the Colleges of Arts & Sciences at the University of South Dakota. As "a predictable response to the pace of change," he also expects an increase in the perceived legitimacy of "academic subjects such as social innovation and intentional disruption."

• **Preparing students for the future, not for the past.** Concern has grown that stringent standards and tests can create a box that actually inhibits the expansive education needed to thrive in a Global Knowledge/Information Age. While having clear expectations is important, "meeting the current standards may not provide kids with all of the conceptual and social skills that will be needed," says Jane Hammond, a veteran school superintendent who served as superintendent-in-residence at the Stupski Foundation. School leaders need to "start with the future conditions our young people will inevitably face and then build back from there," suggests William Spady, president of Change Leaders, a Dillon, Colorado based education consulting firm.[331]

• **Getting students ready for a new economy.** Economic realities are pulling education needs into clearer focus. First, simply being connected to a plethora of social media and having access to a world of information is not an endpoint in itself. Even though we are moving beyond the Industrial Age, we still need to be industrious and put what we know to work in the economy and civil society. Those who are not prepared for an economy based on social and intellectual capital and who have not developed economic, financial, and civic literacy could be among "the new disadvantaged."

Being prepared will mean a full complement of basic management and entrepreneurial skills, the ability to collaborate and work in teams, and the knack of separating truth from fiction and exploring conflicting information with an open mind. Add a high level of curiosity and persistence, creative and critical thinking, an interest in the creative ideas of others, and a profound need to understand and respect people of cultural backgrounds that may be different from their own. Students, all of us, in fact, will need to understand and practice the principles of democracy and value public engagement. Across the board, they will also need to be keenly aware of and sensitive to the ethical dimensions of their discoveries, conclusions, and actions.

• **Insisting that students and others in the community and country are economically and financially literate.** Just as many countries invest too little in civic literacy, they also pay too little attention to economic literacy. Starting at an early age, everyone should be prepared to make rational decisions about needs vs. wants and grasp the idea of dealing with profit and loss, earning interest, borrowing and loaning money, balancing a checkbook, and handling a credit card. Perhaps as important, all must have a deep enough understanding of economic and financial issues to elect or appoint public officials who have well-founded ideas about how to constructively build and support a sound economy within the reality of an Information Age.

• **Serving as a prime source of intellectual leadership for the community.** A credible intellectual leader sees things in context, understands both the big and little pictures, engages in both critical and creative thinking, and helps people understand not just what is happening but why it is meaningful and important.

Every educator should make a commitment to becoming an intellectual leader. We shouldn't even have to say it. Too often, people in education and other fields try to freeze the status quo rather than model behavior that stimulates creative thinking.

A good place to start is engaging staff and community in an informed discussion about how our education system and curriculum need to be shaped to effectively prepare students and the community for a new economy. This book and its invitation to think outside the trenches is designed to stimulate those spirited conversations.

Additional implications of the economics trend include:

• **Offering hands-on internships and real-world experiences in the marketplace.** Having an opportunity to work beside people who hold various positions and learn from actual experience can help students sort through the best fit for their interests, talents, abilities, and motivations. Some will even discover talents they may not have known they had.

• **Ensuring that preparation and professional development programs challenge habits and mindsets.** Our hope is to move the education system and other institutions from Industrial Age mentalities toward Global Knowledge/Information Age realities. That means we'll need to make sure that those on the front lines are prepared to keep us on the cutting edge and in sight of the big picture, both inside and outside the trenches. At the same time, research should be translated into usable knowledge.

• **Developing a knowledge-based economy.** Communities will need to support their education system in making or reinforcing the transition to a knowledge-based economy. In addition to more traditional knowledge and skills, the education system will develop thinking and reasoning, problem-solving, teamwork, creativity, imagination, invention, innovation, and entrepreneurial skills.

• **Creating a new knowledge/information-based model for schools.** Schools and colleges will need to accelerate their move from systems based on an Industrial Age model toward preparing students for a fast-changing world.

The Trickling of the Economy

What is it about *the economy* and *water?* Think about it. We hear about "trickle down" and "trickle up" economic theories. When people get behind on their debts, they're "over their heads." When home values fall below the amount of a mortgage, we are quick to say that those loans are "underwater." When debates erupted about dealing with a "fiscal cliff," we assume that falling off would put us into "hot water." If things just don't work out, we've "missed the boat." On the other hand, most of us admit that we're "waiting for our ship to come in." Fiscal and monetary policies eventually focus on "liquidity."

Trickle down, for example, assumes that if "the income of those at the top increases, it will trickle down to those at the bottom."[332] Sometimes known as "supply-side" economics, it was largely touted by economist Milton Friedman. On the other hand, followers of John Maynard Keynes might call for some government intervention, such as a stimulus, during recessions. When approaches are encouraged for those at the lower end of the economy, the hope is that benefits will "trickle up" to nearly everyone.

Economy

Questions and Activities

1. Prepare a one-page paper with your opinions about why education should be part of our essential infrastructure. If you wish, you may also express why not.

2. What specific demands on schools and colleges have emerged from economic downturns, such as the Long Depression, Great Depression, and Great Recession?

3. Identify ten aspects of society that have been or will be impacted by our transition beyond the Industrial Age to a Global Knowledge/Information Age.

4. What is "intellectual capital?" "Social capital?"

5. What are the characteristics of knowledge workers? Why is growth in the need for these workers important for education?

6. View a presentation by Richard Florida, "The Rise of the Creative Class," based on his book of the same title, UCTV, program (59:23) presented in San Diego.[333] http://www.youtube.com/watch?v=iLstkIZ5t8g,[334] In a paragraph, describe what Florida believes should come first in attracting economic development—the organization/company *or* community.

7. Identify six important things you believe a school or college should do to prepare students for life in a Global Knowledge/Information Age.

8. How can we teach students to become more entrepreneurial in the for-profit and nonprofit sectors?

9. What must happen for education institutions to become more entrepreneurial in creating their futures?

Readings and Programs

1. Florida, R. (2010-2011). *The Great Reset,* Harper Collins, NY.
2. "Economic Report of the President," Transmitted to Congress, Feb. 2012, plus Annual Report of the Council of Economic Advisers, http://www.nber.org/erp/ERP_2012_Complete.pdf. [335]
3. Rifkin, Jeremy. (2011). *The Third Industrial Revolution*. Palgrave Macmillan, NY.
4. View a presentation by Jeremy Rifkin, *TEDx Talk* (11:06), "Leading the Way to *The Third Industrial Revolution,* June 8, 2012, http://www.youtube.com/watch?v=snsb3Pc_C4M.[336]
5. View "The Federal Reserve and You," video, (23.34), produced by the Federal Reserve, http://2www.phil.frb.org/education/federal-reserve-and-you/.[337]
6. "A New Economic Analysis of Infrastructure Investment," A Report Prepared by the Department of the Treasury with the Council of Economic Advisers, March 23, 2012, http://www.treasury.gov/resource-center/economic-policy/Documents/20120323InfrastructureReport.pdf.[338]
7. "Report Card for American Infrastructure," American Society of Civil Engineers (ASCE). View and order: http://www.infrastructurereportcard.org/report-cards and http://www.asce.org/reportcard/. [339]
8. Gibson, R. (2002). *Rethinking The Future.* London: Nicholas Brealey.
9. Putnam, R. (2001). *Bowling Alone.* New York: Simon & Schuster.

Chapter 7

Jobs and Careers
Any Openings?

Chapter 7. Jobs and Careers. *Any openings?*

Trend: Pressure will grow for society to prepare people for jobs and careers that may not currently exist.
Career Preparation ↔ Employability and Career Adaptability

Job Openings

Octogenarian Service Provider
Wind Turbine Repair Tech
Privacy Manager

Preparation and Experience
Required.
Prefer team players with strong
problem-solving and personal
communication skills, plus
good judgment.
Must be available 24/7.

Do you have any openings?
I'm looking for a career, but I'll settle for a job.

The race is on. Across the developed and developing world, communities and countries are trying to put people to work. Unemployment and underemployment, wherever they exist, can increase instability, not only in our homes but also in our communities and nations. At its very heart, a strong and stable economy depends on the opportunity for people to work—to find suitable jobs.

Moving from an Industrial Age to a Global Knowledge/Information Age has thrown everyone for a loop. Globalization has led to outsourcing and offshoring. Traditional manufacturing jobs have a way of shifting from one part of the world to another. Streamlined production, automation, and technology have put a crimp in the number of people needed to get some things done. Additive manufacturing, such as 3D printing, could decentralize manufacturing to devices in our own homes. The demand for quarterly profits has trumped loyalty to seasoned employees. In a fast-changing world, new generations of products, even processes, tend to knock old standbys off the shelf.

As early in our lives as most of us can remember, people have asked us, "What do you want to do?" Another question each of us should ask is more far-reaching and transcends the job. That question is, "What do you want to become?"

> **"It's a recession when your neighbor loses his job; it's a depression when you lose yours."**
> *Harry S. Truman, Former U.S. President*

Creativity May Also Mean...
Creation of New Industries and Jobs

Reality is setting in. We will probably not be able to ride our way into the future. We'll have to invent our way into the future. That means every person, community, country, and economy can never stop developing new industries, new careers, and new jobs. It also means that we'll have to constantly create new generations of ideas, technologies, and processes that will turn out an array of products and services capable of measuring up to 21st century expectations. Think of the opportunity: We'll be able to jump start whole new industries that will employ people and create value, whether it's in the for-profit or non-profit sector. Sadly, some existing industries and clusters of jobs may not make the cut.

Thomas Frey, jobs editor for *The Futurist Magazine*, published by the World Future Society, raises another flag. He tells us, "As a rule of thumb, *60 percent of jobs ten years from now haven't been created yet.*"[340] Internationally respected forecaster Faith Popcorn, founder of Faith Popcorn's Brain Reserve, has said that while those types of projections "might seem overly dramatic" they do "project the current rate of change." Popcorn ventures, "Jobs that are commonplace today will become museum pieces, along with buggy whip manufacturers and typewriter repair people."

All of this is important to us, as individuals, businesses, professions, education institutions, and as countries concerned about our economic futures. A change in careers hits close to home. Sometimes, it's downright traumatic. The rub is that we often tend to identify ourselves, at least in part, by what we do for a living. Popcorn puts it this way, "Job descriptions are the subtitles of the culture."[341]

How many jobs and careers will we hold during our working lives?

The Bureau of Labor Statistics (BLS) estimated the number of jobs that people born during the period of 1957 to 1964 held between the ages of 18 and 44. All would be among younger Baby Boomers. The answer? An average of 11 jobs, about 11.4 for men and 10.7 for women. A job is defined as "an uninterrupted period of work with a particular employer."

As for word on the street that today's young people may go through 5 to 7 career changes during their lives, the BLS declares that, in reality, it "has never attempted to estimate it." However, BLS did estimate that, on average, young people will deal with an average of 7.7 spells of unemployment if they drop out of high school; 5.4 if they are high school graduates; and 3.9 if they are college graduates.[342]

Creating Brain-Gain Communities
Blue Collar, White Collar, Gold Collar, No Collar...All Welcome

Unemployment that came with the 2008 Great Recession proved to be a stickler for numerous freshly-minted college graduates searching for jobs. Yet, the *education dividend* held fairly firm. For example, in the third-quarter of 2012, people over 25 in the U.S. who were high school graduates, on average, made approximately $9,500 more per year than those without that diploma, according to the Bureau of Labor Statistics (BLS). People with a Bachelor's degree or higher, again on average, made between $42,700 to $155,000 a year *more* than the person without a high school diploma.[343]

It is a fact that education and training generally pay lifetime dividends for people in every demographic group, even though equity remains a continuing challenge. Equal opportunity is essential across all diversities.

Intellectual capital, fueled by education, is a centerpiece of a sound economy. That being said, we need to consider Paul Krugman's observations in his article, "White Collars Turning Blue." As the flow of information has become decentralized, people have become "supremely efficient at processing routine information; that is why the traditional white-collar worker has virtually disappeared from the scene."[344]

Brain-Gain Communities/Cities/Metros/Countries. Whatever our line of work, intellectual capital (what we know) and social capital (relationships) will be defining factors in how well we do and whether we even get on the list for a job. Consider this. Brain-gain cities are in fierce competition for knowledge workers to lift their economies toward a more promising future. They know that, while we lead our institutions today, we need to create the ones we'll need tomorrow. Imagination, invention, innovation, and an entrepreneurial spirit will be at the heart of progress for our economy and civil society.

"Creative class" knowledge workers, according to economist and author Richard Florida, represent "a socioeconomic class of professionals who are the driving economic force behind modern growing cities, a group whose jobs require applied intelligence and decision making." In a 2012 article for *The Atlantic*, Florida highlighted 20 metro areas with "the largest concentration of the creative class," based on data from the Bureau of Labor Statistics (BLS). Durham, NC, led the pack with 48.4 percent creative class workers, followed by San Jose-Sunnyvale-Santa Clara, CA, 46.9 percent; Washington-Arlington-Alexandria, DC, VA, MD, WV, 46.8 percent; Ithaca, NY, 44.6 percent; Boulder, CO, 44.4 percent; Trenton-Ewing, NJ, 42.9 percent; Huntsville, AL, 42.7 percent; Corvallis, OR, 41.7 percent; Boston-Cambridge-Quincy, MA, NH, 41.6 percent; and Ann Arbor, MI, 41.3 percent.[345]

"Creatives" earn more, generally pay more in taxes, support cultural institutions, and build overall economies. That reality has stimulated a version of urban warfare among brain gain cities to attract and keep even more of them. Prior to the Great Recession, in 2006, creatives had a 1.9 percent unemployment rate, which escalated to 4.1 in the years immediately following 2008, substantially under the national unemployment rate at the time, Florida adds.[346]

The intensifying demand for knowledge workers poses a growing challenge for the education system and society at large. If a community doesn't meet the education need (grow its own), people from other parts of the nation and world will line up for those local jobs. If a community or country doesn't produce jobs for knowledge workers, then its people will likely migrate and look for other places on the globe where they can profit from their hard work and/or creative genius.

That takes us right back to education. Communities and countries worldwide depend on their education systems to take the lead in discovering and developing the skills, talents, and creative genius of people. *Imagination, creativity, invention, and innovation* are keys to moving the economy forward and creating jobs of the future.

Urban Warfare . . . American Cities Fight for Talent....
"Brain-Gain Cities Attract Educated Youth"
   That was a front-page headline in the November 9, 2003, *Washington Post*. Writer Blaine Harden wasted no time in making his point: "In a Darwinian fight for survival, American cities are scheming to steal each other's young. They want ambitious young people with graduate degrees in such fields as genome science, bioinformatics, and entrepreneurial management skills." Harden observed, "Migrants on the move to winner-take-all cities are most accurately identified by education and ambition, rather than by skin color or country of birth. They are part of a striving class of young Americans for whom race, ethnicity, and geographic origin tend to be less meaningful than professional achievement, business connections, and income."[347]

Shifts in Employment by Economic Sector

As the focus of our local and national economies has shifted, we've had to deal with a scourge of unemployment and underemployment. It's not what we'd like, but it's a reality that needs to be addressed squarely. Pointing fingers and purveying popular political platitudes might stir a shot of adrenalin but that approach is too cheap and easy and generally ineffective. We sometimes are loath to acknowledge a few other realities about boosting the job market.

One of those realities: Innovation is always essential. *A second:* Whole new industries will need to be hatched and grow exponentially if we hope to reach and hold higher social and economic ground. *Let's try a third:* Some people are telling us that many jobs, at least as we've known them, will disappear. Welcome automation, robots, co-working, super- or quantum computers, the Results Only Work Environment (ROWE), and just plain dropping out of the workforce! At a World Future Society Conference, Jared Weiner of Weiner, Edrich, and Brown, a futures consulting firm, suggested terms such as "off-peopling" and "othersourcing." He emphasized that some jobs "are not going to return—they can be done more efficiently and error-free by intelligent software."[348]

The Bureau of Labor Statistics (BLS) keeps a tab on employment in major sectors of the economy. The three that get primary attention are agriculture, industry, and services. Take a look at trend lines reflected in Figure 7.1.

Figure 7.1
Shares of Economic Sectors in the Labor Force (Percent)

Economic Sector	1840	1900	1950	2000	2010	2020
Agricultural	69	37	12	1.6	1.5	1.2
Industrial	15	30	35	16.8	12.4	11.9
Services	17	33	53	73.8	78.8	79.9
Other Combinations	N.A.	N.A.	N.A.	7.8	7.4	6.9
Total	101	100	100	100	100.1	99.9

Percentages for 1840 through 1950 drawn from Historical Statistics of the United States, Bureau of the Census, 1960, http://ageconsearch.umn.edu/bitstream/17629/1/ar610017.pdf, Robert E. Gallman, "Trends in the Location of Population, Industry, and Employment," Ohio State University, Table 1 Those for 2000 and 2010 plus projections for 2020 drawn from "Employment by Major Industry Sector, U.S. Bureau of Labor Statistics, Table 2.1, http://data.bls.gov/cgi-bin/print.pl/emp/ep_table_201.htm. "Other Combinations" includes secondary agriculture, private household, family worker, or self-employed. Details may not total 100 percent due to rounding.

Agricultural Sector. This sector stretches to include forestry, fishing, and hunting. Actual employment in agriculture has gone down precipitously, offset by off-the-charts increases in productivity. Case in point. In 1940, it is estimated that one U.S. farmer could feed 19 others. By the turn of the second decade of the 21st century, that number moved to 143.[351] Agricultural labor efficiency increased dramatically, from 27.5 acres per worker in 1890 to 740 acres per worker in 1990, while the average age of farmers has also gone up.[352] In 1840, agriculture accounted for 69 percent of the labor force.[353] By 2010, it had dropped to about 1.5

percent.[354] Big question: Who will be farming our land in the future?

Industrial Sector. The industrial sector, according to BLS, encompasses manufacturing, mining, construction, and production of goods, excluding agriculture. The Industrial Revolution got a toehold in the early 1800s with the invention of the steam engine and cotton gin. Did it ever! The number of spindles in cotton mills rose from 8,000 in 1807 to 500,000 in 1815. By then those mills employed 76,000 tenders, who were called "industrial workers." Canals and later railroads helped overcome the overwhelming odds of transporting goods and people upstream or along paths and trails rather than roads.[355]

Systemic innovation flourished as people like Henry Ford not only introduced affordable cars but also produced demand for better roads, autoworkers, mechanics, accessory manufacturers, and frequent fueling stations. Thomas Edison harnessed electricity, which led to development of systems for power generation, transmission, and metering. As the 21st century got underway, renewable power generating systems were being proposed, constructed, and coming on line, and a smart grid was in planning stages. A broad range of machinery and appliances—from gigantic earthmovers to pop-up toasters and smartphones—emerged from the foresight of just these two inventors and those who followed.

In 2010, the U.S. maintained its overall world leadership in manufacturing, followed closely by China.[356] A number of efficient and high tech industrial firms are being created across the world. Many new and longstanding firms are reviving domestic manufacturing, using streamlined production techniques and automation/robotics. Those companies often employ fewer but more highly educated and trained people. Molecular manufacturing and 3D printing (additive manufacturing) have also been coming on stream. *Smithsonian Magazine*, in a May 2013 article titled, "The Printed World," declared that "3D printing will democratize design and free us from the hegemony of mass manufacturing." President Barack Obama said it "has the potential to revolutionize the way we make almost anything."[357]

Services Sector. Services represent the fastest growing sector of the economy.

Jobs in this sector fall into several categories: utilities, wholesale and retail trade, transportation and warehousing, information, financial activities, professional and business services, health care and social assistance, leisure and hospitality, federal government, state and local government, and other fields. That includes the military. In services and every other sector, these jobs ride on the shoulders of an increasingly high tech, global economy. We ignore this striking turn of events only at our own peril.

Richard Florida, in *The Great Reset*, points out that, when manufacturing was king, we turned those jobs into good ones that allowed people to feel a sense of identity and make enough money to move into the middle class. Quality circles tapped the ideas and experiences of people on the factory floor. Florida argues that the time has come to turn growing numbers of service jobs into really good ones that pay decently and provide learning and growth opportunities. It's a value-add for any organization and boosts esprit-de-corps. People want to be honestly asked for their suggestions and stand a chance that they will be heard.[358] Better pay can also lead to more support for education.

Outsourcing and Insourcing

Some organizations employ people who are not on staff to perform certain functions, occasionally in the hope of saving money and often because the firm has certain work to do but not enough of it to justify hiring someone full-time. That's generally called *outsourcing*. Let's face it. Some companies buy expertise and production or marketing capacity that they don't have themselves. Occasionally, organizations acquire or merge with others to get that know-how. A number of manufacturing companies, for example, might outsource the production of component parts. In fact, many of the things we buy are partially or fully made and assembled under contract by a company whose name doesn't appear on the label or logo. When that work is done in another country, we often call it offshoring. Occasionally, people call it, "shipping jobs overseas."

Now, the U.S. may be experiencing an "Insourcing Boom," according to Charles Fishman, writing for *The Atlantic*. One example is GE Appliance Park, once, during the manufacturing heyday, employing 16,000 workers. Then, manufacturing headed overseas. At the beginning of the second decade of the 21st century, many of those manufacturing jobs were headed back. GE has also used a "big room" philosophy, "getting manufacturing engineers, line workers, staff from marketing and sales" together to tackle crucial problems, Fishman adds.[359]

What's driving this insourcing movement? One is seamless, labor-saving technologies to streamline the process. Another is that the cost of labor in some countries where jobs had been offshored is going up. "Oil prices are three times what they were in 2000, making cargo-ship fuel much more expensive," Fishman points out, and "U.S. labor productivity has continued its long march upward." Some companies are concerned about the security of their designs and technologies and are guarding against rip-offs. Among many companies beginning insourcing or in-shoring are Ford, Otis Elevator, Master Lock, and Caterpillar.[360]

The Global Dimension

While a chapter of this book is devoted to globalization, we need to acknowledge that people are working and looking for jobs across international boundaries. Expats live in most countries of the world, and, for them and all of us, immigration policies, appropriate pay, language skills, an understanding of cultures, and an opportunity to grow in a job, are all-important. Some work for international/global organizations. Some are posted in another country to provide diplomatic or military service. Some are cosmopolitans who simply want to live, work, and better understand a highly diverse world. A 2012 Deloitte UK survey found that "three quarters of participating companies anticipate that the total number of globally mobile employees (would) *increase* or *increase significantly*" during the second decade of the 21st century. The rate of those increases will tend to vary, depending on national and regional economies and the condition of various industries.[361]

"Creativity, critical thinking, problem solving, innovativeness, communication, collaboration, and citizenship will emerge as more vital than natural resources, such as coal, oil, natural gas, and lumber," foresees Damian LaCroix, a member of Futures Council 21. That, he suggests, "will cause a shift in power toward countries that esteem educators and invest in education." LaCroix adds that countries such as South

Korea and Singapore, where "teachers are regarded as 'nation builders,'" will become increasingly competitive internationally and will "transform their nations from the inside out." He sees prestigious U.S.-based universities being recruited by other nations to expand their presence internationally (both physically and virtually), making those schools "for-profit centers for international learning."

Fastest Growing and Fastest Declining Occupations[362]

The U.S. Bureau of Labor Statistics (BLS) maintains mountains of information about the past, present, and future of our labor force. BLS noted that it was expecting the total number of jobs to grow by about 20.5 million between 2010 and 2020, from about 143 million to 163.5 million.

At the same time, the labor force will "become older and more diverse." Aging and the fact that people are living longer will have a dramatic impact on growth in health care occupations.[363] The stark reality? Baby Boomers will be retiring at the rate of more than 10,000 a day for the better part of two decades, 79 million of them. Growing concerns include: "401(k) accounts have been drained by the recession, pension systems are strained, and Social Security coffers are being drained of money," according to *The Fiscal Times*. That same story, carried by *msn. com*, also reported that "a quarter of middle class Americans plan to delay retirement until they are at least 80 years old," compared to an average U.S. life expectancy of 78.2.[364]

Any labor force and set of occupations is sensitive to a number of factors, such as the state of the economy and where in the world those jobs are actually done. How many jobs there will be depends on demand for certain goods and services; improvements in technology and productivity, which can lead to increases in opportunities in some areas and declines in others; education and training programs; and the creativity of people in developing new lines of products and services, some beyond our imaginations. Another factor is whether people are prepared for jobs that are already there. *CNN Money* reported in 2011 that "nearly 3 million job openings in the U.S. are going unfilled" because of "a mismatch between the qualifications employers are looking for and the skills job hunters have."[365] The following is a sample of what BLS considers the fastest growing and fastest declining of mostly existing occupations between 2010 and 2020 (Figure 7.2)

Figure 7.2
Fastest Growing and Fastest Declining Occupations
2010-2020[362]
U.S. Bureau of Labor Statistics

Fastest Growing Occupations (A Sampling)	Percentage Increase 2010-2020
Personal Care Aides	70
Home Health Aides	69
Registered Nurses	26
Receptionists and Information Clerks	24
Heavy and Tractor-Trailer Truck Drivers	21
Landscaping and Groundskeeping Workers	21
Construction Workers	21
Nurses Aides, Orderlies, and Attendants	20
Childcare Workers	20
Retail Salespersons	17
Office Clerks, General	17
Postsecondary Teachers	17
Elementary School Teachers, Except Special Education	17
Sales Representatives, Wholesale and Manufacturing, Except Technical and Scientific Products	16
Food Preparation and Serving Workers, Including Fast Food	15
Fastest Declining Occupations (A Sampling)	Percentage Decrease 2010-2020
Shoe Machine Operators and Tenders	-53
Postal Service Mail Sorters, Processors, and Processing Machine Operators	-49
Postal Service Clerks	-48
Fabric and Apparel Patternmakers	-36
Postmasters and Mail Superintendents	-28
Switchboard Operators, Including Answering Services	-26
Textile Cutting Machine Setters, Operators, and Tenders	-22
Textile Knitting/Weaving Machine Setters/Operators/Tenders	-18
Semiconductor Processors	-18
Telephone Operators	-17

Jobs Growth by Level of Education

The *education benefit* is alive and well. Among jobs listed as fastest growing or fastest declining between 2010 and 2020, the only ones requiring *less than a high school education* were some positions in food service, health aides, laborers, cashiers, janitors, landscape workers, sewing machine operators, and some construction laborers. Job growth will be most robust among those with *some postsecondary education*. For example, job opportunities for people with doctorate or professional degrees are likely to rise 20 percent; master's degrees 22 percent; bachelor's degrees 17 percent; associate's degrees 18 percent; postsecondary non-degree awards 17 percent; some college with no degree 18 percent, a high school diploma or equivalent 12 percent; and less than high school 14 percent.

Considering new jobs, BLS is projecting 7.5 million of them for those with a high school diploma or equivalent, 1.4 million for those with associate's degrees, 3.65 million for those with bachelor's degrees, 431,000 for those with master's degrees, and 877,000 for those with doctorates. Of course, the lines of work available to people in each of those categories will vary widely, as will compensation, depending on education, experience, demonstrated talent, personal skills, and an entrepreneurial spirit.[367]

College of Arts & Sciences Dean Matthew Moen raises a caution for policymakers who "continue to drive universities to become engines of economic development rather than repositories of wisdom." He adds, "Students may ultimately pull back somewhat from the current focus on careerism, professionalism, and vocationalism as it becomes evident that a liberal education provides a better foundation in a rapidly changing world." Ed Gordon, president of Imperial Consulting in Chicago, a noted workforce development expert and author of *Future Jobs*, reinforces the need for a foundation in the liberal arts and an education-talent creation system.[368]

What Creates Job Openings?
Growth and Replacement

We know that the number of job openings goes up and down based on the economy, efficiency, and the creation of new industries, jobs, and professions. Of course many openings are created because people leave for greener pastures, such as retirement, go back to school, just make a switch, or leave the job for a while to take care of family responsibilities. Some might have to deal with an illness. Others might have seen their jobs disappear and are having a tough time finding the next one.

Considering all of the above, growth and replacement needs during 2010-2020 will likely lead to 7.45 million job openings in office and administrative support. Sales openings will likely total 6.45 million; healthcare 5.6 million; food preparation and serving 5.1 million; transportation and material moving 3.6 million; education, training and library services 3.4 million; construction and extraction 2.76 million; management 2.57 million; business and financial 2.55 million; computer and information technology 1.37 million; protective services 1.19 million; architecture and engineering 798,000; life, physical, and social sciences 546,000; media and communication 417,000; legal 344,000; entertainment and sports 340,000; arts and design 309,000; and farming, fishing, and forestry 291,000.[369]

Welcome to the Home Office. You're right, *home office* has taken on another meaning. On the one hand, it could be the headquarters of the company. On another, it could just be an office in your home that you use to telecommute. Writing for *Associations Now*, Katie Cascuas observed that, "While *opponents* of teleworking argue that working outside the office can reduce collaboration, innovation, and productivity, *proponents* point to business benefits such as reduced operational or real estate costs, improved continuity of service, and reduced absenteeism."[370]

Conundrum: "High productivity can catch the admiring attention of superiors, but we also know that productive employees are often 'rewarded' with more of the same kind of work, and the resentment of less-productive workers."
Mark Athitakis, Associations Now[371]

Hot Prospects for the Future
New Wrinkles

"One of the easiest ways to begin thinking about future careers is to focus on what may be a problem in the future and invent a job that solves it," suggests Cynthia Wagner, editor of the World Future Society's *Futurist Magazine*.[372]

Try this. We can divine possible jobs for the future by studying each of the trends included in this book and ask, "What kind of jobs will we need to create to deal with it?" Then, look at relationships across trends we've included. For example, what products and services can the technology industry create for members of an aging society? Assistive Device Developer comes to mind. Or, what types of skills and bodies of knowledge will people need to succeed in a global economy? Think of diplomatic, language, open-mindedness, and strategic thinking skills, coupled with a broad understanding of what etiquette means in various parts of the world. Missouri's Parkway Schools Superintendent Keith Marty observes, "Jobs will no longer be defined by hours, nor a place, rather by the outcome and/or product. That means teaching and learning may also no longer be defined by hours and a place called a schoolhouse."

Try these obvious and promising career or job possibilities, again drawn from various chapters of *Twenty-One Trends*.

• **Data analytics** will likely see explosive growth as we pile layer upon layer of new data into our supercomputers and the cloud. It's not enough to have the data, or "intelligence," if we can't keep up with figuring out how it comes together to create new ideas and reveal oncoming problems, even fend off terrorist attacks. The demand will multiply as quantum computers seriously come on stream.

• **Neuroscience**, the study of the brain and nervous system, is considered one of the most exciting areas of biomedical research, according to the Society of Neuroscience, whose membership had grown from 500 in 1969 to 40,000 by 2010.[373] How our brains develop and our minds impact behavior has implications for everything from figuring out the best way to design a car to how we personalize education.

• **Superconducting Technologist, Electro-Chemist.** These are among the people who will help us further conceive of, build, monitor, and maintain a smart grid for our power distribution system. They will also be among the folks who build our energy storage (battery) capacity, which will be essential for keeping our smartphone alive, our electric car taking you more miles with less charge, and our renewable power generating systems keeping the lights on when the wind isn't blowing or the sun isn't shining.

• **Energy Inventors, Entrepreneurs, and Technicians.** As demand increases and the environment becomes even more threatening because of a buildup of greenhouse gases, growing numbers of people will be involved as inventors, entrepreneurs, and technicians in fields such as energy generation, storage, distribution, efficiency, and disposal or use of waste. In the field of renewables, the title, *energy harvester*, will become more common. Hydrogen fuel cells also fit into this category.

• **Robotics Engineers, Inventors, Technicians, and Ethicists.** Like it or not, automation of various functions is on the increase. Robots are delivering medications, and searching for survivors of hurricanes and earthquakes. They are flying over highways to check traffic and over clandestine areas in search of terrorists. Every new technology, from GPS information that is stored in the cloud to "Wikileaks," other clandestine release of security information, and drone strikes, raises ethical concerns. *Ethicists*, along with legal experts, will be needed to sort through these issues.

• **Nano-, Bio-, and Forensic Scientists and Technologists.** Nanotechnology, which can involve moving atoms around within a molecule, is becoming mainstream, leading to many job opportunities, according to the Institute of Electrical and Electronics Engineers (IEEE). *TryNano.org* was, in fact, developed by IBM and the New York Hall of Science to connect the possibilities with parents, teachers, school counselors, and others. Among those jobs is development of new generations of semiconductors. Another cluster is development and fabrication of nanotools to perfectly position nano materials. Interest is growing in molecular manufacturing.[374] As in nanotechnology, careers in biotechnology also seem somewhat endless, ranging from actual clinical research and quality control to information systems, regulatory affairs, manufacturing and production, marketing, and sales as well as administration.[375] Television programs such as *Crime Scene Investigator (CSI)* helped trigger growing interest in forensic science careers. While forensics can cover a vast number of industries, just the CSI area alone can involve tasks such as: reconstructing crime scenes, collecting and analyzing DNA samples, taking fingerprints, interpreting laboratory finds, and so on.[376]

More Jobs of the Future: "Whenever a column is written about the best paying jobs of the future, jobs like civil engineers, registered nurses, and computer system analysts, they are the jobs that currently exist," observes futurist Thomas Frey. In "55 Jobs of the Future," one of Frey's articles, he declares, "As data becomes cheaper, faster, and more pervasive, the nature of our work begins to change as well." We'll share a few of Frey's thoughts about future jobs, starting with some we might expect before 2020:

• **Waste Data Managers.** We have so many "redundancies that have been built into the system" that we'll need these "de-duplication specialists who can rid our data centers of needless copies and frivolous clutter."

• **3D Printing Engineers.** This technology and the people who operate it will likely revolutionize "virtually every field of manufacturing, stemming the tide of outsourcing." Machines that do this type of additive manufacturing have been placed in many high schools and colleges, certainly in industries, and are being used to produce everything from auto parts to body parts. Watch for **Food Printing Engineers.** What you make depends on design and also what you put into the print cartridges.

• **Privacy Managers.** In a world where everything seems to be recorded and much of it shows up under our name in "the cloud," these professionals will help us sort out our "preferred privacy-transparency balance."

• **Elevated Tube Transport Engineers.** In theory, a massive "tube transport system" might be developed. These vacuum tubes would be equipped with maglev tracks and would be able to carry people and cargo at high speed and low cost.

• **Octogenarian Service Providers.** Barring some calamity, the numbers of people in their 80s, 90s, and 100s will grow to record numbers. Industries are likely to be created or will expand to provide services for people in this seasoned demographic group.

• **Competition Producers.** Following the lead of Peter Diamandis and his Ansari X-Prize, organizations, communities, and nations may hire these producers to stage competitions as they seek possible solutions to big problems.

• **Augmented Reality Architects.** With augmented reality, students are able to immediately see a simulation of an historic event or science experiment they are studying. Home designers can show us how a new house will fit into an established neighborhood or give us commentary on what we see as we walk down Main Street. Augmented reality can range from how we perceive pain to how we see paint. *Google Glass* is a head-mounted wearable computer device, much like glasses, that can respond to voice commands to display the information or images you want to see.[377]

• **Dismantlers.** With rapid change, many organizations will need to dismantle old policies and infrastructure and replace them with something more sustainable. These professionals can provide needed counsel and help.

Beyond 2030, Frey suggests a number of other job and career possibilities, such as **clone ranchers**, **memory augmentation therapists**, **earthquake forecasters**, and **executioners for clandestine virus-builders**.[378] These possibilities for future jobs and careers are not meant to be the final list but to stir entrepreneurial imagination.

"Who am I anyway? Am I my resume?"
Broadway musical, A Chorus Line, music by Marvin Hamlisch and Edward Kleban[379]

So What Are…Employability Skills?

Many jobs, in both the for-profit and nonprofit sectors, require preparation or experience, sometimes both. On-the-job training helps bring most people up to speed on at least some of what we need to know and be able to do to take on the task at hand. Some are immediately qualified for certain jobs and some are not. Whatever the case, employers are generally looking for particular employability skills that they consider basic to becoming part of the organization. Some test for those qualities and skills. Others observe them as they review resumes, do interviews, and conduct evaluations. Another thing: Even if we're self-employed, having these skills will likely improve our shot at success.

SCANS. The classic and rather timeless *SCANS Report*, issued in 1991 by the U.S. Department of Labor, is consistent with employability skills that were being sought more than 20 years later. SCANS, which stands for the Secretary's Commission on Achieving Necessary Skills, clusters those skills into four areas. They include:

- **Basic Skills:** reading, writing, arithmetic, mathematics, listening, and speaking.
- **Thinking Skills:** creative thinking, decision-making, problem solving, visualization, knows/learns, reasoning.
- **Personal Qualities:** responsibility, self-worth, sociability, self-management, and integrity/honesty.
- **Work Competencies:** utilizing resources, interpersonal skills, utilizing information, using systems, and using technology.[380]

Forbes Magazine. In an article titled, "What You Don't Know Will Hurt You," written by contributor Kathy Caprino, *Forbes*, listed eight essential skills for professionals. Among them: communication, building relationships, decision-making, leadership, the ability to advocate and negotiate for yourself and your causes, planning and management, having a sense of work-life balance, and boundary enforcement (knowing yourself, your needs and wants, and what, for you, is non-negotiable).[381]

Conference Board of Canada. This business organization identified *Employability Skills 2000+*. "These are skills needed to enter, stay in, and progress in the world of work, whether you are on your own or part of a team," the Board declares. Its list highlights fundamental, personal management, and teamwork skills.[382] Clearly, collaboration is fast becoming central to employability.

The White House. As we began the second decade of the 21st century, the White House emphasized a STEM Initiative, aimed at developing student strengths in science, technology, engineering, and math. The White House also called for innovation skills that would "promote continued economic growth and prosperity."[383]

Career Education
What am I gonna do?

While all education, including life experiences, might be considered preparation for a career, sorting out the multitude of possibilities and matching them to our talents, abilities, and interests can be tough. That's why career education is vitally important. Counselors are often on the front line in guiding students toward their interests, especially as they think about a career or simply what to study at a postsecondary school.

Most agree on some grouping of knowledge, skills, and behaviors that are basic to career education. Examples include: career awareness and preparation, attitude development, career exploration, internships, career acquisition (getting a job), career retention (keeping a job), advancement, and entrepreneurship.[384]

Career Clusters. As early as possible in our lives, we should be introduced to career clusters. Knowing what they are, we can connect "school-based learning with the knowledge and skills required for success in the workplace," according the National Association of State Directors of Career Technical Education Consortium (CTE). Those 16 career clusters include: agriculture, food, and natural resources; architecture and construction; arts, audiovisual technology, and communication; business management and administration; education and training; finance; government and public administration; health science; hospitality and tourism; human services; information technology; law, public safety, corrections, and security; manufacturing; marketing; science, technology, engineering, and mathematics; and transportation, distribution, and logistics. These clusters can be tweaked to address emerging needs.[385]

Educators at all levels can help students sort through growing career possibilities. Rather than simply talk about existing jobs, we might ask students to think about what they would like to accomplish, then consider a set of goals to get from here to there. Check out existing jobs that contribute to reaching those goals. A step that is often left out: consider jobs, careers, or industries you may want or need to create. Throughout the process, students should be honest with themselves about their strengths and what might need strengthening. All should carry a deep understanding of the importance of lifelong learning, which will not only make life more interesting but also prepare students for their next jobs, careers, and avocations.

Implications for Education and Society
Jobs and Careers

• **Education systems at all levels will face growing demand to produce good members of civil society who are employable and prepared for jobs and careers in a 21st century global economy.** Educators and community leaders will face the daunting task of balancing a chorus of demands from both the public and private sectors for people who are "college and career ready." All will understand that every student, every potential worker, will require the breadth and depth of education and the flexibility to become a contributing member of civil society and a fast-changing national and world economy. Society will insist on exemplary education systems to serve as magnets as we compete for talent and resources on a local, regional, national, or an international stage. Those expectations need to be coupled with financial and moral support. Montana Superintendent Lauri Barron believes "career pathways" that allow students to make choices within their education program to "ensure they are ready for college and careers" can be helpful. Chicago educator Debra Hill calls for discussion of assessment to determine student college and career readiness.

• **Communities and countries will need to become and remain competitive for talent and resources.** We may want to grow our local or national economies, but we will always be faced with some tough questions. One, "Is this a place I'd want to live?" Two, "Can I make enough money to sustain myself and my family?" Whether we like it or not, egos aside, we're in competition for both talent and resources.

• **Fresh approaches will be needed to teach career and entrepreneurial skills.** Career and vocational education will constantly take on a greater dynamic as it responds to realities and possibilities in both the for-profit and not-for-profit world. In addition to being employable, potential workers will generally need entrepreneurial skills to succeed, whether in an existing job or career or one they invent. Getting experience, through an internship or working as a volunteer, can help. A seasoned mentor can offer wisdom and counsel based on life experience. Of course, knowing how to develop a compelling and accurate resume and presenting well in an interview can also make a big difference.

• **Educators will be among those who are deeply involved in discussions about economic development and quality of life in every community.** That, of course, means that educators will stay in constant touch with the world outside the trenches. Constantly scanning the environment and anticipating possibilities for social and economic development is baseline. The status quo is unsustainable. Sensitivity to present and future needs should be pervasive and should invigorate and unite educators and communities in a sense that "we're all in this together" and "the future depends on all of us." Among the most sought-after educators will likely be those who can not only prepare students for employability but who can also counsel a community on how to capitalize on emerging industries or careers. That counsel will likely stimulate job creation, and the forward thinking will increase chances for attracting existing firms or encouraging the development of new ones.

• **Education and training programs and systems will need to reflect changes in industries and careers—and be able and willing to adapt.** Emphasizing the importance of understanding history, Gary Rowe, president of Atlanta-based Rowe, Inc., points out, "entire railroads disappeared, heavy industries turned to rust, and new technologies transformed workplaces." For high school students, "the expectation of change must be a part of their view of the world," he emphasizes. "Seat time in the classroom is no guarantor of the skills needed for modern vocations and careers," Rowe adds. "The best way for students to absorb how to be entrepreneurial, how to manage, and how to be flexible comes with 'up-out-of-the-seat' experiential learning."

Careers that capture the interest of students might change frequently in response to the job market and employment opportunities. As part of that responsiveness, schools and colleges will continue to be centers for continuing education, training and retraining, elder care, and other services, possibly all under one roof.

A few additional implications for consideration: Job security will require flexibility that comes from a broad education, not just training for a skill. Pressure will grow to produce people with the education and training to perform as members of an international/global economy.

Words of Wisdom on Jobs and Careers
Futures Council 21

Members of Futures Council 21 shared a wealth of wisdom for *Twenty-One Trends*. Here are just a few of their comments:

• *Patrick Newell* of the Tokyo International School sees "a rise in so-cial entrepreneurship." He adds, "People who want to make a difference may not work for profit machines," but may want to instead focus on preparing themselves "to meet the spiritual needs of people to share and give." Newell also suggests that education needs a greater focus on "mak-ing and designing things," to further provoke creativity, critical thinking, and collaboration. In fact, the *maker movement* is encouraging students to put together materials, tools, creativity, and skills to do that.

• *Concepcion Olavarrieta*, president of El Proyecto del Milenio in Mexico, suggests, "People need opportunities for new jobs that will in-crease their incomes." She calls for "local, regional, and global integra-tion that will help others."

• *David Pearce Snyder* observes, "Post-secondary educational insti-tutions are producing degreed graduates that the Labor Department projects will exceed workplace requirements by 2-1/2 times." He adds, "Overproduction of workers with degrees is economically rational since employers routinely pay college graduates 25 to 75 percent wage pre-miums…as tacit evidence they have acquired advanced cognitive skills simply by matriculating." Snyder points out that the new curriculum will "incorporate real-world tasks, either virtual or actual, as essential components, (plus) lifelong learning."

Jobs and Careers
Questions and Activities

1. What qualities and skills do people need to be employable, whatever jobs they hold?

2. You've just received an assignment to provide counsel on how an education system can get its students ready for careers that do not currently exist? Develop a brief, one-page, ten-point recommendation.

3. Read "The Insourcing Boom" by Charles Fishman in the December 2012 issue of *The Atlantic*, http://www.theatlantic.com/magazine/print/2012/12/the-insourcing-boom/309166/.[386] Develop a five-minute PowerPoint presentation addressing "five ways our community might attract jobs that have been outsourced or offshored." Apply this activity to your country, wherever you are.

4. Develop six suggestions for how your local education system or your college or university might become even better connected to the economic infrastructure of your community. Your two aims will be to make the system more aware of work force needs and to provide counsel for the community on possibilities for economic growth that might exist now or in the future.

5. View "How Do We Prepare Students for Jobs That Don't Exist Yet?" Edudemic animated presentation by Sir Kenneth Robinson, (11:41), http://edudemic.com/2011/10/students-of-the-future/.[387] What five changes do you believe your community needs to make...to get students ready for the future?

6. Identify two multidisciplinary industries or careers not mentioned in this chapter. The possible industries or careers you identify do not have to currently exist. Examples would be biotechnologist, nanotechnologist, and astrogeologist. Feel free to use your imagination.

Readings and Programs

1. "Hard at Work in a Jobless Future," Lee, James H., World Future Society, *The Futurist*, March-April 2012, http://www.wfs.org/book/export/html/2477.[388]

2. U.S. Department of Labor, Bureau of Labor Statistics. "Fastest Growing Occupations, 2002-2012" and "Occupations with the Largest Job Declines." http://www.bls.gov.

3. Check out http://trynano.org/ to learn more about nanotechnology and discover related careers.[389]

4. View "Hot Jobs of the Future," Good to Know, *ABC News* Video, (13:41), featuring an interview with Jordan Goldman, founder and CEO of *Unigo*.

com, Dec. 21, 2009, http://abcnews.go.com/Business/video/hot-jobs-future-9396730.[390] Consider that data reflect the time it was taped.

5. Explore *The Futurist* magazine, World Future Society, Bethesda, MD. Check http://www.wfs.org.

6. Review annual issues of world almanacs.

"I owe my soul to the company store."
Sixteen Tons[391], Merle Travis, George Davis, Tennessee Ernie Ford

Chapter 8

Energy
It's a Power Game

Chapter 8: Energy. *It's a Power Game.*

Trend: The need to develop new sources of affordable and accessible energy will lead to intensified scientific invention and political tension.
Energy Affordability, Accessibility, Efficiency ↔ Invention, Investment, and Political Tension.

> **"Energy is the life blood of advanced civilizations."**
> *World Changing, Edited by Alex Steffen*[392]

Energy: What Do We Have in Common?

Where do you get your energy? That's a question we often ask people who get a lot done. Steve Jobs, Bill Gates, and Mark Zuckerberg created devices, software, and networks that have revolutionized technology. In the early days of the automotive industry, Henry Ford came up with the Model T, a car that nearly anyone could afford to buy. The late, great musical genius Duke Ellington admitted that, "I merely took the energy it takes to pout and wrote some blues."

Just to be clear, one of the greatest sources of energy is the ingenuity of people. In fact, that creative energy is essential if we hope to meet the challenge of powering our economy and our civil society as we move toward a more sustainable future. If there is one thing we all have in common, it is our need for energy.

Among the basics? *Energy literacy.* That urgent need, in itself, has significant implications for education. It's imperative that we prepare and encourage people who are capable of scientific discovery, invention, and innovation…people who are able to develop sources of affordable, accessible, and sustainable energy. Then add *investment* and *political will. This chapter includes just a few things we need to know as we pursue our energy future.*

What Is Energy? What Forms Does It Take?

Energy is basic to our existence, no matter who we are or where we live. In fact, it is so pervasive that seldom, if ever, do we try to define it. The U.S. Energy Information Agency (EIA) tells us succinctly that, "Energy is the ability to do work."[393] *About.com/Physics* confirms it: "The capacity of a physical system to perform work."[394]

It's clear that our work is cut out for us. In everyday conversation, the definition of energy might emerge as: "We haven't been able to do anything since the storm snapped the utility lines." "We just got a new fridge that saves quite a bit of energy." "Those kids have so much energy we can't keep up with them."

OK, it's a quiz show. How many *forms of energy* can you name? You have thirty seconds. "Let's see, there's *heat* (thermal energy), *light* (radiant energy), and *motion* (kinetic energy), as well as *electrical, chemical, nuclear,* and *gravitational* energy." Congratulations, you nailed the answer.[395]

Energy: Back to Basics
Energy Dependent or Energy Junkies?

For the most part, our energy originates from the sun. Some of it is stored in the ground. We find concentrations of energy in oil, gas, coal, and uranium. With the exception of some species that thrive primarily on chemical reactions, most plant life thrives on sunlight. Move up the energy chain and it becomes obvious that each of us is dependent on that medium-sized star at the center of our solar system.

As humanity has moved through its phases, from hunting and gathering toward Agricultural and Industrial Ages, we have become increasingly dependent on an even greater supply of energy. In fact, our voracious and expanding appetite for energy has turned us into captives. Cut the power supply and we have big trouble, with a capital T. Some might even see it as catastrophic when the batteries run down at a critical time.

Constant, clean, reliable sources of energy are basic to our existence. Our economy and civil society are, to say the least, *energy-dependent.* Some even tell us that we're *addicted to oil*—that we've become *energy junkies.*

Sustaining relatively equal access to a continuous, efficient, and affordable energy supply and managing demand are among the critical challenges nations and the world will face during much of the 21st century.

Where Do We Get Our Energy?
Renewable, Nonrenewable, Conservation, R&D, a Sense of
Urgency, and Political Will

Let's zero in on sources of energy that keep our lights on, home or office air-conditioned, smartphone operating, batteries charged, and car running. That stream of energy is also basic to agriculture, manufacturing, construction, and health and social services.[396]

Backing off to the big picture, there are basically two *types* of energy: *nonrenewable* and *renewable*, which could also be described as *finite* and *infinite*. According to a U.S. Department of Energy Education and Workforce Development presentation, *nonrenewables* will "eventually dwindle, becoming too expensive or too environmentally damaging to retrieve." *Renewables* or infinite sources "do not use up natural resources or harm the environment and can be replaced in a short period of time."

Nonrenewables. Among those sources of energy that might, over time, be depleted are fossil fuels that resulted from "the remains of plants and other organisms that were buried in the earth's crust and altered by heat and pressure over millions of years." While generally thriving, existing extraction industries understand the need to develop an even more extensive energy mix.

• **Coal** is often said to be the most abundant of those fossil fuels with reserves that could last 200 years. About 68 percent of it is extracted from surface mines, while the remainder comes from underground mines. Coal is generally either anthracite or bituminous.

• **Petroleum,** also a fossil fuel, is generally found in "geologic faults and folds in the earth's crust."[397] Most often, oil wells are deep. Offshore drilling has become a source of some controversy, magnified by the 2010 *Deepwater Horizon* platform blowout in the Gulf of Mexico. Other sources? Oil sands or tar sands, sometimes called bituminous sands, are found largely in Canada, Kazakhstan, and Russia, as well as some other places.[398] Shale oil reserves, found in a number of locations, including the Williston, Uinta, Green River, Great Divide, Washakie, and Piceance Basins, as well as other locations, have been estimated at four or more times the reserves of Saudi Arabia.[399] However, the fracking process used to retrieve it from organic sedimentary materials called *kerogen*, is a source of some controversy.

- **Natural Gas.** Although other theories exist for how it developed, *natural gas* is generally considered a fossil fuel.[400] Usually extracted from wells, natural gas is, for the most part, made up of methane—and lesser portions of ethane, propane, butane, carbon dioxide, and other gases.[401] Fracking (fracturing) is being used more often to release natural gas and oil from geological formations deep underground by "blasting pressurized water, industrial additives, and sand into wells." Environmental concerns have been raised about hydraulic fracking and horizontal drilling, which require high pressure.[402]

- **Uranium.** While still a nonrenewable source of energy, *uranium* was likely produced "in one or more supernovae over six billion years ago," according to cosmochemists. In short, we've inherited it from our solar system.[403]

Renewables. Sources of energy generally described as *renewables*, sometimes as *clean energy* or *alternatives*, include solar, wind, hydrogen, hydroelectric (water), geothermal, and biomass. The scientists, technicians, and a host of others who pursue these sources might be called *energy harvesters*.

- **Solar.** Rooftops have been sprouting solar panels. Solar farms have been created in various parts of the nation and world to capture the sun's rays and turn them into heat or electricity.

- **Wind.** Wind turbines and entire wind farms are becoming more abundant.

- **Hydrogen.** This gas is often on the list because it is a prolific element and can be produced with renewables for use in fuel cells.

- **Water.** Water plays a central role in energy generation. Most commonly, it turns hydroelectric turbines at dam sites around the world. Water power is also getting more attention as we use buoys to capture the motion and energy of waves rising and sinking. Other ways water can be converted into energy include using tides to turn turbines or through capturing energy in thermal variations.

- **Geothermal.** Geothermal wells are frequently used to tap heat within the earth to provide electricity, heating, and cooling.[404] Water that seeps into these wells produces steam that can be used to power turbines. Somewhat simpler ground-source systems for heating and cooling are becoming common. According to the Union of Concerned Scientists (UCS), "the U.S. has more geothermal capacity than any other country, with more than 3,000 megawatts in eights states." While widespread, hot spots are found in areas near active or geologically young volcanoes and along fault lines. UCS notes "the amount of heat within about 33,000 feet of Earth's surface contains 50,000 times more energy than all the oil and natural gas resources in the world."[405]

• **Biomass.**[406] An array of bio-based energy sources command substantial public attention. What is called *biomass* can produce liquid or gaseous *biofuels*, such as ethanol (primarily from corn) and biodiesel.[407] *Bipower, co-firing*,[408] *anaerobic digestion*,[409] and *production of bio-based products*[410] are part of the biomass playbook.

In an article by Chad Wocken of the Energy and Environmental Research Center (EERC), *Biomass Magazine* highlights *algae* as "a sustainable biomass source for energy and fuels." When grown, it is capable of producing "1,000 to 5,000 gallons of oil per acre," Wocken declares.[411]

Political Will, R&D, Conservation, and a Sense of Urgency. Energy has become a perennial hot political issue. Its outsized impact rears its head in sustained arguments about laws, policies, and regulations. Some get into heated tiffs about the type of public and/or private incentives that might be appropriate to develop a sustainable energy future. Case in point: Various companies and politicians sometimes see movement toward greater development of renewable sources of energy as competition for their current products or as counter to various groups in their political constituencies.

While conservation is a key to extending our energy supply and helping rescue the environment, it is occasionally seen as a limitation by certain vested interests as they try to get the last dollar from traditional energy sources. Despite this perennial dust-up, some enlightened energy companies, politicians, and entrepreneurs see the sense of urgency for conservation, efficiency, and renewables, even confronting the need head-on. The times demand increased investment in *research and devel-opment (R&D)*, and that must be coupled with *political will*, if we hope to dig out of our energy trenches.

Energy Production and Consumption...by the Numbers

While some people rave, accuse, and occasionally bend numbers to make a case for their special interest, whatever it might be, here are a few basic facts about energy production and consumption from the U.S. Energy Information Administration (EIA) and other sources.

• **Energy Independence, Dependence, Interdependence.** In 2011, domestic energy production met about 80 percent of U.S. energy demand. That meant that, as the second decade of the 21st century got under way, the country was relying on imports for approximately 20 percent of energy needs. That reality has introduced issues such as *energy dependence*, reaching *energy independence*, or pursuing *energy interdependence*. However with increases in domestic production, the International Energy Agency (IEA) speculated that the U.S. would close that gap and become self-sufficient in oil production by about 2035. The projection has put the spotlight on a number of immediate and future environmental, economic, and foreign policy questions.

Figure 8.1.
Energy Sources, Types, and Uses (2012)
Energy Information Administration
Not all totals equal 100 percent.

Energy Production Source Percent		Energy Consumption Type Percent		Energy How It Is Consumed Use Percent		Energy Educational Bldgs. Use Percent	
Natural Gas	30	Petroleum	36	Transportation	26.9	Space Heating	41
Coal	28	Natural Gas	25	Industrial Uses	20.4	Cooling	6
Petroleum	19	Coal	20	Residential/		Water Heating	22
Renewables	12	Renewables	9	Commercial	11.0	Lighting	20
Nuclear	11	Nuclear	8	Electrical Power		Other Purposes[413]	11
		Other	2	Generation[412]	41.0		

- **Measuring Energy.** Various energy sources are physically measured in different ways. Examples include barrels of oil, cubic feet of natural gas, tons of coal, and kilowatt-hours of electricity. The BTU(British thermal unit) is generally used to measure heat energy. In 2011, primary energy use in the U.S. amounted to approximately 98 quadrillion (one thousand trillion) BTUs. In the energy business, that would likely be called "98 quads."[414]

Oil and the Hubbert Peak

Petroleum has been an old reliable. It powers our planes, trains, and automobiles; helps heat our homes; yields hundreds of derivative products; and lubricates the wheels of industry. Oil has become the catalytic agent for the world economy. Yet, until about 200 years ago, we hardly knew it existed.

In about 1956, long before most people even considered a future shortage of oil, geophysicist Marion King Hubbert developed what we now call Hubbert's Peak. As an employee of Shell Oil Co., he conceived of a graph to anticipate when total usage of fossil fuels would outstrip the amount of these fuels stored in the earth. We want to believe that the flow of oil will continue unabated, at a price we can afford, but it won't.

David Goodstein, in his book, *Out of Gas*, says he believes we should develop new sources of energy and propulsion with the same level of energy we devoted to making our first trip to the moon. "There were formidable technological obstacles to overcome, but we are very, very good at overcoming that kind of obstacle when we put our minds to it," he says. "The energy problem is exactly of that nature."[415]

Emerging Possibilities
Harvesting Renewables

The population of the world will soar to around nine billion by 2050. It's not much of a stretch to figure out that demand for energy will grow. Some of it will come from more or expanded generation facilities, greater efficiency in transmission and use, and advances in energy storage. In filling the gap, a lot depends on our energy infrastructure. Getting that infrastructure up to speed will require invention and innovation by talented scientists and technicians as well as economic and social entrepreneurs. The need is formidable. The sense of urgency is growing exponentially.

Consider a few possibilities for renewable energy harvesting. Some have been up and running and are expanding. Others are just ideas or in the development stage.

• **Solar Cells.** These photovoltaic cells (PVCs), which can turn light into electricity, are assembled into panels and often clustered into arrays that can be found in many places, from rooftops to solar farms. Scientists point out that absorption of incoming radiation by the atmosphere and reflection back into space by clouds and other surfaces can cut the actual light that strikes the PVCs by about 50 percent.[416] A report from the European Union's Joint Research Centre indicated, "the vast majority of solar panels installed worldwide in 2011 were placed in Europe." The most rapid growth had been in Asia, and China was the largest producer of solar panels.[417] Advancement of the industry in the U.S. had been sometimes held up by political wrangling.

• **Space-Based Solar Collectors.** A fascinating idea is the possibility of orbiting clusters of *space-based solar collectors*. Up there in space, those cells could more directly capture rays of the sun, oblivious to day and night, weather conditions, seasons, and the earth's tilt on its axis. Challenges would include the initial investment, microwaving the energy back to earth, and the ability to store that much power.[418] Some who are sold on the idea tell us that the end run around more traditional sources might "help prevent wars, because we would be less dependent on foreign oil."

• **Lunar-Based Solar Collectors.** A hybrid of the big idea of orbiting PVCs is the possibility of *lunar-based solar collectors*. In concept, these solar farms might be placed along the western and eastern edges of the Moon, as seen from earth. They would then "beam power via relay satellites" to our own land-based collectors. In theory, those stations could be made from "materials already available in the lunar soil," such as "silicon, aluminum, and iron." Again, challenges include cost and development of suitable technologies.[419]

• **Other Ways to Harvest Energy. Heat vacuums** might suck up heat from something like asphalt parking lots, roads, or sidewalks and use it to create steam to turn turbines. **Embedded PVCs.** Some have considered embedding photovoltaic cells along similar public spaces or in roofing materials to capture the light of the sun. **Floor Generators.** Try this idea. How about *floor generators* that would capture energy we use walking across a room to turn turbines. In theory, these generators could even convert vibrations along roadways and railroad tracks. **The Bloom Box** could become the equivalent of an individual home or neighborhood power plant fueled by natural gas. A similar idea is based on developing fuel cell technology. Firms like Capstone Turbine were considering development of microturbines.[420] We've already discussed **wind turbines**, **wave generators,** and other ways of harvesting renewable energy.

Personal and Public Transportation

Driven by inevitably higher energy costs and the environmental impact of CO_2 and other emissions, consumers are looking for more efficient trains, planes, automobiles, and other vehicles. Aircraft manufacturers, for example, have made fuel efficiency a major target. Auto manufacturers have been turning out more hybrid and electric cars. Railroads tout their tons of freight hauled per unit of energy burned.

"From Columbus, Georgia, to Batesville, Arkansas, to Brownstown, Michigan, our investments in advanced batteries and other electric vehicle components are putting Americans to work and helping make our country more competitive," according to the U.S. Department of Energy.[421]

While people often choose their vehicles for "speed, flow, and operation," along with a sense of "freedom or...added control of our lives,"[422] we have, perhaps, become too connected to the shape, size, and even the infrastructure that we already have. We might want to clean the slate for fresh brainstorming on both what we need and what might make sense for personal and public transportation in the future

---

The Energy Infrastructure
A Look at Transmission/Distribution, Efficiency, and Storage

We've briefly discussed energy generation and harvesting. Now let's look at what happens after that energy has been captured. Of course, we all know that distributing energy is big business. Our highways and railroads are teeming with tankers. Gas stations and a growing number of charging stations are laced along roadsides. Coal trains work their way across the landscape. Energy companies often have to deal with landowners to get the right-of-way for their power lines or pipelines, whether they are above or below ground. Sub-stations serve as distribution points. Cable is frequently strung on poles or buried. Some transmission lines droop from tall towers that meander across the countryside.

From the beginning, even as Thomas Edison built his first system for generating electricity, he had to deal with distribution of that power to those who needed and wanted it. The expanding enterprise became a model for *systemic innovation*.

• **The Distribution, Efficiency, and Reliability Dilemma.** Most infrastructure, in many parts of the world, simply must be updated to meet the needs of an emerging high-tech economy and growing demand from civil society. That includes the need for energy efficiency and reliability. Much of the existing infrastructure is aging, in need of repair, sometimes dangerous, and, too often, inefficient. Words to describe what we might hope for are: reliability, redundancy, resilience, and sustainability, all at a good price.

The Energy Information Administration (EIA) makes clear that "annual...transmission and distribution losses average about seven percent of the electricity that is transmitted in the U.S."[423] Taking that point a bit further, Neil deGrasse Tyson, a respected scientist and host of NOVA on PBS, points out that "more than half our energy is lost in the way we produce, transmit, and use it."[424] Cutting down on energy loss could lead to an even better price for all of us.

Energy is one of the biggest industries on earth—which makes even small improvements huge. CNN reported "improving the efficiency of the national electricity grid by (only) five percent would be the equivalent of eliminating the fuel use and carbon emissions of 53 million cars."[425]

Power outages bring convoys of bucket trucks to communities hit by weather events, from hurricanes to blizzards. Many of them bring down the lines. Those incidents put a spotlight on a transmission system that has been in place for generations and has, in some cases, become out-of-date and inefficient. Conversation is intensifying about the need for a "smart grid" that incorporates state-of-the-art technologies.

"I am a lineman for the county, and I drive the main
road... Searchin' in the sun for another overload..."
Wichita Lineman, Composer Jimmy Webb, Singer Glen Campbell, 1968[426]

• **Smart Grid.** Improving and in some cases replacing a big slice of our
energy infrastructure by moving to a *smart grid* is no small matter. Replacing 215,000 miles of high-voltage power lines in the U.S. could cost an
estimated $1.5 trillion over 20 years, according to *The Daily Beast*. That
article emerged in response to massive and costly power outages caused by a
devastating hurricane that had struck the U.S. east coast.[427]

What makes up a smart grid? "Like the Internet, the smart grid
would consist of controls, computers, power lines, automation, and
new technologies and equipment working together. Sensors would be
placed along the lines to quickly detect outages and increase chances for reliability. Those technologies would respond digitally to our
quickly changing electricity demand," according to *SmartGrid.com*.

Benefits of a smart grid might include: "more efficient transmission, quicker restoration of electricity if it goes down, reduced operating costs, reduced peak demand, increased integration of large-scale
renewable energy systems, and better integration of customer-owned
power generation systems, including renewable energy."

According to advocates, the system would be more resilient and use
computer power to avoid large-scale blackouts—just in case a portion
of the system is disrupted by anything from a solar flare, earthquake,
tsunami, or weather event to sabotage or a terrorist attack. A *USA Today* editorial put it this way, "Thanks to computers and smartphones,
Americans are more dependent than ever on electricity. But the nation's
20th century power grid is incompatible with its 21st century economy
and increasingly extreme weather."[428]

Imagine yourself with a *smart metering* system that lets you know, in
real time, how much energy you are using for various purposes, gives
you a tally on the cost, and contributes to conservation. You might couple all of that with smart appliances and heating/air conditioning systems, equipped with wireless communication, in what many are calling

"a smart home." Consider less resistance in power lines. That would mean greater efficiency, less wasted energy, and possibly a better price.

With a smart grid, if you have your own wind turbine or solar panels and have extra power, you might be able to sell it back into the grid.[429] There's much more, but that's a brief introduction to this commanding idea. Watch this spot.

Renewing and Creating Energy Infrastructure

From power generation through the whirring of a blender when it's plugged into a socket, the entire energy system is part of our infrastructure. *OilPrice.com*, an energy newsletter, notes, "It's easy to forget that every piece of the current infrastructure—roads, rails, runways, bridges, industrial plants, housing—was built with a certain temperature range in mind." Temperature and rainfall have been changing.

On the up side, new and more efficient technologies have been developed and others are in the pipeline, if we have the will to use them. We'd better take the issue seriously, since demand continues to grow as we move into a high tech economy. Expectations for our energy infrastructure among members of civil society are increasing exponentially.

In that same July 2012 *OilPrice.com* article, writer Kurt Cobb makes no bones about it. When we think of renewing infrastructure, we need to fold in "our electrical generating system (including dams, nuclear power stations, and conventional thermal electric plants which burn coal and natural gas)."[430]

• **Energy Storage.** A 21st century nightmare? *Losing battery power without the time or place to plug in.* We feel cut off from each other and the world. When we're traveling, the dilemma sends us scrambling in search of a charging station or an airport or train station plug in. The demand for smaller, more powerful, longer-lasting batteries seems to grow exponentially in direct proportion to the number of smartphones we're carrying or hybrid, all-electric, or solar-powered cars we're driving.

Relentless research and development are under way to develop advanced batteries. In many cases, that means "grid-scale," "utility-scale" storage. That would allow for, among other things, "transmission and distribution deferral," smoothing out the power supply as it is needed on the grid.

A study reported by *Business Wire* forecast "the market for advanced batteries will roughly double each year between 2012 and 2017, when the market is expected to reach $7.6 billion. That same study anticipated that the market for these advanced batteries would reach $29.8 billion by 2022.[431]

Think about industries, jobs, and careers. Electrochemists, superconducting technologists, and people involved in photonics, among others, will be working to increase battery power and further develop the smart grid. Think of the significant contributions they and their work will make to local, regional, and national economies and our quality of life. One look the Electrochemical Society's publications can help spell out the possibilities. Whether we are interested or not, most of us in today's and tomorrow's world are sure to benefit from their cutting-edge work.[432]

• **Waste.** Leftovers from energy production are hard to escape. From CO_2, methane, and other gases in the atmosphere to spent nuclear material, and slag from mining operations, every nation is faced with the possible consequences of waste that is a natural byproduct of extracting, refining, and using various types of energy, especially fossil fuels. Some are using that waste to generate more energy.

Energy…A Worldwide Challenge

"Over 1.3 billion people are still without access to electricity worldwide, almost all of whom live in developing countries," according to the World Bank's energy program. In a report focused on energy aims by 2030, the Bank points out that "about 2.5 billion people use solid fuels—wood, charcoal, and dung—for cooking and heating. Every year fumes and smoke from open cooking fires kill approximately 1.6 million people, mostly women and children, from emphysema and other respiratory diseases." Economic development is also inhibited by a shortfall in energy, and approximately 26 percent of households, or 550 million people, mostly Africans, were without access to electricity. "At present, more than 80 percent of energy consumed comes from burning fossil fuels, which produces greenhouse gases that cause climate change."

The World Bank Group, in its *Sustainable Energy for All (SEFA)* initiative, sets three objectives for 2030: "universal access to energy, including electricity and modern cooking and heating fuels; double the renewable share of power produced and consumed from 15 percent to 30 percent; and double the energy efficiency improvement rate."[433]

The top ten energy consuming countries during 2011 were, in order: China, United States, India, Russia, Japan, Germany, Brazil, Canada, South Korea, and France.[434]

Energy Literacy

Energy is important to our quality of life, even our existence. That's reason enough for all of us to be energy literate. We're not necessarily talking about another course in school. However, we are being challenged to address this big question: *What do all of us need to know about energy?* The U.S. Department of Energy, working with the American Association for the Advancement of Science (AAAS) and numerous other organizations, has developed a booklet, available online, titled *Energy Literacy...Essential Principles and Fundamental Concepts for Energy Education.*

Bottom line, "The energy-literate person...

- Can trace energy flows and think in terms of energy systems.
- Knows how much energy he or she uses, for what, and where the energy comes from.
- Can assess the credibility of information about energy.
- Can communicate about energy and energy use in meaningful ways.
- Is able to make informed energy and energy use decisions based on an understanding of impacts and consequences.
- Continues to learn about energy throughout his or her life."

How should we teach those concepts and how can we learn more about them ourselves? A good starting point is to read this chapter and review *Energy Literacy* at http://www1.eere.energy.gov/education/pdfs/energy_literacy_1_0_low_res.pdf.[435]

A series of lesson plans, also available from the Department of Energy, is located at http://www1.eere.energy.gov/education/lessonplans/default.aspx. *(Click on topics.)*[436]

Implications for Society
Energy

• **Be ready for changes in lifestyles.** Some will want to live nearer work to reduce transportation costs and help address environmental concerns. Demand will grow for expansion of mass transit.

• **Expect demand for energy education.** Worldwide, people will expect schools and colleges to help students understand and be able to address energy challenges.

• **Consider new energy-related industries and their job-creating potential.** Every challenge brings an opportunity. Renewable energy will become a growth industry and a source of jobs and careers.

• **Insist that current sources of energy are efficient, broadly available, and people-friendly.** To ensure quality of life, each person must have access to affordable clean energy. Be prepared to address the tension between short-term profits and long-term health, even survival, issues that have become more acute across the energy spectrum. Start conversations about the quality of life we might expect moving forward.

• **Promote sustainable energy for your community.** As a community, look beyond the moment. Consider future needs for energy. Do short-, medium-, and longer-term plans to provide the energy you'll need in the future?

• **Consider science first; then develop political positions.** Scientific knowledge continues to expand. A massive body of that work has been devoted to studying energy and the environment. To the extent possible, we should try to build our political positions on top of scientific realities, not preconceived opinions.

• **Ask the tough questions.** Ask office-holders and candidates about energy challenges and what they recommend we do to deal with them. Consider their answers.

• **Be ready to consider and support research and development.** Unfortunately, we invest too little in R&D. When legitimate opportunities arise to support it, we should give those proposals our serious consideration. Then, we should strongly consider investments we need to make in improving our energy future. The smart grid, fast rail, and recycling centers for lithium carbonate batteries are just a hint of what might be on the agenda. Our discussions should likely focus on *benefits* and *possible unmet needs* rather than the too often polarizing *yes* and *no*.

Implications for Education
Energy

• **Energy Literacy will become a significant aim for every student and educator.** In some cases, energy becomes the elephant in the room, so big and pervasive that we don't recognize it. An aim might be to make every educator a part of the energy team and to offer key elements of energy literacy across disciplines. Each of us should understand the urgency of our energy challenges and the roles we might play in responding to them. Students will discover growing career opportunities in energy-related industries. Dean Matthew Moen urges campuses to "accelerate their efforts to be green. He anticipates a growth in the study of petroleum engineering, which he believes "will become more interdisciplinary as new techniques to extract oil will require additional input from other academic disciplines." Impacts of extraction are often accompanied by environmental, health, economic, social, and cultural concerns.

• **Education systems at every level will have frequently updated plans for energy use and conservation.** In the process of developing and implementing plans, educators might involve members of the community and staff as well as students. The broader the involvement, the more ideas and experiences we'll have for consideration and the more real-life lessons students will be able to learn. Keith Marty, superintendent of the Parkway Schools in Missouri hopes that "public and private incentives will lead to a new surge of American genius and new ideas" devoted to dealing with energy concerns. He hopes to see "political will to push for change" rather than "giving in to special interests and those who don't want to take the risk." Marty points out that his school district had installed solar panels on nearly all of its schools and planned to purchase 30 compressed natural gas (CNG) buses, as well as set up a CNG filling station on campus.

• **Schools and colleges/universities will set an example for their communities.** Education institutions will continue to take leadership in employing local, renewable sources of energy to partially offset rising energy costs. What they do will provide an authentic example for their communities. They will also become champions of reasonable energy efficiency.

• **Education systems, along with other institutions, will likely face rising energy costs.** What schools and colleges have to invest in actual instruction could be profoundly impacted by the rising cost of powering electronic technologies, transporting students, keeping lights on, and providing for heating and cooling. Budgets and priorities will be caught in the middle. Unfortunately, the cost of energy for those who live in the community may, in some cases, compete with the level of support they are willing to provide for the education system. Growing numbers of educators and communities will discuss creative ways to get the most for what they invest in energy through conservation and actually reconfigure how the system has traditionally operated.

• **Students can practice their thinking and reasoning skills as they consider energy challenges.** Learning about capturing and conserving energy will energize classroom discussions. Teachers will have an opportunity to ask students to use their creativity and imaginations in conceiving of answers for energy concerns and opportunities. Among the many areas they might consider are energy harvesting; generation; transmission; storage; efficiency; safety, reliability; and waste.

• **Policies and procedures as well as distributed leadership will be needed to deal with energy-related concerns.** While frequently updated policies and procedures are needed to guide education systems in dealing with energy concerns, they must allow for distributed leadership. Ideas should be encouraged from across the system for everything from dealing with increases in the cost of field trips and fuel for buses to figuring out how to balance temperature and humidity controls for people with a variety of preferences ranging from "too hot" and "too cold" to "just right."

Energy

Questions and Activities

1. Review *Energy Literacy…Essential Principles and Fundamental Concepts for Energy Education…A Framework for Energy Education for Learners of All Ages*. Meet with a small group to discuss what this booklet has to say, and identify five things we should do now to increase the energy literacy of students, educators, and our community. You'll likely find it at http://www1. eere.energy.gov/education/pdfs/energy_literacy_1_0_low_res.pdf.[437]

2. Review "Lesson Plans, Teach and Learn, Energy Education & Workforce Development," U.S. Department of Energy, http://www1.eere.energy.gov/ education/lessonplans/default.aspx.[438]

3. Review "Energy Independence and Security Act of 2007: Summary of Provisions." If you were a Secretary or Minister of Education in your country, what provisions would you add or change? http://www.eia.gov/oiaf/aeo/otheranalysis/aeo_2008analysispapers/eisa. html.[439]

4. View "Smart Grid," *NOVA Science Now*, PBS Video. Write a one-page summary addressing benefits and concerns in moving toward a smart grid. http://video.pbs.org/video/1801235533/ (8:53).[440]

5. What is Hubbert's Peak?

6. On a percentage basis, identify how various sources of energy (petroleum, etc.) have recently been consumed in the U.S. or in your country.

Readings and Programs

1. Review the Energy Information Agency's "EIA Technical Review Guidelines: Energy Generation and Transmission, Volume II, Appendices," and consider maintaining it as a reference. http://www.epa.gov/international/ regions/lac/eia-guidelines/energyvol2.pdf, July 2011 or updates as they occur.[441]

2. Consider *The Big Flatline…Oil and the No-Growth Economy*, Rubin, Jeff, Pallgrave Macmillan, NY, 2012 (print or Kindle). http://www.amazon. com/The-Big-Flatline-No-Growth- ebook/dp/B008PBX102/ref=zg_ bsnr_16233441_5#_.

3. Explore the Electric Power Research Institute's "Research Portfolio," 2013, http://portfolio.epri.com/ (2:39).

4. "Solar Power," Video, National Geographic Society, http://video.national- geographic.com/video/environment/energy-environment/solar-power/.[442]

5. "Great Energy Challenge," National Geographic Society, [443] http://vid- eo.nationalgeographic.com/video/environment/energy-environment/

great-energy-challenge/ (2:00).

6. "U.S. Energy Facts Explained, Consumption and Production," U.S. Energy Information Agency, http://www.eia.gov/energyexplained/ print.cfm?page=us_energy_home.[444]

7. "Smart Grid: A Beginner's Guide," NIST (National Institute for Standards and Technology), http://www.nist.gov/smartgrid/beginnersguide.cfm, post created July 17, 2012.[445]

Chapter 9

Environmental and Planetary Security
Where is Paul Revere?

Chapter 9: Environmental and Planetary Security.
Where is Paul Revere?

Trend: Common opportunities and threats will intensify a worldwide demand for planetary security.
Personal Security/Self Interest ↔ Planetary Security
Common Threats ↔ Common Opportunities

> "We haven't inherited this planet from our parents,
> we've borrowed it from our children."
> *Jane Goodall, British Primologist and Anthropologist*[446]

Where is Paul Revere?

On the 18th of April, 1775, the legendary Boston silversmith Paul Revere mounted his horse and spread the warning to village and farm—"To arms. The British are coming." The American Revolutionary War was under way. Now, in a world faced with what some believe are climate changes that could have a devastating impact on our planet, some are asking, "Where is today's Paul Revere?"

A Balancing Act for Survival

If you're like most of us, you're deeply committed to personal security. Some of us, in fact, are even willing to trade some of our basic freedoms for what we hope will make and keep us safe. Many students of human behavior tell us we're driven by self-interest, certainly by self-preservation. Some organizations fight for their corporate interest. What can override all of these interests and yearnings for safety and security? The answer is *planetary security*. Everything around us exists in the environment. Destroy that…and all bets are off.

During 2012, scientists issued a sort of report card. Scientific instruments detected that carbon dioxide levels in the atmosphere had reached 400 parts per million, a level not seen on earth for three million years, long before the roughly 8,000 years that relatively civilized humans have occupied the planet.[447]

How well we take care of our environment, how well we balance our need for economic growth and development with planetary security, how deeply we grasp issues ranging from the need for fresh water to the melting of ice caps and climate change, have deep implications for us and for whether humanity will actually have a future.

Are we capable of adopting public policies that cut some slack for the planet that sustains us? Are we will willing to invest in new technologies or even consider lifestyles that would be friendlier to the biosphere? Do we understand the benefits and consequences of our actions for our long-term survival? Any guarantees about the future are largely connected to our behavior, not just on the playground but also as custodians of our environment. We will always be in the middle of a balancing act for survival.

Other Big Security Questions. How prepared are we to deal with *human-caused problems*, ranging from litter in the streets to shootings, wars, nuclear catastrophes, and acts of terrorism? How about *natural disasters*, such as tornados, earthquakes, tsunamis, floods, mudslides, fires, drought, and even meteor or asteroid strikes? Are we willing to concede that some of those *natural disasters* might be provoked by *human activity*?

> "The Earth has been through a lot in its 4.5 billion years of existence, but human beings are its biggest challenge yet. An explosively growing human population, already 7 billion plus, is using up natural resources, stripping land of its forests, and polluting the air and water."
>
> *Bryan Walsh, Time, Global Warming: The Causes, The Perils, The Solutions*[448]

Excuse Me. Is This on the Test?

"Look. I know you're well intentioned, but I'm busy trying to get my kids ready to pass their math tests. Who has time for big issues like threats to the environment? Besides, we've had training on handling threats. We call 911, and in a few minutes a SWAT team might come through the door."

Yes, we're busy. Yes, math and the test are both important. Yes, it's good to call 911 when we face an immediate physical threat. However, what do we do when our health, well-being, and even our planet are in peril? At some point, we need to stop and smell the hydrocarbons.

Because the atmosphere has no political boundaries and the sea laps against every shore, what happens to our planet has a profound impact on every one of us. Learning about the environment and acting on what we learn can help us become members of another kind of team, one that could rescue the biosphere for future generations.

While we're turning out good citizens for our communities and our nations, we also need to be sure we're producing good citizens of the world, or the world will start closing in on us. Among other things, we need to understand that a sense of safety and security is a prerequisite for most people to pursue other aspects of their lives. The whole concept isn't that complicated. We just need to "do the math."

Maslow . . . and the Hierarchy of Needs

Psychologist Abraham Maslow identified a now-famous list of human needs that are firmly attached to our basic instincts:

- **Physiological:** the need for oxygen, food, water, and a relatively constant body temperature.
- **Safety and security:** the need for security balanced against insecurity, the need to stay out of danger.
- **Belongingness and love:** the need to affiliate with others, overcome feelings of loneliness and alienation, and be accepted.
- **Esteem:** the need to achieve, to be competent, to gain approval or recognition, the need for a stable, firmly based, high level of self-respect and respect from others, a sense of self-confidence.
- **Self-actualization:** the need to be problem-focused, appreciate life, engage in personal growth, develop the ability to have peak experiences, and pursue what we were "born to do." The poet wants to write. The musician wants to make music.

Physiological as well as safety and security needs are a foundation for everything else. Unless our physical needs are met and we are safe and secure, most of us will feel basically imprisoned, unable to pursue other needs and an interesting, satisfying life.[449]

Let's face it. It will be the students in our schools and colleges today who will develop the ideas, techniques, technologies, behaviors, and lifestyles that will help us sustain this planet. Unless they have some understanding of the issues they confront, they may have a tough time dealing with them…or even grasping the growing sense of urgency. They may be missing one or several of the fundamentals for our very survival. *Climate change fatigue is not an option.*

What Can We Endure?
Taking the Long-Term Temperature

Since 1992, a group of countries has gotten together "to cooperatively consider what they could do to limit average global temperature increases and the resulting climate change, and to cope with the possible impacts." The effort is called the United Nations Framework Convention on Climate Change (UNFCCC).

In 1997, several UNFCCC nations met in Japan where they issued the Kyoto Protocol, an attempt to set greenhouse gas emission reduction targets for the period of 2008 to 2012. Basic to the Protocol was reduction in emissions (largely carbon dioxide) to help ensure that "global temperature increases are limited to below 2 degrees Celsius (3.6 degrees Fahrenheit) higher than preindustrial levels." In 2011, a platform was developed in Durban, South Africa, for possible extension of the Protocol for another five to eight years. During a 2012 meeting in Doha, Qatar, delegates agreed on a framework for negotiating a pact that would take effect in 2020." At the time, the world's two biggest greenhouse gas emitters and most vibrant economies, the United States and China, had historically not ratified the Protocol.[450]

The consequences are deadly serious. In 2012, a UNFCCC background report pointed out: "Climate change is a complex problem, which, although environmental in nature, has consequences for all spheres of existence on our planet. It either impacts on—or is impacted by—global issues, including poverty, economic development, population growth, sustainable development, and resource management."[451]

> **"Mount Kilimanjaro will be losing the last of its snow and ice as temperatures rise, leaving the entire (African) continent ice-free for the first time in at least 11,000 years."**
> *Mark Lynas, Author of Six Degrees: Our Future on a Hotter Planet.*

A Growing Sense of Urgency
Climate Change

A *National Geographic* documentary, *About Six Degrees Could Change the World*, attempted to explain the impact of climate change on temperatures. Author, journalist, and environmentalist Mark Lynas, who wrote *Six Degrees*, followed scientists who explained what to expect with various increases in the average world temperature.

- We're on track for a .5 to 1 degree C (0.9 to 1.8 degrees F) increase in temperature. We've seen extreme weather and melting ice in glaciers and polar regions.
- A 2 degrees C (3.6 degrees F) increase would lead to the disappearance of glaciers and some lower lying islands.
- "At 3 degrees C (5.4 degrees F) higher, the Arctic would be ice free all summer; the Amazon rainforests would begin to dry out; and extreme weather patterns would become the norm."
- With a 4 degrees C (7.2 degrees F) increase, the oceans would rise drastically."
- At a 5 degrees C (9 degrees F) increase, "part of once temperate regions could become uninhabitable."
- A 6 degrees C (10.8 degrees F) increase would be "a doomsday scenario as oceans become marine wastelands, deserts expand, and catastrophic events become more common." [452]

Author, educator, and environmentalist Bill McKibben wrote a hard-hitting July 19, 2012, climate change article for *Rolling Stone*. "So far," McKibben said, "we've raised the average temperature of the planet just under 0.8 degrees Celsius (±1.44 degrees F), and that has caused far more damage than most scientists expected," such as the loss of "a third of summer sea ice in the Arctic" and "oceans that are 30 percent more acidic."[453]

The National Center for Atmospheric Research (NCAR), in a November 2012 *Science Magazine* article by researchers John Fasullo and Kevin Trenbert, considered factors such as moisture and clouds, as well as heat, in developing their observations about temperature. They were looking at models ranging from a 3 degree to an 8 degree Fahrenheit increase. Fasullo and Trenbert noted, "At current rates of global emissions, doubling (of carbon dioxide in the atmosphere) will occur well before 2100."[454]

The National Snow and Ice Center's Randolph Glacier Inventory tracks "surface areas for most of the world's estimated 160,000 glaciers as well as locations of many of the world's smaller glaciers and ice caps. While massive melting of glaciers and polar ice sheets in Greenland and at the north and south poles often get attention, less is focused on smaller glaciers in mountainous, mid-altitude, and tropic areas that have locked-up and then dispensed water for eons. At imminent risk is water for drinking, agriculture, and hydropower. Climates are drying. Andean nations are already feeling the pinch, as are other parts of the world. The International Panel on Climate Change (IPCC) has declared "widespread mass losses from glaciers and reductions in snow cover over recent decades are projected to accelerate throughout the 21st century."[455]

In 2005, a Millennium Ecosystem Assessment, "the most comprehensive analysis ever conducted of how the world's oceans, dry lands, forests, and species interact and depend on one another" reached a dramatic conclusion. A total of 1,300 authors from 95 countries participated in the study. Conclusion: "Human actions are depleting earth's natural capital, putting such strain on the environment that the ability of the planet's ecosystems to sustain future generations can no longer be taken for granted."[456]

The Intergovernmental Panel on Climate Change (IPCC) has declared, "Climate change is happening even faster than previously estimated; global CO_2 emissions since 2000 have been higher than even the highest predictions. The need for urgent action to address climate change is now indisputable." Representatives of science academies in

Brazil, Canada, China, France, Germany, India, Italy, Japan, Mexico, Russia, South Africa, the United Kingdom, and the United States, who signed the IPCC paper, declared, "Education and public awareness programmes will be essential as we pursue this agenda. We must build on the current enthusiasm and engagement of a younger generation."[457]

The Climate Landscape
Recognition of Climate Change Increases, Skeptics Express Concerns

Take a look at the landscape for dealing with climate change. It's all uphill. That's in part because some people and institutions will support science only when it reinforces their politics or their short-term bottom line. An argument among those who believe the climate is changing is whether it is simply a natural phenomenon or whether human activity is contributing to it. A relatively small minority denies the problem exists or sees attempts to solve it as a conspiracy to impose world government. Others say reducing climate-altering gases will, in turn, reduce the number of jobs or adversely impact the economy—possibly make electricity and fuel more expensive.[458]

Scientifically, some skeptics say the trend is part of a 1,500-year cycle that is to be expected, that CO_2 is good plant food, or that Greenland was meant to be green, not covered with ice.[459] Scientists ask each other hard questions as they conduct and review research. When some of those questions were discovered in emails, they were used to claim dissent in the scientific community rather than confirm the rigor of ongoing work.[460] Some want *good news* rather than *facts* about environmental concerns. The good news would be that we are committed to dealing effectively with our problems.

A March 2012 study titled *Six Americas*, conducted by the Yale Project on Climate Change Communication and George Mason University Center for Climate Change Communication, aimed at determining Americans' perceptions about global warming. Among findings: 38 percent believe that global warming is happening due mostly to human activities; 19 percent believe that global warming is happening due mostly to natural causes; 14 percent don't believe that global warming is happening; and 29 percent either don't know or haven't made up their minds.[461] In that study, 13 percent said they were *alarmed* about the issue, 26 percent *concerned*, 6 percent *disengaged*, 15 percent *doubtful*, and 10 percent *dismissive*.

An ongoing question is, "Why has the U.S. signed but not ratified the Kyoto Protocol?" In fact, it has never been presented to the U.S. Senate for ratification, likely because of a sense that it would be rejected. In part, it's a matter of political will. With most of U.S. energy coming from fossil fuels, economic concerns have been raised. Some have declared that global warming is a hoax; the fossil fuel lobby is strong; and the media have been compliant, according to *Time*'s Bryan Walsh. Add this. Developing nations, such as India and China, which were not included in the original Protocol, "now account for more than 50 percent of global greenhouse gas emissions." Note: *Time* also reports, "Developed nations that ratified the deal have largely been able to make the required cuts." [462]

An unavoidable conclusion is that our personal security depends on the security of our planet. We are past the point that any one group, wherever it is, can survive only at the expense of others. At some point, "United We Stand," may become a rallying cry to preserve life on earth. What happens to one of us happens to all of us. We're all in this together. We have nowhere else to go. Optimistically, our common threats may be common opportunities.

> **"When we try to pick out anything by itself, we find it hitched to everything else in the Universe."**
> *John Muir, Scottish-American Environmentalist and Writer*

Observations, Warning Flags
What's Up? What's at Stake?

In his compelling book, *The Fate of the Species*, Fred Guterl, executive editor of *Scientific American*, notes that scientists are trying to identify the "early-warning signals for climate that precede abrupt transitions." We could be reaching tipping points that will, in a dynamic way, impact each other. Among those systems is El Niño, originating in the Pacific but "causing weather changes across wide swaths of the globe." Guterl adds that the build-up of carbon dioxide in the atmosphere will likely impact the West African and Indian monsoons, possibly bringing drought to some areas and flooding rains to others, as patterns change. "Other tipping points could be the melting of sea ice, glaciers, and ice sheets in the Arctic, Greenland, and Antarctic and the rise in sea levels, coupled with destruction of the Amazon rain forest and Canada's boreal forests." Guterl suggests that policy makers pay more attention to *dynamical-system theory* (systems influence each other in a more-or-less chain reaction).[463]

Guterl highlights the work of James Hansen, a seasoned and outspoken climate scientist who retired from NASA's Goddard Institute for Space Studies. Drawing from Hansen's book, *Storms of My Grandchildren*, Guterl writes, "If we burn all the available coal and oil, the Venus effect is likely, and if we also burn other sources of fossil fuels, such as tar sands and tar shale, I believe the Venus syndrome is a dead certainty."[464] (Venus' atmosphere is 96.5 percent carbon dioxide. The planet is shrouded in clouds of sulphuric acid, with a temperature of about 867 degrees F, hot enough to melt lead.)[465]

Technologist and author Ramez Naam, a senior associate at the Foresight Institute in Seattle, shared some startling information with a 2012 World Future Society (WFS) audience in Toronto. During recent history, he observed, forests covered two-thirds of the planet; now it is one-third. As higher temperatures thaw Arctic permafrost, CO_2 and methane are released into the atmosphere. *On the planet*, we are now using about 1½ earths of resources. *In North America*, it amounts to about five earths.[466]

Zhouying Jin, director, Center for Technology Information and Strategic Studies, Chinese Academy of Social Sciences, Beijing, China, also spoke at that WFS Conference. She suggested Gross Domestic Product (GDP) does not necessarily reflect the quality of economic growth across a society. Another measure, she said, might be a Genuine Progress Indicator (GPI), which would include a balance of economic and environmental factors in addition to social progress and use of natural resources. Jin concluded that she would like to see a *Future 500*, not just a *Fortune 500*.[467]

Former chair of the Council of Colleges of Arts & Sciences Matthew Moen suggested conversation about whether climate change is occurring would shift to actual planning for how to adapt. He added, "Debate, with winners and losers, will shift from whether this is just a cyclical event to what we're going to do about it." Moen noted that scientists are planning for "a world in which the temperature is actually higher."

An Energy Strategy for the Future
A U.S. Approach

Blueprint for a Secure Energy Future[468] is a 20-page, 2012 White House report that lays out a proposed way forward in approaching solutions to energy and environmental concerns. In it, President Barack Obama, addressed issues such as: securing American energy (reducing dependence on oil, promoting energy efficiency, and investing in clean energy); the environment (protecting the nation's precious natural resources); climate change (reducing emissions that contribute to climate change and pollute our air and water); and gas prices.

The President called for deeper fuel economy standards for heavy-duty vehicles as well as cars and trucks, with an aim of an average 54.5 miles per gallon by 2025. Also included was a goal of putting more plug-in hybrid and electric vehicles on the road, which bolsters the need for advanced battery technologies, and the further development of cellulosic biofuels. The President also wanted to see expansion of high-speed rail and modernization of the aviation sector to reduce fuel consumption, reduce carbon dioxide emissions and improve air space management. (Corporate Average Fuel Economy "CAFÉ" standards were first enacted in 1975 and continue to aim toward increasing the fuel economy of cars and trucks.[469])

High on the agenda were clean, renewable energy technologies, elimination of fossil fuel subsidies, and generation of clean energy on public lands. The President reinforced the need for energy efficiency ratings and support for home energy-saving retrofits. His *Blueprint* supported setting new standards for residential and commercial appliances, a better buildings initiative, and a more efficient electrical grid.[470]

Cap and Trade would allow companies—that achieve emissions that are under certain limits—to sell credits to those that might exceed them. This proposed way of controlling greenhouse gas emissions would help control pollution of the atmosphere. As of 2012, the approach was held up in a highly polarized Congress.

The U.S. Environmental Protection Agency (EPA) is among the world's major sources of information about environmental issues, www.epa.gov.[471]

> "It's not true that we can't solve big problems through technology; we can. We must. But all these elements must be present: political leaders and the public must care to solve a problem, our institutions must support its solution, it must really be a technological problem, and we must understand it."
>
> *Jason Pontin, Editor in Chief and Publisher, MIT Technology Review[472]*

Personal and Planetary Security . . .
Making the Connection

You've heard it—"Why should I be concerned about other people? I've got enough on my plate just handling my own problems." That attitude has driven the behaviors of people embedded in hundreds of generations over thousands of years. It reflects a level of self-centeredness, a sense that we have no problem, even if our physical well being, our security, comes at the expense of others, including future generations.

However, the number of us on our planet is growing exponentially. Between 2000 and 2050, for example, the earth's population is expected to grow 50 percent, from 6 billion to about 9 billion people. Each of those 9 billion will be born with a hierarchy of human needs to be satisfied. Battles will likely rage over access to food. Global warming might change the climates of various regions of the world, turning rich farmland into desert and triggering wars over access to fresh water.

Intensifying polarization could provoke extremism, even fanaticism, and set off actual use of chemical, biological, radiological, or nuclear weapons, often referred to as weapons of mass destruction. Of course, conspiracy theories abound.

The good news is that these life-threatening scenarios may be avoidable. After all, everyone on earth has something in common. We live on the same small planet, rotating at about 1,000 miles per hour, circling the sun every 365 days. The fact is that we are the first generations of people who have the capability to destroy the world, and we may be the last generations who can save it. Much depends on how we educate our people. Contemporary philosopher, Ken Wilber, in his book *A Brief History of Everything*, makes clear that if we destroy the biosphere, we destroy ourselves.

Perhaps our greatest dilemmas are much more mundane, generated by an unwillingness to understand what motivates the values, opinions, attitudes, customs, beliefs, and perceived needs of people who are different from ourselves.

One Thing Leads to Another

Pick up a balloon. Press it at one spot. It will expand somewhere else. To some extent, that same principle applies to the growing interconnectedness of our planet.

For example, the price of a barrel of oil impacts the price of food, construction, chemicals, pharmaceuticals, transportation, heat, and light. A 2010 Deepwater Horizon Gulf oil spill created havoc for people and nature. The impact can become geometric, quantum, exponential. A further price could be unsustainable contamination of the environment, a tipping point that some might call a point of no return.

How will we be remembered? In *The Third Industrial Revolution*, Jeremy Rifkin asks us to think about what our age might be called, if humanity survives. He speculates that we might be called "the fossil fuel people and the period (might be referred to) as the Carbon Era, just as we have referred to past periods as the Bronze and Iron Ages."[473]

Common Threats and Common Opportunities
Natural and Human Caused Environmental Threats

Nothing galvanizes us more than common threats or common opportunities. Unfortunately, it has too often taken natural or human-caused disasters to rally our attention. Opportunities never seem as compelling as threats.

Perhaps that is why we are constantly threatened—because we fail to recognize and pursue the opportunities that surround us. Damage control has become the new norm, but will there come a time when we ultimately do damage that is beyond repair? Mass extinction events, natural or human-caused, are getting more attention, from the loss of species and recent killer viruses to our interest in disappearance of dinosaurs.

Natural Threats

Massive changes in the ecology of our planet have likely been caused by events of nature, such as volcanic activity, earthquakes, or direct hits by meteorites, asteroids, or comets, according to scientific consensus.

Other natural disasters, such as those caused by floods, hurricanes, typhoons, droughts, heat waves and extreme cold, snowstorms and blizzards, desertification, and tsunamis have largely been thought to be somewhat predictable, minimally controllable, and largely unpreventable. All can cause death, injuries, displacement of people, and heavy economic and social consequences, as we've witnessed time and again.

As concern grows about the impact of greenhouse gases on climate change, scientists suspect changes in the ecosystem may be intensifying some weather events, making them even more violent. *Accuweather* identified what it considers the world's five most expensive natural disasters in recent history, including: the 2011 9.0 magnitude earthquake and tsunami in Japan that resulted in a nuclear disaster at the Fukushima Power Plant near Sendai; the 1995 6.8 magnitude Kobe earthquake, also in Japan; the 2005 Hurricane Katrina that hit New Orleans and the U.S. Gulf Coast; the 1994 6.7 magnitude Northridge earthquake in California; and the 2008 8.0 magnitude Sichuan earthquake in China.[474]

In 2012 alone, Hurricane Sandy devastated parts of the New Jersey shore and inundated sections of New York City; a sudden *Derecho*, a flash-windstorm, struck the U.S. east coast; forest firestorms burned homes and forests across Colorado, Texas, and other states; and extreme drought conditions destroyed crops in the Midwest.

Asteroids...From Yucatans to Tunguskas
Natural Disasters Waiting to Happen

At Capital Science 2004, a March conference organized by the Washington Academy of Sciences, Martin Schwab of the Homeplanet Defense Institute presented scenarios for possible threats from wayward asteroids. He placed them in three categories: kilometer-sized "Yucatan Class," which he described as "earth killers;" 200- to 800-meter "Eros Class," the most common threat; and 50-plus meter "Tunguska Class" asteroids, which he described as "city killers."

While scientists track asteroids, there are so many of these bodies that they are almost too numerous to follow. Solid plans and technologies to meet the threat of an asteroid or comet colliding with the planet are generally not in place and could prove to be astronomically expensive.[475]

Since 2001, the U.N. has hosted a Science and Technical Subcommittee that includes a Near-Earth Object Working Group.

A potential mission to an asteroid has been mentioned for 2025. In August 2012, U.S. President Barack Obama said, "As we continue work with the International Space Station, we are focused on a potential mission to (an) asteroid as a prelude to a manned Mars flight."[476] Among likely benefits: development of new technologies, better understanding of planetary security, the possibility of a wealth of resources, building greater interest in science, and taking a strategic step forward in human space flight.

Human-Caused Threats

Instability can lead to threats, which often have worldwide implications. Just a few of those threats include:

• **Terrorism.** The same technologies designed to improve our lives have been used to disable and destroy. Growing numbers of incidents girdle the globe. Weapons of mass destruction could slip from control by nations into the hands of non-state, sub-national groups with a reach that ranges from local to international. Those weapons, coupled with a variety of tactics, are used to intimidate; to cause death, injury, and destruction; and to make a case for a group's particular point of view. While substantial attention is focused on biological, chemical, radiological, and nuclear weapons, other concerns include clandestine bombers and cyber attacks.

• **The Economic Divide.** Extreme income disparities exist both among and within nations. In 2012, the country with the estimated highest GDP per capita was Qatar, with an average of $106,000 per person, while the poorest was the Democratic Republic of Congo, with less than $365 per person.[477] Between 1960 and 1995, the disparity in per capita income between the world's 20 richest and 20 poorest nations more than doubled, from 18 to 1 to 37 to 1, according to the World Bank. The disparity has caused suffering and unrest that has led to misunderstanding, disagreements, and attacks.[478]

• **Refugees and Genocide.** Experts at an annual meeting of the American Association for the Advancement of Science (AAAS) warned that, "In 2020, the UN has projected that we will have 50 million environmental refugees." Growing numbers of people are forced to move because "if you have no water from a drought, have no food due to flooding, or if your home is quite simply underwater, what other options do you have but to flee?" Unbreathable air, plant diseases, multiplying numbers of pests, and water-borne and other diseases are among other major factors in the great migration of people seeking refuge.[479]

• **Carbon Emissions.** In *Vital Signs*, the Worldwatch Institute noted that, in 2007, U.S. carbon emissions per person amounted to 5.4 tons, compared to a worldwide average of 1.2 tons. "CO_2, a natural component of earth's atmosphere, traps heat that would otherwise radiate into outer space, thereby keeping earth's temperature within a habitable range. Human activity, including the use of fossil fuels by an increasing energy-hungry population, has put more CO_2 into the atmosphere, trapping more heat, raising the average global temperature, and changing the climate." [480]

• **Poverty and Armed Conflict.** Wars have continued to deepen poverty, ruin economies, destroy private and public infrastructure, and displace millions. As noted earlier, refugees are often trapped in camps, subjected to the will of warlords, and made susceptible to disease or exploitation. Some crowd the borders of other countries, their land or other property destroyed or peppered with mines, and their productivity put on indefinite hold. Wars and the spread of arms reduce investments that could strengthen economic growth for individual people and nations.

The Stockholm International Peace Research Institute (SIPRI) reported that world military expenditures amounted to approximately $1.74 trillion in 2011, up from $839 billion in 2001.[481] In 2013, without Sequestration, the United States accounted for approximately 40.2 percent of the total. China came in at approximately 7.9 percent; the U.K, 3.88 percent; Japan, 3.7 percent; Russia, 3.7 percent; Saudi Arabia, 3.2 percent; India, 3.0 percent; Germany, 2.8 percent; France, 2.5 percent; and Brazil, 2.2 percent. All others accounted for about 26.8 percent, according to *Time*.[482]

• **Natural Resources and Drug Trade Used to Support Combat and Suppression:** Warlords and organizations specializing in terror have captured the resources of some countries to support their interests. Those resources include but are not limited to diamonds, lapis lazuli, emeralds, timber, oil, opium, cocaine,[483] and rare earths. Illicit drugs have also led to ongoing smuggling, gang wars, underworld activity, health dilemmas, and international incidents.

Health Concerns and the Environment

"As much as 24 percent of global disease is caused by environmental exposures which can be averted," according to the World Health Organization (WHO). Among those ailments are *diarrhea*, largely from unsafe water, sanitation, and hygiene; *lower respiratory infections*, often caused by indoor and outdoor air pollution; *malaria*, as a result of poor water resources and land use management; and *Chronic Obstructive Pulmonary Disease (COPD)*, largely a result of exposure to workplace dusts and fumes and other forms of indoor and outdoor air pollution. The environment is also believed to impact *cardiovascular disease* and *cancers*.[484] *Cholera* outbreaks are often related to contamination of water and sewer systems (or the lack of them). *Salmonella* is generally spread through contaminated food products. *Asthma* and other allergies can be stimulated by environmental factors, ranging from secondhand smoke, dust mites, and molds, to chemical irritants, pests, pets, wood smoke, outdoor air pollution, nitrogen dioxide,[485] and pollen. *Stress-related ailments* and *obesity* are taking a growing toll, in some cases encouraged by economic, social, and other forms of tension or uncertainty.

Concern has grown about antibiotic resistant bacteria; *sepsis*, sometimes contracted in hospitals; and bacteria or viruses that might be released with climate change or by someone who creates new strains of them synthetically. Still another: Many have been injured and warned of dangers posed by the road, rail, and pipeline distribution of hazardous and flammable chemicals, fuels, and other materials.

Environmental factors are being studied to determine if they increase or decrease autism risk, according to an Autism Speaks Environmental Factors in Autism research project.[486] In 2008, approximately 1 in 88 children (1 in 54 boys and 1 in 252 girls) had been identified with an autism spectrum disorder (ASD), according to estimates from the Centers for Disease Control's (CDC's) Autism and Developmental Disabilities Monitoring Network (ADDM).[487] The 1 in 88 compares to 1 in 150 in 2000.

Water

A Natural Resource Vital to Life As We Know It

Nations and economies turn themselves upside down in their quest for oil and other forms of energy. At the same time, supplies of clean water, an essential for life as we know it, is moving front and center on every agenda. While oil and water may not mix, they are becoming the two most volatile liquids on our planet

In its 2004 special issue devoted to "The Future of Earth . . . A Planet Challenged from the Arctic to the Amazon," *U.S. News* reported, "In a nation where abundant, clear, and cheap drinking water has been taken for granted, it is hard to imagine residents of a major city adjusting to life without it." Gary Wolff, an economist for Oakland's Pacific Institute for Studies in Development, Environment, and Security, observes, "The idea of water as an economic and social good, and who controls this water, and whether it is clean enough to drink, are going to be major issues in the country."[488] Water quality and quantity will be critical environmental issues in the 21st century.[489]

Common Opportunities

One thing we know with some certainty. The earth's species have been adapting to life on this planet for millions of years. We do not have evidence of a nearby planet that could readily sustain us. In short, we have nowhere else to go. That should be incentive enough to see problems as opportunities in disguise.

To the extent we've been trying to ignore environmental problems, that cost of neglect has passed its due date. We are surrounded by problems that must be addressed, and that should bring a tingling to the nerve endings of students who are searching for future economic and civic opportunities. Think, for example, about employment opportunities that could be created by seriously pursuing any or all of these possibilities:

- Harvesting renewable sources of energy and increasing battery capacity.
- Conceiving of treatments, cures, and prevention strategies for diseases. Creating technologies and treaties that will guarantee

adequate fresh water.

- Pursuing careers in diplomacy.
- Educating and nurturing future scientists who will be needed to probe the depths of the atom and the outer reaches of the universe.
- Becoming one of a new generation of economists capable of further developing economic theories, policies, and strategies that consider quality of life as part of the bottom line.
- Considering a future as an enlightened politician, office-holder, or public servant.
- Promoting and supporting even better environmental education. Why is education listed last? Because education is the bottom line. It will be our schools and colleges, whether they meet in a fancy building or under a tree in some remote part of the world, that will spell the difference between opportunity and disaster. Yes, education is that serious.

"Man shapes himself through decisions that shape his environment."

Rene Dubos, French-American Microbiologist

Spaceship Earth
The Chick and the Egg

Internationally renowned master of science and design, philosopher, and author R. Buckminster Fuller, in his 1969 classic *Operating Manual for Spaceship Earth*, suggested giving greater consideration to the world human beings inherited.

We were blessed with an "abundance of immediately consumable, obviously desirable, or utterly essential resources," he said. Those resources, "have allowed us to carry on despite our ignorance. Being eventually exhaustible and spoilable, they have been adequate only up to this critical moment."

Fuller compared those resources to what a bird has available when it's still inside the egg; they only last a certain length of time. "The nutriment is exhausted at just the time when the chick is large enough to locomote on its own legs," he wrote. "And so as the chick pecks at the shell seeking more nutriment, it inadvertently breaks the shell." Then,

"Stepping forth from its initial sanctuary, the young bird must now forage on its own legs and wings to discover the next phase of its regenerative sustenance."[490]

Fuller warned that we must prepare for that day when our sustaining nutriment is exhausted. We will likely come to the stark realization that much of what we thought was infinite—such as clean air and water—are finite, unless we intervene to protect them. The myth of our traditional definition of "infinite supply" might very well be shattered.

Implications for Education and the Whole of Society
Personal/Planetary Security

The following implications have repercussions for both education and the whole of society as we work together to address our individual and planetary security.

• **Balancing economic development and environmental sustainability.** As we make progress in our communities, in our nation, and globally, we need to measure what we do against its impact on our environment and the legacy we leave for future generations. Let's put it this way: In the short-run, an unsustainable activity might make a few people much richer. In the long-run, it might make all of us much poorer. "Our grandchildren and their children will be left trying to clean up the environmental mess we are leaving behind," observes Ted Blaesing, a longtime Minnesota educator. "All ages of schooling will have an increased focus on the science and ethics of environmental issues," he predicts.[491]

• **Insisting on environmental education.** The evidence is multiplying that our environment is in peril, and so are all of us who occupy the planet. To solve these problems, we must overcome the gridlock as we make the case for a sense of urgency. Environmental education in our schools and communities is one important step. According to EPA, "Environmental education increases public awareness and knowledge about environmental issues or problems. It also energizes people and gives them the knowledge, skills, and experience to take responsible action. A primary desired outcome of environmental education programs is environmental literacy."[492]

Tokyo International School Vision Navigator Patrick Newell declared a sense of urgency: "The time has come to reconnect with nature. A survey recently found that 16 to 22-year-olds preferred having a connection with digital tools over smell." Meera Balachandran, director of the Education Quality Foundation of India and a member of Futures Council 21, pointed out the need for educators worldwide to do their part in an effort aimed at "awareness and prevention of environmental devastation."

• **Making study of the environment an interdisciplinary springboard for creativity, problem solving, media literacy, an understanding of globalization, and other areas of study.** This multidisciplinary concern can be addressed in courses or units in science, math, engineering, technology, business, economics, social sciences, civic education, law-related education, international education, government, media literacy, communication, and environmental studies, to name a few. For example, science and technology students might be asked to conceive of creative, renewable, clean sources of energy. Community and education leaders should work together with business and government to build even greater understanding of corporate citizenship and social responsibility. All should have a working knowledge of *Maslow's Hierarchy of Needs*.

• **Creating environmentally friendly schools and other buildings.** Buildings provide an "envelope" where people generally gather to work or learn, often both. They are also symbols for how we think and feel about the environment. Indoor Environmental Quality (IEQ) should be considered essential. In fact, if teachers and students are exposed to allergens, pollutants, and chemicals…if they encounter poor ventilation, lighting, acoustics, and temperature control…they may not be able to do their best teaching, learning, or just plain functioning. Some symptoms that indoor environmental health might be lacking include: respiratory irritation, sore throats, drowsiness, headaches, asthma attacks, and the inability to concentrate, according to the U.S. Environmental Protection Agency (EPA).

For schools and colleges, maintaining "green" buildings can provide a living laboratory for students who are learning about the environment. On a broader scale, to be sustainable, all structures should reflect good siting, design, construction, operation, and maintenance.[493]

Additional implications of this environmental trend include:

• **Promoting school and community commitment to the environment.** What do we want to be known for? If we hope to attract and keep the people we need to create a future, then we may want to pursue a "green" identity.

• **Creating and encouraging environmental entrepreneurs.** Consider an ongoing campaign or movement to produce and support environmental entrepreneurs.

• **Expecting schools, colleges, businesses, governments, and others to demonstrate their commitment to the environment.** What we say is important. However, the message is more powerful if we practice what we preach.

• **Offering futures courses.** As students study futures, they develop techniques for better understanding the social, political, economic, technological, demographic, and environmental forces that drive us today and the potential impact our decisions will have on those forces tomorrow.

• **Developing a philosophy of possibility.** How we feel about *possibility* will play a huge role in charting our future. At its best, education and community leaders should instill that sense of possibility.

Our Legacy:
We are the first generations of people who have the capacity to destroy the world . . . and we may be the last generations capable of saving it. What happens will depend largely on how well we educate our people.

Environment

Questions and Activities

1. How does Abraham Maslow's "Hierarchy of Needs" support the need for us to pay more attention to planetary security?

2. Tell someone you know Buckminster Fuller's story about the chicken and the egg. Prepare a five- to six-minute PowerPoint presentation focused on five ways you believe human activities can have a profound impact on the natural world.

3. Review "Teacher Resources and Lesson Plans" for teaching about the environment. These materials were developed by the U.S. Environmental Protection Agency (EPA). Review teaching materials/lessons. Write a one-page paper or blog-post emphasizing three ideas from that resource that "I'd like to see taught in my classes or school." http://www.epa.gov/students/teachers. html.[494]

4. View "This Bulb," National Geographic Society, http://video.nationalgeographic.com/video/kids/green-kids/this-bulb-kids/ (1:45).[495]

5. "Learning Center: Climate Change Courses & Consequences," information plus teaching resources, National Center for Atmospheric Research, http://spark.ucar.edu/climate-change-causes-and-consequences.[496] Review and consider ideas.

6. Review "Alex Steffen: The Route to a Sustainable Future." Steffan co-founded *Worldchanging.com* and is author of books such as *Carbon Zero*, 2005, Oxford, U.K., http://climatechange.worldbank.org/sites/default/files/Turn_Down_the_heat_Why_a_4_degree_centrigrade_warmer_world_must_be_avoided.pdf. Identify five things you like or dislike about his ideas.[497]

7. What should schools and colleges teach students about environmental opportunities and threats facing people across all political boundaries?

8. Prepare a paper or presentation indicating how our personal security is connected to planetary security. Conclude the work with at least five to 10 ways we can teach students about those connections.

Readings and Programs

1. Review and keep as a reference: *Turn Down the Heat...Why a 4 degree C. Warmer World Must Be Avoided*, World Bank Report by the Potsdam Institute for Climate Impact Research and Climate Analytics, World Bank, Washington, D.C., Nov. 2012, http://climatechange.worldbank.org/sites/default/files/Turn_Down_the_heat_Why_a_4_degree_centrigrade_warmer_world_must_be_avoided.pdf.[498]

2. Walsh, Bryan. (2012). *Global Warming: The Causes, The Perils, The Solutions*, Time Home Entertainment, Inc. TIME Books, NY[499]

3. Read transcript or view the program, Jane Goodall, interviewed on *Bill Moyers Journal*, PBS, Nov. 27, 2009, http://www.pbs.org/moyers/journal/11272009/watch.html.[500]

4. View "Global Warming, A Way Forward: Facing Climate Change," National Geographic Society Video, Washington, D.C., with support from the U.N.Foundation, http://video.nationalgeographic.com/video/environment/global-warming-environment/way-forward-climate/ (7:40).[501]

5. Check out environmental videos at MIT Video, Environment, http://video.mit.edu/channel/environment/.[502]

6. Fuller, R.B. (1969). *Operating manual for spaceship earth.* New York: South Illinois University Press, Touchstone Book by Simon and Schuster.[503]

7. Guterl, Fred. (2012). *The Fate of the Species*, Bloomsbury USA, NY[504]

8. Wilber, K. (2000). *A Brief History of Everything.* Boston: Shambhala Publications.[505]

"We are either headed for breakthroughs or breakdowns."
Ervin Laszlo, Founder, Club of Budapest International

Chapter 10

Sustainability
Bearable, Viable, and Equitable

Chapter 10: Sustainability. *Bearable, Viable, and Equitable*

Trend: Sustainability will depend on adaptability and resilience in a fast-changing, at-risk world.
Short-Term Advantage ↔ Long-Term Survival
Wants of the Present ↔ Needs in the Future

> "What's the use of a fine house if you haven't got a tolerable planet to put it on?"
> *Henry David Thoreau*[506]

Great Idea! But is it Sustainable?

That's a question that pops up virtually everywhere. Generally understood as the *ability to endure*, sustainability has become a flagship concern for the environment and the economy. While its implications are spread across those and other chapters of this book, the urgency of sustainability is ingrained in every problem we solve and every issue we address in nearly every walk of life. That's why we're briefly setting it apart and putting it in the spotlight.

Sustainability is another elephant in the room, so big that we sometimes don't even see it. In fact, in a polarized society, where honest discussion becomes buried in political rhetoric, that herd of elephants in the room seems to be increasing exponentially. In an either/or, right/wrong, my way/no way society, civil discussions about sustainability are too often reduced to something like a multiple choice test. We can fill in the bubble for one of two false choices, *a dark cloud* or *a silver lining*.

Sustainability
Back to Basics

Let's start with the environment and call it our *natural capital*. If the environment provides the ingredients for life, we'll be able to exist and possibly even become self-sustaining. At some point, we might develop a society that thrives because of the contributions of *human capital*. Based on the foundation of natural and human capital, various forms of industry might develop. We'll call that *economic capital*. As simple as it seems, the whole thing is truly profound. Fundamentally, the environment and society don't exist because of the economy, but the economy exists on a foundation of natural and human capital.

Of course, social, environmental, and economic spheres overlap. At the intersections of those forces, we have an opportunity to think about whether our dreams and plans will be *bearable, viable,* and *equitable*.[507] If what we're considering doesn't pass the test, it may simply not be sustainable.

Nature. The Industrial Revolution left the impression that we could do whatever we pleased with our natural environment and it wouldn't fight back. We were convinced that it would adapt to us, however we chose to use it. With urbanization, most of humanity became cloistered in cities, and virtually billions of people lost touch with and ended up with little *empathy* for the natural world. Among the consequences? If we continue to consume as we have been, we'll likely need four or five planets to sustain us.

Social thinker, author, and senior lecturer at the Wharton School of Business Jeremy Rifkin, reminds us: "To empathize is to civilize." The philosophical debate about whether those trees and animals live in our world or we live in theirs may have to move to a later agenda as we realize that we are likely interdependent. Rifkin, drawing on observations from the late British philosopher Owen Barfield, reminds us that, "for more than 90 percent of our existence on earth, we lived as foragers and hunters." Those ancients, Rifkin notes in his book, *The Third Industrial Revolution*, were utterly dependent on the "goodwill" of "Mother Earth." He strongly suggests that parents, schools, and society firm up a relationship with the natural world.[508]

Striking a sustainable balance with nature is getting increased attention as global temperatures rise and life on earth, at least as we've known it, is threatened. Being able to endure may depend on our willingness to understand that our thin environment is only resilient to a point. Sustainability may depend on whether we can adapt to scientific realities. Consider this premise: While we might follow the environmental arguments of our favorite political commentators, nature keeps right on following the laws of science.

Sustainable: (a). relating to, or being a method of harvesting or using a resource so that the resource is not depleted or permanently damaged. (b). of or relating to a lifestyle involving the use of sustainable methods. *Synonyms:* defendable, defensible, justifiable, maintainable, supportable, tenable. *Antonyms:* indefensible, insupportable, unjustifiable, unsustainable, untenable. *Merriam-Webster Dictionary Online.*[509] Additional definitions include the ability to endure, prolong, keep up, exist, and to nourish.

The Expanding World of Sustainability

In a fast-changing world, the *ability to endure* does not mean simply *maintaining the status quo.* Among the many essentials for sustainability are: adaptability, resilience, transparency, staying in touch, educating and training people, recognizing talent, creating new knowledge, cultivating creativity and imagination, balancing political and market forces with the common good, adding value, avoiding obsolescence, respecting civil and human rights, and demonstrating courage in facing the possible longer-term benefits and consequences of our decisions.

Education and Infrastructure,
A National Strategic Narrative, **Mr. Y**

What does it take to be sustainable? In 2011, two active military officers, using the pseudonym, Mr.Y, developed what they called, "A National Strategic Narrative." Among publishers was the Woodrow Wilson Center.

These thoughtful leaders called for a shift from a longstanding military and foreign policy strategy focused on *containment* toward one that gives greater attention to *sustainability.* Our ability to compete in the world depends on our "willingness to invest in skills, education, energy sources, and infrastructure." That investment should include "national resources that can be sustained, such as our youth and our natural resources (ranging from crops, livestock, and potable water to sources of energy and materials for industry)." They added, "We think that we are an exceptional nation, but a core part of that exceptionalism is a commitment to universal values—to the equality of all human beings, not just within the borders of the United States but around the world."

"Our first investment priority," they declare, "is intellectual capital and a sustainable infrastructure of education, health, and social services to provide for the continuing development and growth of America's youth." These officers call for "a system that demands adaptability and innovation." They add, "Only by developing internal strength through smart growth at home and smart power abroad, applied with strategic agility, can we muster the credible influence to remain a world leader." All of that and more will be needed if we hope to sustain "a pathway of promise and a beacon of hope, in an ever changing world."[510]

The Mr. Y report was not published as official policy but as a cluster of big ideas to consider in framing the nation's position in the world and building on its strengths. It's the kind of narrative we might consider developing for our own communities.

The Urgency of Sustainability on Many Frontiers

While a lion's share of discussions about sustainability focus on the environment, the need extends to nearly every frontier of our existence. While our commentary doesn't pretend to be inclusive of every need, our hope is that it will expand the conversation. We'll make brief stops as we travel the road from community planning and homes to finance and lifestyle.

• **Community Planning.** The Sustainable Cities Institute considers land use planning that involves "decisions on crosscutting and multi-layered issues that affect air quality, water quality, access to transportation options, economic vitality, and quality of life." Rather than placing a massive commercial storage facility next to an historic monument, they recommend residential, commercial, civic, industrial, open space, and agricultural land zones. They suggest paying attention to sustainable water resources and a strong sense of place. The organization also emphasizes the need for energy-saving retrofits and minimizing waste.[511]

• **Cities.** In 2011, the City of Chicago announced that it would "retrofit 100 of the city's buildings to increase their energy efficiency." Set to be included, according to SustainableBusiness.com's Andrew Burr, were upgrades in lighting, mechanical equipment, and water conservation technology involving about 6.5 million square feet of space. The initiative was expected to "save taxpayers around $4 million to $5.7 million a year, reduce the city's carbon footprint, and create

375 direct jobs and 1,100 manufacturing and related jobs.[512] The city was also promoting electric vehicles and trying to make its own fleet more energy efficient.[513] The City of Toronto launched an Energy Retrofit Program following a Smog Summit in 2004.[514] In fact, numerous cities nationwide and worldwide have undertaken *energy* and *resilience* building retrofits. Examples include Chicago's Willis Tower (formerly the Sears Tower)[515] and the Empire State Building in New York City.[516]

• **Businesses.** There are many sides to sustainability for businesses. Justifying their existence generally involves more than profit. These institutions also need to be aware of how they honestly reflect their corporate citizenship and social responsibility. Their triple bottom line should include "profits, people, and planet," according to *Financial Times*. Of course, attracting investors, securing talent, developing systems, and producing a viable return on investment are very much a part of their survival mix. Visited a sustainable restaurant lately? The Sustainable Restaurant Association (SRA) gauges an eating establishment's approaches to "society, environment, and sourcing."[517]

Writing for the January 2011 issue of *Harvard Business Review*, Michael Porter and Mark Kramer observe that businesses are often "trapped in an outdated approach to value creation that has emerged over the past few decades. They continue to view value creation narrowly, optimizing short-term financial performance in a bubble while missing the most important customer needs and ignoring the broader influences that determine success." They ask, "How else could companies overlook the well-being of their customers, the depletion of natural resources vital to their businesses, the viability of key suppliers, or the economic distress of the communities in which they produce and sell?" Porter and Kramer recommend that businesses focus on "creating shared value." They add, "Companies must take the lead in bringing business and society back together."[518]

The World Council for Economic Development points out that business sustainability involves "meeting the needs of the present without compromising the ability of future generations to meet their own needs." That means attention to "economic efficiency (innovation, prosperity, productivity), social equity (poverty, community, health and wellness, and human rights), and environmental accountability (climate change, land use, and biodiversity)." Using a model of life cycle analysis, "Those organizations wanting to take a large leap forward should systematically analyze the environmental and social impact of the products they use and produce.[519]

• **Education Institutions.** Schools and colleges play multiple roles in sustainability. First, they model it. Second, they teach it. Third, they pay a great deal of attention to whether what they teach and how they teach it matches the needs of a fast-changing world that is outside the trenches. All the while, they try to provide a safe and stimulating learning environment and set a good example.

When schools and colleges build or upgrade buildings, growing numbers of them are using guidelines developed by the U.S. Green Buildings Council. Those guidelines are part of the Leadership in Energy and Environmental Design (LEED) program. They zero in on five key areas of environmental and human health: energy efficiency, indoor environmental quality, materials selection, sustainable site development, and water savings. Students are involved in these projects and learn firsthand about how what they are learning applies in the real world.[520]

Since colleges and universities are often like small cities, they are "some of the best incubators of sustainable solutions to myriad problems, and they are molding and inspiring the bright minds that will inherit the environment from the current generation of leaders," according to *The Daily Green.* Like many elementary and secondary schools, colleges and universities are likely to feature organic gardens, green roofs, solar panels, wind turbines, and ground-source energy systems.[521] Add low-flow toilets using rainwater and skylights to the mix. Writer Brian Clark Howard, in a separate article about cafeterias, added that food service programs are a key part of the sustainability effort, "buying local, composting waste, and saving energy."[522]

The University of South Dakota is among postsecondary institutions offering a degree program in sustainability with an aim of strengthening a workforce that will contribute to "wind energy, hydroelectric power, geothermal energy technologies, green construction techniques, and biofuels." The university sees in the program "a dimension of sustainability (that will be) an essential part of the state's economy for the foreseeable future." USD Dean of the College of Arts & Sciences Matthew Moen expects to see sustainability programs sprouting up on a number of college campuses.[523]

Southern Methodist University (SMU) in Dallas has a degree program plus a Sustainability Committee, "a group of faculty, students, and staff who sponsor activities and events on campus that encourage lifelong environmentally conscious behavior." Also in Texas, Lady Bird Johnson Middle School in Irving gives students "hands-on experiences with environmental stewardship, energy conservation, and such topics as geothermal science, rainwater collection, solar panel usage, and turbine efficiency." [524]

It doesn't stop there. In a broader context, sustainability for schools and colleges involves developing a clear and established identity, a distinctive culture, and an excellent product, according to Ian Symmonds & Associates, a research and strategy firm serving independent schools, colleges, universities, and nonprofts.[525]

Schools are center-stage in shaping the future for a community, country, or the world. Their curriculum and instructional methods, plus the technologies they use, the support they get from their boards and legislative bodies, and their ability to respond to the fast-changing needs of society are essential to their sustainability.

• **Buildings in General.** When you think about it, a building is a bit like an envelope, a container that puts a roof over our heads and helps rather than hinders us in getting things done. That could range from healing the sick in a hospital, or educating a child in a school, to providing a home base for a hotel, restaurant, office, or the place you live. Building sciences are important for us for many reasons because, among other things, we spend much of our lives in a *built environment*.

Buildings should *and increasingly must* be sustainable. An aim is a zero energy footprint, which means that a building doesn't contribute to greenhouse gases. The road to that goal is likely paved with renewables, such as solar, wind, and ground source energy, coupled with ramped up efficiency. Not only is a high-energy efficiency rating for equipment important, but so is insulation. James Woods, a noted indoor environments consultant, emphasizes the importance of air quality; thermal issues connected with heating, ventilation, and air conditioning; humidity control; lighting; acoustics; and a host of other related issues. Allergens are a significant concern. So are antibiotic-resistant bacteria that occasionally catch up with people in hospitals.

Architects, engineers, energy/environmental scientists, and contractors make sure that any structure they design and construct complies with the latest codes. Then there are building engineers who keep mechanical, electrical, and environmental systems running at peak efficiency. Everyone involved tries to create a comfortable, healthy, and productive working and learning environment.

After the 9/11 terrorist attacks, a group of engineers, scientists, security professionals, and others formed an organization called TISP, The Infrastructure Security Partnership. Members of this group, on a grand scale, are working to make the built environment even more sustainable and resilient in the face of *natural hazards*, such as earthquakes, tornados, hurricanes, and tsunamis, and *human-caused hazards*, such as an accident or a deliberate attack. In addition to buildings, TISP pays a great deal of attention to the protection and resilience of transportation systems, such as air, land, and seaports, as well as dam and levee systems.

• **Social Systems.** Programs such as Social Security, Medicare, and Medicaid are sustainable only if society provides support, generally through taxes. As Baby Boomers retire in droves and life expectancies expand, these "safety nets" face actuarial realities that have been known for decades. The same is true for other programs that serve people at many levels of society. Most people have contributed to these programs during significant portions of their lives and consider them part of a social contract. Whether we maintain that contract becomes a measure of our sense of community, our empathy, and our reliability. Sustainability depends on whether society, in concert with the people it elects, can adapt to emerging realities and ensure reliability across generations. Some argue that multi-generational social contracts should not be held hostage to short-term political advantage.

• **Lifestyle.** How do we identify the good life? Do we get a tingle

of satisfaction from consuming healthy foods and doing what we can
to save energy to help sustain the environment? Or do we get that sense
of satisfaction from yanking the handle eight times on a paper towel
dispenser in a public restroom, only to dry our hands in three seconds
and drag what might have held the equivalent of a Dead Sea Scroll
into a trash can? Which is a true symbol of *the good life*? The answer?
We don't know yet. But more people are interested in clothes made of
sustainable fabric and cars that spew less carbon into the atmosphere.
Largely, the survival of our species may depend on our lifestyles, which
often reflect an insistence on overpowering the earth's natural capability to restore itself. Perhaps there are more sustainable ways to demonstrate our power.

> **"Every profession bears the responsibility to understand
> the circumstances that enable its existence."**
> *Robert Gutman*[526]

Big Questions, The Sustainability Landscape

Our goal in this chapter is not to cover every aspect of sustainability,
but to raise the issue, provide some background, and provoke further
discussion. With that goal in mind, let's consider a few questions about
sustainability in a few more arenas.

• **Financial Investments.** How important is a diversified portfolio
in sustaining what we hope will be the longer-term viability of our investments? How can we be sure that those investments, whether they
are in bank accounts, real estate, commodities, precious metals, stocks,
bonds, or other instruments, are reasonably secure? When should we
buy and sell? What rules and regulations might be needed to ensure
that investments are safe? What lessons about sustainability have we
learned from the dot.com bubble of the early 1990s and a housing bubble inflated by mysterious derivatives that served as a building block for
the Great Recession that began in 2007-08? What questions should we
ask in selecting a financial advisor?

• **Energy.** What are the differences between finite and renewable
sources of energy? Should we invest in solar, wind, bio, ground source,

and other renewable energy technologies now or wait for prices of finite sources to increase dramatically, creating a rolling scarcity for those who increasingly won't be able to afford them? What impact will use of fossil fuels at current or accelerated levels mean for the planetary environment, the economy, and our quality of life—even our survival? Thinking ahead, what narrative should describe our approach to energy sustainability?

• **Transportation and Logistics.** Should we be reinventing personal and public transportation to make them more sustainable? Considering substantial increases in population worldwide, especially urban areas, how can we logically transport people, safely and affordably, from one place to another? Is our transportation system adequate for moving goods from producer to market? Are our transportation and logistics systems sustainable in the longer-term? How could we lessen their environmental footprint? How should we define a "sustainable car?" Will it be a hybrid, a plug-in, or operate with a hydrogen cell? Will we have smart cars and roads with built-in guidance systems?

The Concorde. Why is the supersonic Concorde no longer flying? Bottom line, it probably wasn't the gashed tire and fuel tank explosion in 2000. More likely, it was the law of diminishing returns. How much is it worth to cross the Atlantic in 3 rather than 7½ hours? Now, some are wondering if we are ready for hypersonic flights—3,125 mph, altitude 20 miles, Paris to New York in 90 minutes?[527] What will be sustainable? When?

• **Technology.** Have we become enamored with using computers and their mobile cousins to find quick information and stay in constant touch? Should we be spending some of that time considering and maybe inventing new generations of those technologies that are both viable and affordable, possibly even improving the human condition? How do we define sustainability as we consider buying a computer, knowing that computing power will double in 18 months, making whatever we buy at least marginally obsolete? How do we ensure that the manufacture of technological devices, including batteries, and their disposal, do not adversely impact the environment?

• **Agriculture.** Are current methods of tilling soil, planting, fertilizing, and harvesting crops and raising livestock sustainable? What impact does irrigation have on the fresh water supply? How susceptible is agriculture to extreme changes in climate or weather patterns? What forms of bio-fuels, often refined from agricultural products, have least impact on the foodchain? What is agriculture's impact on greenhouse gases? How could *anaerobic digestion* help us break down animal waste and capture biofuels? What are the upsides and downsides of the rise in often-massive corporations that dramatically influence the inputs and outputs of agriculture? What is the future of organic farming and vertical farming? Who owns rights to seed grain? Are policies needed to ensure the viability of genetically altered species? Will we likely see more development of protein from stem cells? How can we ensure that knowledgeable and experienced people will tend to farms and ranches in the future? Do we, as people, understand the importance of producing food and fiber to the sustainability of life itself?

Our Assignment, If We Accept It...

Legendary biologist, author, and philosopher Edward O. Wilson, in his book, *Letters to a Young Scientist* urges us to find an unsolved scientific problem and imagine ways to solve it. His counsel may be a key to sustainability.[528]

Implications for Society, including Education
Sustainability

• Leaders should expect to be constantly challenged with maintaining the sustainability of their schools or colleges and the future of their communities.

• Demand will grow for more sustainable buildings and an infrastructure that is not only built to endure but also subject to ongoing maintenance and repair.

• Sustainability should be part of the discussion whenever we consider world, national, community, and personal decisions.

• Courses and degree programs in sustainability are likely to increase substantially at colleges and universities in response to growing demand and the urgency of societal need. Study units will likely be developed for all levels of education, from kindergarten through professional development in every field.

• For-profit and non-profit organizations can be expected to grow, aimed at enhancing sustainability. Related careers will likely see an upswing.

• Since our short- and longer-term future is at stake, astute educators and other leaders will likely see sustainability as a concept for uniting people across disciplines and walks-of-life. As diverse groups of people are drawn together to discuss these issues, they will gain greater understanding of demographic realities, developments in technology, economic needs and opportunities, ethics, and scores of related issues.

• With limited funds, new ideas and initiatives to maintain the sustainability of the education system will sometimes "be funded by the quick abandonment of old practices," foresees Keith Marty, a member of Futures Council 21 and a school superintendent. "Schools are slow to change," he observes. In a fast-changing world, education systems with scarce resources may need to make changes that are "more systematic" and may seem "sudden for many people."

• In areas of the curriculum such as economics, social studies, environmental studies, physical and biological sciences, and other subjects, students will be expected to learn how they can apply the principles of sustainability. The need to address the issue of sustainability will spur class and student discussions across disciplines.

• To further enliven some classes, students will be asked whether certain inventions or strategies in the past, present, and possibly the future, have been or will be sustainable. As a spur to their thinking and reasoning skills, they might be asked *why* or *why not*. Applying active learning and learning through inquiry, students will be challenged to conceive of what might be *even more sustainable*.

• In the hands of creative educators, students, and others in the community, discussions of this concept can help build a more sustainable future for our education system, economy, and civil society.

Sustainability

Questions and Activities

1. Briefly, based on what we've shared in this chapter and your own creative genius, write a brief definition of sustainability that you would recommend for an upcoming dictionary.

2. Identify a local community project or institution. In no more than one page, share your thoughts on how the one you've chosen will contribute to sustainability and whether it will, itself, be sustainable.

3. What five recommendations would you make to your school, college, university, or community to ensure its longer-term sustainability?

4. Read "A Framework for Sustainability, Indicators at EPA," produced by the Environmental Protection Agency. Prepare a five-minute PowerPoint presentation devoted to an aspect of that report, http://www.epa.gov/sustainability/docs/framework-for-sustainability-indicators-at-epa.pdf.[529]

5. Review "Sustainability Education Resources," Portland State University Graduate School of Education, numerous resources that include teaching materials and videos, http://www.pdx.edu/elp/sustainability-education-resources.[530]

Readings and Programs

1. "A National Strategic Narrative," Mr. Y, published by the Woodrow Wilson Center, Washington, D.C., 2011, http://www.wilsoncenter.org/sites/default/files/A%20National%20Strategic%20Narrative.pdf.[531]

2. Explore *Science Daily*'s web site, including its informational and teaching materials, such as videos, http://www.sciencedaily.com/videos/earth_climate/.[532]

3. View "Ecosystem Markets Task Force Video: A New Lens for Business; A New Economy," March 2013, University of Cambridge, U.K. Business executives discuss a sustainable economy, (5 min.), http://www.cpsl.cam.ac.uk/Business-Platforms/Natural-Capital-Leaders-Platform.aspx?#fragment-2.[533]

4. Wilson, E.O., *Letters to a Young Scientist*, (2013), Liveright, W.W. Norton & Company, NY, http://www.amazon.com/Letters-Young-Scientist-Edward-Wilson/dp/0871403773.[534]

Chapter 11

International/Global
We're all in this together

Chapter 11: International/Global. *We're all in this together.*

Trend: International learning, including relationships, cultural understanding, languages, and diplomatic skills, will become basic.

<u>Sub-trend</u>: *To earn respect in an interdependent world, nations will be expected to demonstrate their reliability and tolerance.*
Isolationist Independence ↔ Interdependence

> **Did you know?**
> **The facts are stunning. According to the U.N. Population Division, for every 100 people who live on the planet, 60 percent live in *Asia*, 15 percent in *Africa*, 11 percent in Europe, 9 percent in *Latin America and the Caribbean*, 1 percent in *Oceania*, and 5 percent in *Northern America*.** [535]

The World
How Small Is It?

The earth has somehow gotten smaller. If Greece catches a cold, people in other parts of the world start to sneeze. If a nuclear reactor goes off kilter in Asia, nations an ocean away start testing for radiation. Even though the road has not been smooth, we are moving from isolationist independence toward a realization that the people of this relatively small blue planet are interdependent.

Actually, our earth is about the same size it was when I was a kid— 25,000 miles around, roughly 7,926 miles in diameter at the equator and 7,901 miles pole to pole. We're an *ellipsoid*, slightly wider than we are tall.[536] Even though it might seem very big when you're standing in one spot, the earth, for all practical purposes, seems to be shrinking. Instant communication and high-speed air transportation can bring us together in anything from microseconds to a few hours. We carry on conversations with people around the world without breaking a sweat. In a sense, everyone is our neighbor.

A headline might read, "Small Planet Spawns Big Tensions." Rather than sensing the common blessing of a place that can harbor life such as ours in the vastness of space, we fight over territory, power, authority, religion, economic advantage, traditional energy resources, water, and a host of other massive issues. We fight over whether we should protect our environment and even over whether we should fight. A 1991 Los Angeles riot survivor, Rodney King, asked, "Can we all get along?"[537] The answer, whatever it is...may foretell our future.

Former U.S. Attorney General Elliot Richardson once noted, "The human beings who make up every ethnic group and every political entity share common bonds with all other human beings on this planet. All of us want a safer, more orderly, and more humane world, not simply because such a world is better for ourselves, but because we recognize our kinship—however distant—with others."[538]

The World
How Flat is It?

The World is Flat. That's the title of *New York Times* columnist and author Thomas Friedman's now classic book, first published in 2005. So what is he getting at? While you can read the book and draw your own conclusions, most of us likely agree on a few basics. *One*, the playing field is being leveled. *Two*, a long list of technologies, including many that will continue to emerge, make it possible for people to work, in essence, side-by-side, even if they are physically thousands of miles apart. It's a convergence, a connectedness, and a part of the foundation for an increasingly global society. Friedman wonders if the world has gotten too small and too fast for human beings and their political systems to adjust in a stable manner. He points out, "The faster and broader the transition to a new era, the greater the potential for disruption."[539]

Joseph Hairston, President and CEO of Vision Unlimited in Maryland, foresees "the flat world becoming evident in our schools. Multicultural environments will expand in our classrooms as learning becomes an *anywhere anytime* experience." As for interdependence, Frank Kwan, director of communication services for the Los Angeles County Office of Education, called it "a forced choice for America to remain globally competitive. The challenge is for educational institutions that are not traditionally agile to incorporate changes that serve students in meeting the needs of a global economy."

Globalization. How would you Define it?

We know that globalization has intensified with boundary-jumping advances in transportation and communication. We now realize that we are increasingly interdependent. Consider these observations that could help us define this phenomenon.

"Globalization is the accelerating traffic of people, capital, and cultural products around the world. (It) embodies opportunities and risks for individuals and societies worldwide." *Introductory materials, 2012 Harvard Future of Learning Institute.*[540]

"Globalization—whether economic, political, cultural, or environmental—is defined by increasing levels of interdependence over vast distances." *Foreign Policy Magazine, January/February 2001.*[541]

Globalization is "the broad movement among economies and societies... that is knitting the world closer together and affecting capital markets, technology, and the exchange of information." *Business consultant, author, and lecturer Daniel Pink, in an interview for the State Department's February 2006 issue of eJournal USA.*[542]

In their Globalization Index, A.T. Kearney and *Foreign Policy Magazine* consider economic integration, technological connectivity, personal contact, and political engagement.[543]

Certainly, globalization has helped us gain a broader perspective about ourselves and those who live on the other side of rivers, mountains, deserts, oceans, and political borders, often thousands of miles away. We have the capacity to communicate and collaborate instantly...if we're willing.

Globalization
Educators Weigh In

Globalization is happening in real time, and everyone, everywhere, feels its impact, whether they want to admit it or not. There is nowhere to hide, whether we work in government, business, education, or any other field. We've listened to some seasoned educators who shared observations about this phenomenon.

Debra Hill, an associate professor at Argosy University in Chicago, a longtime K-12 educator, and a former ASCD president, cautions that, "We will not be able to successfully function in a global framework until we recognize that multiple language acquisition is occurring everywhere else across the globe, with the exception of the U.S." She challenges schools to "gear up for successfully teaching other languages." Hill cautions, "A singular focus on test scores and assessments is hampering schools from teaching what students need to know in this global environment."

James Harvey, executive director of the National Superintendents Roundtable, which pays particular attention to international relationships and learning, believes the education system needs "a new emphasis on developing global competency in our graduates—with more attention in the curriculum to history, geography, and languages." Harvey is concerned that the narrow scope of assessments is "likely to shortchange the global competence agenda." He calls support for learning languages "a critical national need." We don't know where the next global emergency will develop, Harvey observes, and "the U.S. will need a broad base in many different languages," among them, people who are adept at languages used in the Middle East and Slavic-speaking countries.

Joseph Cirasuolo, executive director of the Connecticut Association of Public School Superintendents (CAPSS), makes clear that if we do not "educate all children from a global perspective, they will be ill equipped to be productive in a global economy." Unless today's young people are prepared for life on a global stage, they will see a reduction in their lifestyles compared to an older retiring generation. That situation, he warns, could "exacerbate social unrest."

Concepción Olavarrieta, creator of a Millennial Prize for students worldwide, is President of Nodo Mexicano, El Proyecto del Milenio, A.C., in Mexico City. She emphasizes the need "to teach people knowledge and respect for other cultures."

Matthew Moen, a past-president of the national Council of Colleges of Arts & Sciences, sees language instruction proliferating. However, he speculates that much of it will be happening "outside of universities, through mechanisms like computer software and specific language and cultural sensitivity training for people who will be living and working abroad." He calls on language departments to "offer something other than rudimentary language instruction as their bread and butter."

Challenges for Leaders
In an Interconnected World

What once was vertical has become horizontal or lateral. Leadership has been dispersed. Collaboration is in, and yesterday's system of getting things done may not jibe with today's and tomorrow's needs.

Last night, we might have dreamed that everyone streamed into our office and shouted in one voice, "We're all in this together!" While we need to hang on to that dream, most of us have now realized that getting there often means reaching across cultures, habits, histories, ideas, points of view, prejudices, and time zones.

If that complexity weren't enough, we also have to deal with realities of a fast-changing world. Rowan Gibson, founder and chairman of Rethinking Group, reminds us "the future will not be a continuation of the past. It will be a series of discontinuities." He adds, "linear thinking is useless in a non-linear world." [544]

When did we realize the power of emerging technologies to turn us into a global society? While Thomas Jefferson may have traveled to Europe in the 1700s, the voyage took a while. He lived in a four-mile-an-hour world. News of Abraham Lincoln's assassination in 1865 didn't make it into London newspapers until 12 days later. The very next year, the transoceanic cable was up and running. With that cable in place, news about the 1883 eruption of the volcano Krakatoa made the world's major newspapers the very next day. [545] Those who noticed must have realized we had taken a giant leap toward an interconnected, globalized world. Now, we see the aftermath of terrorist attacks, hurricanes and tsunamis, and massive demonstrations in real time. As for the demonstrations, some have led to revolutions, often stirred and directed using some version of social media.

The smartphone that we stuff into a pocket or purse is nearly equivalent to being across the table from people in the far corners of the planet. We can use it to crowdsource an idea, call a meeting, check on a grocery list, or foment an uprising. We can get an idea to someone within microseconds. However, that *someone* can also respond by sharing an objection to our brainstorm, along with a few video clips that support their case, with an expanded list of thousands, in the bat of an eye.

Marcello Suarez-Orozco, now dean of the UCLA Graduate School of Education, recalls that on September 15, 2008, as storm clouds thickened and a fuller reality of the Great Recession was setting in, the global financial firm, Lehman Brothers, filed for bankruptcy protection. With that collapse, "the fragility of the world became more apparent. We discovered how profoundly we were interconnected," Suarez-Orozco remarked. He observed that while it could take the Spanish queen two years to get a message through to Peru back in the 1500s, we get news of an uprising in Somalia within minutes or hours. "There is no here and there anymore."[546]

Growth and Shift
As the World Churns

World Population. In 1960, the world population reached 3 billion. By 2000, it had doubled to 6 billion and passed 7 billion in 2011. That's right. The population doubled in just 40 years. By 2050, the U.S. Census Bureau is expecting the world population to reach about 9.3 billion. That's a 50 percent increase in just 50 years.[547] While the annual planetary population growth rate peaked at slightly more than 2 percent in the early 1960s, it was expected to dip to about 0.5 percent by 2050.[548]

That population growth is far from uniform across regions of the world. For example, between 2000 and 2050, the population of Africa was expected to grow from 803 million to 1.786 billion, a 2.2 times increase. Asia was expected to grow from 3.4 to 4.8 billion, an increase of 41 percent. Northern America was slated to increase from 314 million to 462 million people, up 47 percent. Meanwhile, the population of Europe was slated to decrease from 802 million to 776 million, a drop of 3.2 percent. Even more dramatic, while *less developed countries* were expected to grow by 60 percent, *more developed countries* were projected to grow only 4.3 percent. The challenge is apparent. (See Figure 11.1)

A Measuring Stick for Members of the Global Community
Four Factors[550]

Do thoughtful and legitimate leaders worldwide consider the performance and reputations of countries? Absolutely. In fact, some of their approaches provide an excellent framework for stirring discussions

Figure 11.1
World Population Increase by Regions, 2000 to 2050
Actual and Projected by Region and Development Category[549]
U.S. Census Bureau, International Data Base
Midyear Population in Millions (m) or Billions (b)

Region	2000	2050	Increase/Decrease
World (2050 total varies slightly from later reports)	6.079 b.	9.079 b.	+ 49.3 percent
Less Developed Countries	4.887 b.	7.836 b.	+ 60 percent
More Developed Countries	1.192 b.	1.243 b.	+ 4.3 percent
Africa	803 m.	1.786 b.	+ 122 percent
Near East	171 m.	396 m.	+ 132 percent
Asia	3.435 b.	4.832 b.	+ 41 percent
Latin America & Caribbean	524 m.	782 m.	+ 49 percent
Europe (Western, Eastern, New Indep. States)	802 m.	776 m.	- 3.2 percent
Northern America	314 m.	462 m.	+ 47 percent
Oceania	31 m.	45 m.	+ 45 percent

among students and citizens. Let's take a look at one prime example.

The A.T. Kearney/*Foreign Policy* Globalization Index, a ranking of nations, was conducted and published annually by *Foreign Policy Magazine* from 1999 until 2007. The study rated 62 countries for a number of years, then added ten, making it 72. This seminal work was among the first to thoughtfully rank nations on their progress toward globalization. Those rankings and the scores each earned were based on four factors:

- **Economic Integration**—trade, foreign direct investment, portfolio capital flows, and investment income.
- **Technological Connectivity**—Internet users, Internet hosts, and secure servers.
- **Personal Contact**—international travel and tourism, international telephone traffic, and remittances and personal transfers.

- **Political Engagement**—memberships in international organizations, personnel and financial contributions to U.N. Security Council missions, international treaties ratified, and governmental transfers.

While we urge you to examine the full studies, we've included a few example of how certain nations fared in the globalization rankings published in 1999, 2003, and 2007. We're presenting some at or near the top, in the middle, and toward the bottom.

For example, in 2007, U.S. globalization efforts ranked seventh overall. While the country came in first in technological connectivity, it ranked 71 out of 72 nations in economic integration. Leading the

Figure 11.2
A.T. Kearney/Foreign Policy Globalization Index

—————— 2007 Rank by Category ——————

2007 Ranking (Overall)	2003 Ranking (Overall)[548]	1999 Ranking (Overall)[549]	Country	Economic Integration	Technological Connectivity	Personal Contact	Political Engagement
1	1	1	Singapore	2	15	3	40
2	NI	NI	Hong Kong	1	17	1	71
3	5	5	Netherlands	4	6	16	8
7	11	4	United States	71	1	40	51
8	7	6	Canada	34	2	11	13
9	NI	NI	Jordan	10	50	5	1
13	21	NA	Australia	26	3	39	41
22	17	21	Germany	45	16	34	19
34	24	27	Italy	56	26	38	11
48	NI	NI	Vietnam	19	52	50	57
49	51	42	Mexico	50	41	45	37
54	50	47	Argentina	61	38	63	20
66	53	54	China	43	56	67	65
71	57	61	India	66	63	59	69

Foreign Policy Magazine, November/December 2007, Ranking of 72 Nations[547]
(NI=not included in 2003. NA=not available. Note that in 1999 and 2003, 62 nations were included in the Index and those earlier issues of the magazine, whereas 72 were included in 2007.)

pack on globalization were Singapore, Hong Kong, the Netherlands, and Switzerland. Because of Hong Kong's political status, it gained a number two position despite not being able to sign most international treaties or engage in U.N. peacekeeping operations.

National Reputations . . .
They depend on each of us and all of us.

Aristotle had it right. When he said that our individual reputations—the personalities we project—are based on our competence, good character, and good will, he opened a Pandora's box that will not stay closed. Today, we judge not only individuals by those criteria, but we also judge nations.

The behaviors of governmental and non-governmental organizations, businesses, and individuals can shape national reputations. How people feel about a country depends on more than glowing words in a travel brochure. Policies and actions speak more loudly than words. Over time, national reputations become directly tied to the level of respect any country enjoys as a member of the community of nations. That level of respect could ultimately be a key to a country's success, even its survival.

Let's take a quick look at some of the most basic of those international relationships. Most would agree that all are enhanced or inhibited, depending on the situation, by instant electronic communication and a 24/7-news cycle. Every student who emerges from our schools and colleges should understand these critical criteria.

• **Relationships among governments** are often a reflection of foreign policy, the handling of expected and sometimes unexpected issues or events…and longstanding friendships or rivalries. Issues might range from war and peace to trade agreements, environmental concerns, and human rights. The stage is often *bilateral*, between two nations, and increasingly *multilateral*, involving several nations, as opposed to *unilateral*. Sometimes, these relationships are influenced by regional alliances and world organizations.

• **Business relationships** bring people together to tap natural and human resources; develop, sell, or purchase products or services; and make a profit, sometimes called trade. A growing issue? Intellectual property. Another? Stability. Communities, states, and nations often send delegations to another country in an attempt to attract investments in business or infrastructure. Transactions cross political boundaries at breathtaking speed.

- **Educational, scientific, and other non-governmental relationships** often involve associations, societies, education institutions, and a variety of international organizations and individuals who share information, conduct research, sponsor exchanges, and rally communities of interest around projects or ideas.

- **Personal relationships** are often stimulated by study, various types of correspondence such as social media, and travel across international boundaries.

Exponential growth in online conferences, web sites, blogs, a variety of social media, an expansion of online courses, and speedy but sometimes strenuous transportation, have led to a blossoming of international relationship building and idea sharing. Our ability to convene, however distant our physical locations, has exponentially multiplied the pace of change and has led to nearly instant international response to national or even local decisions. Development of this book, for example, involved Futures Council 21, a special advisory group made up of people worldwide who were willing to share their ideas, knowledge, and experiences across time zones and physical space. Enlightenment often begins with collaboration.

Hot-Button Realities...
For Globally Sensitive Schools, Communities, and Countries

The road to helping students, educators, communities, countries, and continents become more sensitive to international/global issues is getting busier. The traffic is intense because the sense of urgency is increasing exponentially. Thinking is expanding beyond traditional boundaries at the same time the planet seems to be shrinking. With that in mind, here are a few hot-button realities that deserve our attention:

- **Thinking Globally, Acting Locally.** A growing chorus is asking us to *think globally and act locally*. That could mean considering the impact of local actions on the larger world, such as implications for the environment. If we live up-river, we might want to consider the needs of people downstream. In reverse, it could mean considering the needs and wants of the world, or parts of it, as we develop products and services or just attempt to expand our market.

Princeton Professor and author of *Cosmopolitanism: Ethics in a World of Strangers*, Anthony Appiah, asks us to consider being "global citizens with national citizenship."[554] Marcus Newsome, superintendent of the Chesterfield County Public Schools in Virginia, speculates that schools may need to offer "a world curriculum." He foresees "citizenship and service learning that will encourage students to also be global citizens." Meera Balachandran, director of the Education Quality Foundation of India, makes clear that interdependence depends on learning other languages, understanding other cultures, and developing diplomatic skills. She calls for "immediate attention to these issues beginning at a very young age." Without that kind of action, she cautions, we could see growing disturbances in society.

• **Balancing International Competition with International Collaboration.** The hue and cry is to support education, economic policies, and a host of other issues and concerns to make us "more competitive" in the international marketplace. However, we urgently need to include another factor. Somehow, we need to balance our competitive nature with a big dose of collaboration, if we hope to effectively tackle the urgent planetary need to develop renewable sources of energy, protect the environment, and establish economic policies that are workable in a fast-changing, even more heavily populated world.

Consider defense and foreign policy. As the country moved beyond the Cold War, the U.S. faced several possibilities. One was to take on the mantle of *superpower*, becoming the world's policeman. Another was a return to what might be called a *bi-polar world* made up of distinct friends and enemies, similar to the Cold War. A third policy, one that challenges long-held beliefs about power- and geo-politics, might call for each country to become a good member of *a family of nations*, maintaining its sovereignty but, at the same time, addressing planetary concerns. Despite our longings, some believe a hybrid of the three may be the best we can do in the short run.

The tug between *sovereignty and collaboration* won't go away soon. While sovereignty is a legitimate concern, we often sidetrack any progress by simply declaring, "If we have to share an accomplishment with someone else, maybe we'll just forget about it." It's the lonesome highway approach that leaves us stranded in the trenches, victims of the status quo as the world zips by. Yet there is plenty of precedent for legitimate partnerships, collaborations, coalitions, and joint efforts to get things done.

• **Brain Gain vs. Brain Drain.** Unless we provide opportunities for people to use their talents, abilities, and outright genius in our own communities or countries, they will move on. Since pre-history, people have hit the road looking for greener pastures. In our own time, those people include everyone from construction workers and farmhands to nanotechnologists, medical doctors, and research scientists. For example, many students in India attend the elite Indian Institutes of Technology (IIT). Yet, "large numbers of graduates end up in Silicon Valley or on Wall Street," according to a 2011 article in *The Economist*. Some poorer countries that realize they are suffering a brain drain may argue that the remittances their citizens send back home while working in other countries help ease the economic pain.[555]

A benefit for *brain drain* countries and communities could be that they will, once and for all, accept the fact that they will have to offer the incentives required to attract and keep the talent they need in a highly mobile workforce.[556] Some are suggesting a new frame for the brain drain/brain gain concern, instead referring to it as *brain circulation* that encourages scientific exchange.[557]

The most classic stories of brain drain and gain popped up around WWII when Wernher von Braun, a German rocketry scientist, and Albert Einstein, a theoretical physicist who developed his theory of relativity, emigrated to the U.S. On the other hand, Moses declaring, "Let my people go!" before crossing the Red Sea was fairly dramatic in its own right.

• **Seismic Shifts.** The 1990s and 2000s brought seismic economic and demographic shifts. Even though most were generally expected, the world now faces new realities. The global climate outlook is deteriorating with the inability of nations to collaborate on far-reaching agreements to limit greenhouse gases. Developed nations are facing continued aging and some are dealing with population declines. On the other hand, developing nations, on average, have more youthful populations, and they are often growing quickly. While the number of people in Europe was projected to decline by 3.2 percent between 2000 and 2050, the population of Asia was expected to increase by 41 percent. Japan and Russia were expecting serious population declines. Massive adjustments were taking place in the Middle East, accelerated by the "Arab Spring."

"China has emerged as a major global economic and trade power," confirming the rise of Asia. In the early part of the second decade of the 21st century, the country took its place as the "the world's second largest economy following the U.S. China had become the world's largest merchandise exporter, the second largest merchandise importer, the largest manufacturer, the largest holder of foreign exchange reserves, and the largest creditor nation, according to a June 2012 report from the Congressional Research Service (CRS). "Many U.S. firms," the report noted, "use China as the final point of assembly in their global supply chain networks," CRS noted. We're also seeing growth across regions such as Southeast Asia.

• **Global Challenges as Opportunities.** Some challenges facing our planet, if we hope to deal with them, will require the combined efforts of many people and nations working together. As we make a turn onto the road toward interdependence, just a few of those challenges we'll be meeting are: energy, the environment, fresh water, poverty, terrorism, avoiding or dealing with pandemics, and issues of war and peace. Even during the Cold War, nations worked across their political divisions to launch a network of weather satellites. The International Space Station, the Hadron Collider (particle accelerator), and research into finding cures or treatment regimens for threatening diseases, depend on collaboration among nations. A looming question is often, "How much are we willing to contribute to dealing with massive worldwide concerns that have profound implications for our well-being, even our survival, wherever we are?"

• **Jobs and Careers.** The competition for resources, including human resources, will continue, as it has for centuries, even millennia. If we hope to retain or attract people, we need to be attractive. That often means a relatively high quality of life, including an opportunity for a good education and a stimulating job that comes close to matching our individual abilities—plus decent pay.

Workforce mobility is becoming a driving issue for organizations that stretch globally. If we hope to be successful, wherever we are, we probably need to be thinking about how we can draw from a planetary-wide talent pool.[558]

Nature Magazine conducted a survey asking which countries would be "producing the best science in their field by 2020." *Global Talent Strategy* reported in 2013 that "more than 60 percent of respondents in both biological and physical sciences picked China as an option. However, only eight percent said they would be prepared to relocate in China—instead preferring the U.S., Europe, Canada, and Australia." China is very likely taking this finding seriously. The question is whether each of us, in our own communities or countries, face up to a similar challenge. If so, how would our schools as well as our colleges and universities respond?[559]

• **Immigration/Migration, Acceptance, and Language.** Let's start with the premise that all children are our future. During 2012 in the U.S., "nearly 17 million of them ages 17 and under lived at home with at least one immigrant parent, accounting for 24 percent of all children in that age group," according to Migration Information Source. Those children, then in the U.S., had arrived from countries worldwide.[560] Of course, all children, whatever their parentage or social or economic circumstances, will need equal access to a broad, deep, and purposeful education. That includes opportunities for them to fill any gaps and get on a level playing field so that they can become contributing members of the economy and civil society.

All students need to have a grasp of what it will take to work in other parts of the world. They need a grounding in how to get along and get constructive things done, as well as how to work with people whose backgrounds, including their countries of origin, cultures, histories, races/ethnicities, social and economic situations, languages, and other distinctions are different from their own.

The U.S. Center for Immigration Studies (CIS) pored over 2005-2007 census data and found that 45 percent of maids and housekeeping cleaners in the country were *not* born in the U.S. Neither were 44 percent of medical scientists, 34 percent of software engineers, 27 percent of physicians, and 25 percent of chemists.[561] Meanwhile, the European Commission reported in 2008 that, in the United Kingdom (UK), 30 percent of physicians and 13 percent of nursing staff were non-natives.[562]

At a 2012 Harvard Future of Learning Institute, Senior Program Associate for *Facing History,* Laura Tavares, suggested that the flow of immigrants worldwide begs the question, "Can we create an identity that is inclusive?" As educators, Tavares says, we need to ask ourselves how we respond in certain critical moments, from the student wearing the headscarf to one whose language and habits don't mirror our own. "What are the moments we have to teach membership and belonging?" she asks. Acceptance is a need shared by most people.[563]

"I think this calls for a new emphasis on developing global competence in our graduates, with more attention in the curriculum to history, geography, and languages," suggests James Harvey, executive director of the National Superintendents Roundtable.

• **Human Rights.** While civil rights are described in laws, the Universal Declaration of Human Rights is part of the foundation for any nation that hopes to secure respect among its worldwide neighbors. A dedication to these basic rights can also build confidence that all voices are being heard and that progress is built on a firm foundation.

First adopted by the United Nations General Assembly in 1948, the declaration notes that all human beings are born free and equal in dignity and rights; that everyone is entitled to these rights, despite their differences; that everyone has the right to life, liberty, and security; that no one should be held in slavery; that no one should be subjected to torture, or to cruel, inhuman, or degrading treatment or punishment; and that everyone has a right to be recognized as a person before the law. In 30 articles, the declaration describes these and other rights that are bestowed at birth.[564]

This document and others can be used to spark discussions about a measuring stick for a respected nation. Without a concerted effort to deal with expanding inequities within and among nations, the world risks sustained conflict that could bring unprecedented human suffering and death.

• **International Students and Exchanges.** While festivals, food, and field trips can be helpful in introducing young people and adults to an array of cultures, they don't substitute for a grounding in history, geography, and language. Many high schools, colleges, and other organizations in several parts of the world sponsor exchanges, visits, tours, or missions to other countries. Those who have, perhaps, the most in-depth experiences are international students who study or work on projects abroad.

Since its founding in 1919, the Institute for International Education (IIE) has conducted an annual census of those international students. As she joined in releasing the report, Ann Stock, assistant secretary of state for educational and cultural affairs, noted that "international education creates strong, lasting relationships between the U.S. and emerging leaders worldwide. Students return home with new perspectives and a global skill set that will allow them to build more prosperous, stable societies."

IIE's 2012 *Open Doors Report* found that "international students at colleges and universities in the U.S. increased by six percent to a record high of 764,495 in the 2011-12 academic year." U.S. students studying abroad increased by one percent.

Languages, Students, and Exchanges

The realities of student exchanges, international students, and the flow of world languages are startling. Here are a few facts:

• **Most Spoken Languages.** In 2005, the *Chronicle of Higher Education* reported that, "Mandarin was the most spoken language in *the world*, followed by English, Hindi, Spanish, and Russian." Writer Francisco Marmolejo revealed that, *by 2050*, Mandarin will likely continue as the world's most spoken language, followed by Spanish. English and Arabic would be tied for third and fourth places, followed closely by Hindi/Urdu.

• **Language at Home.** "At least 20 percent of *the (U.S.) population age*

5-years and older (spoke) a language other than English at home. After English, the most commonly used languages were: Spanish, Chinese, Tagalog, French, Vietnamese, German, and Korean."[565]

• **Languages on the Web.** A 2011 report on the "Top Ten Languages used in the Web" indicated that English users of the Internet totaled approximately 565 million or 26.8 percent, while Chinese users totaled 509 million or 24.2 percent, followed by Spanish, 164.9 million or 7.8 percent, and Japanese, 99 million or 4.7 percent.[566]

• **U.S. Students Abroad.** During the 2010-11 academic year, according to a report from the Institute for International Education (IIE), an all-time high number of American students studied abroad for academic credit, a total of 273,996. They noted that 14 of the top 25 destinations for those students were outside Europe. The top 15 included: the United Kingdom, Italy, Spain, France, China, Australia, Germany, Costa Rica, Ireland, Argentina, India, South Africa, Mexico, Japan, and Brazil.

• **Studies.** Fields of study for international students in the U.S. included, among others: business and management, 21.8 percent; engineering, 18.5 percent; math and computer science, 9.3 percent; social sciences, 8.7 percent; and physical and life sciences, 8.6 percent. Among several other areas, approximately 2.2 percent studied education and 1.3 percent studied agriculture.[567] Copies of the full *Open Doors Report* are available for purchase from IIE at http://www.iie.org/Research-and-Publications.

• **Contribution to the Economy.** The U.S. Department of Commerce noted that, in addition to relationships and gaining knowledge of the nation, "international students contribute more than $22.7 billion to the U.S. economy." Comparable contributions of varying amounts also benefit other receiving countries.

Worldwide, Multinational, Continental, Regional, Local Organizations

The precedent for working together across national boundaries isn't new. Governments, businesses, associations, the media, non-profit civil society organizations, research groups, those who work toward mutual security, and a host of others are on the front lines of issues. They range from local to international organizations.

Businesses find raw materials, means of production and distribution, and markets worldwide. The clothes we wear were probably designed in one country, manufactured in another, and sold everywhere. In fact, multinational businesses can have operations in a few or several countries. The same is true of **media**, including broadcast, motion picture, Internet, and social media organizations that girdle the globe. **International associations** represent professions or entire industries. **Scientific, cultural, educational, health, and research groups** share information that can lead to broader understanding and sometimes even breakthroughs in everything from avoiding a pandemic to making education available to women. **Communities, states, regions, and countries** send missions to other nations to promote tourism and attract industries that will help them grow their economies.

Worldwide Organizations, such as the *United Nations, World Trade Organization (WTO)*, the *Organization of Petroleum Exporting Countries (OPEC)*, and the *G20* are examples of ongoing international groups that address a range of issues. There are also **security alliances** such as the *North Atlantic Treaty Organization (NATO)*, and *ANZUS*, involving Australia, New Zealand, and the U.S. In **continental and regional organizations**, rarely do nations cede their sovereignty but consider it in their common interest to collaborate. Examples of **continentalism** include: the *European Union*,[568] and the *North American Free Trade Agreement (NAFTA)*.[569] An *Association of Southeast Asian Nations (ASEAN)* was established in 1967 "to accelerate the economic growth, social progress, and cultural development of the region."[570] [571] *Mercosur*, the *Andean Community*, and the *Union of South American Nations (UNASUR)*,[572] are among South America's strategic continental organizations. The *African Union (AU) is an* organization that brings together nearly 30 countries from across that continent.[573] While not a continental organization, the *Arab League* brings a number of countries in North Africa and the Middle East together to consider issues they might have in common.

Going Urban
The Massive Growth of Mega-Cities
"By 2030, the world's urban population alone will have grown to almost 5 billion, with growth centering on cities as rural area populations shrink," according to *World Politics Review*. "Some 60 percent of

those urban dwellers will be under the age of 18," with "urban areas around the globe currently swelling at more than 1.2 million people a week."[574] A quick look at *metropolitan areas* worldwide, sometimes called *agglomerates* or *mega-cities*, can help put what we're hearing into a broader perspective. In 1970, only two metro areas, Tokyo and New York, had populations of more than 10 million. By *1990*, it was 13; and by *2011*, 23. Projections put the number of mega-cities of 10 million+ at 37 by *2025*, with 21 of those cities in Asia. Three are expected to be in the U.S., one in Mexico, five in South America, two in western Europe, and three in Africa, plus Moscow and Istanbul.

The world population is expected to reach 8.3 billion by mid-year of 2030.[576] At an August 2012 meeting of the World Health Forum, a U.N. event, Ramona Vijeyarasa, head of ActionAid International, warned that "we're seeing a real urbanization of poverty." She added, "Urbanization has two sides. It can bring marginalization, violence, and sub-standard housing for the poor, while at the same time offering opportunities to grow and advance."[577]

Globalized Cities
How the Top 15 Rated on the Globalization Index

In 2010, using *the same criteria for countries and adding cultural experience*, A.T. Kearney and *Foreign Policy Magazine*, scored some of the most globalized cities in the world. New York, London, and Tokyo ranked among the top ten cities in all five categories: *business activity, human capital, information exchange, cultural experience,* and *political engagement.* Washington, D.C., New York, and Brussels led the field in *political engagement.* London, Paris, and New York were among the top three in *cultural experiences.* New York, Geneva, London, Brussels, and Paris were among top cities in *information exchange* scoring. New York, London, Chicago, and Los Angeles were among leaders for *human capital.* New York, Tokyo, and Paris were the top cities for *business activity* (Khanna, Parag, "Metropolis Now" and "2010 Global Cities Index," A.T. Kearney and *Foreign Policy Magazine,* Sept.-Oct., 2010, http://www.atkearney.com/news-media/news-releases/news-release/-/asset_publisher/00OIL7Jc67KL/content/2010-global-cities-index-ranks-new-york-london-tokyo-and-paris-as-top-global-cities/10192,).

Figure 11.3
Mega-Cities with More Than 10 Million Inhabitants
1970, 1990, 2011, 2025
Population of Urban Agglomerates, World Ubanization Prospects
United Nations Department of Economic and Social Affairs/Population Division[575]

Rank	Urban Agglomerate, Mega-City	Population (millions)	Rank	Urban Agglomerate, Mega-City	Population (millions)
1970			**2025**		
1	Tokyo, Japan	23.3	1	Tokyo, Japan	38.7
2	New York-Newark, USA	16.2	2	Delhi, India	32.9
1990			3	Shanghai, China	28.4
1	Tokyo, Japan	32.5	4	Mumbai (Bombay), India	26.6
2	New York-Newark, USA	16.1	5	Mexico City, Mexico	24.6
3	Mexico City, Mexico	15.3	6	New York-Newark, USA	23.6
4	Sao Paulo, Brazil	14.8	7	Sao Paulo, Brazil	23.2
5	Mumbai (Bombay), India	12.4	8	Dhaka, Bangladesh	22.9
6	Osaka-Kobe, Japan	11.0	9	Beijing, China	22.6
7	Kolkata (Calcutta), India	10.9	10	Karachi, Pakistan	20.2
8	Los Angeles Metro, USA	10.9	11	Lagos, Nigeria	18.9
9	Seoul, Rep. of Korea	10.5	12	Kolkata (Calcutta), India	18.7
10	Buenos Aires, Argentina	10.5	13	Manila, Philippines	16.3
2011			14	Los Angeles Metro, USA	15.7
1	Tokyo, Japan	37.2	15	Shenzhen, China	15.5
2	Delhi, India	22.7	16	Buenos Aires, Argentina	15.5
3	Mexico City, Mexico	20.4	17	Guangzhou, Guangdong, China	15.5
4	New York-Newark, USA	20.4	18	Istanbul, Turkey	14.9
5	Shanghai, China	20.2	19	Cairo, Egypt	14.7
6	Sao Paulo, Brazil	19.9	20	Kinshasa, Dem. Rep. of Congo	14.5
7	Mumbai (Bombay), India	19.7	21	Chongqing, China	13.6
8	Beijing, China	15.6	22	Rio de Janeiro, Brazil	13.6
9	Dhaka, Bangladesh	15.4	23	Bangalore, India	13.2
10	Kolkata (Calcutta), India	14.4	24	Jakarta, Indonesia	12.8
11	Karachi, Pakistan	13.9	25	Chennai (Madras), India	12.8
12	Buenos Aires, Argentina	13.5	26	Wuhan, China	12.7
13	Los Angeles Metro, USA	13.4	27	Moscow, Russian Fed.	12.6
14	Rio de Janeiro, Brazil	12.0	28	Paris, France	12.2
15	Manila, Philippines	11.9	29	Osaka-Kobe, Japan	12.0
16	Moscow, Russian Fed.	11.6	30	Tianjin, China	11.9
17	Osaka-Kobe, Japan	11.5	31	Hyderabad, India	11.6
18	Istanbul, Turkey	11.3	32	Lima, Peru	11.5
19	Lagos, Nigeria	11.2	33	Chicago, USA	11.4
20	Cairo, Egypt	11.2	34	Bogota, Colombia	11.4
21	Guangzhou, Guangdong, China	10.8	35	Bangkok, Thailand	11.2
22	Shenzhen, China	10.6	36	Lahore, Pakistan	11.2
23	Paris, France	10.6	37	London, U.K.	10.3

Primer from Lagarde

A Global Economy for a New Generation

International Monetary Fund (IMF) Managing Director Christine Lagarde, addressing a conference in Davos, Switzerland, during January of 2013, explored *A New Global Economy for a New Generation*. "The burning question," she asked, is "how we can make sure that all regions grow strongly, converge rapidly, and succeed in meeting the aspirations of their people."

What did Legarde tell us about "the groundwork for future success?" She suggested "embracing some of the emerging values of this new generation," which has been "weaned on immediacy, democracy, and global reach of the social media." Three of those values include: greater openness, stronger inclusion, and better accountability.

"In an era of globalization, cooperation needs to be hardwired into the psyche of policymakers," Lagarde declared. She added, "As we saw clearly during the crisis (Great Recession), this is a world where economic jitters in one region or market can have instant repercussions all across the globe. In a flat world, there is no room for silos." However, she admitted, "Old instincts die hard."[578]

International Testing and Assessment

A variety of assessments attempt to provide comparative data on how students are doing from one country to the next. While rankings of participating countries might vary from year to year, and some question how they can be used to improve education, the studies provide a scorecard and observations that can stimulate discussion about a range of education issues in a global context. Among the more prominent are:

—TIMSS (Trends in International Mathematics and Science Study*)* has been conducted at four-year intervals since 1995. It focuses on mathematics and science achievement at the fourth- and eighth-grade levels. A total of 63 countries and 14 benchmarking entities participated in TIMSS2011. *Both TIMSS and PIRLS are under direction of the International Study Center at Boston College.*

—PIRLS (Progress in International Reading Literacy Study) measures trends in reading comprehension among students at the fourth-grade level in each of 49 countries and nine benchmarking entities. First offered during 2001, PIRLS is on a five-year cycle.[579]

—**PISA (Program for International Student Assessment)** has been conducted by the OECD (Organization for Economic Cooperation and Development) every three years since 1997. An aim is to assess 15-year-olds' competence in reading, mathematics, and science. More than 70 countries and economies from around the world participate in the PISA process. In addition to global rankings, the study produces information such as "whether education systems are offering students the best training for entering the workforce of tomorrow, and why."[580] *PISA's offices are at OECD Headquarters in Paris.*[581]

Guideposts for International/Global Education

The ball is squarely in our court. The world is changing before our very eyes. As educators, what are we doing to get our students, our whole societies, in sync with tectonic shifts and quantum leaps into an interconnected future?

Fernando Reimers is the Ford Foundation Professor of International Education, Director of Global Education, and Director of the International Education Policy Program at Harvard. Writing for ASCD's September 2009 issue of *Educational Leadership*, Reimers remarked:

"Good educators know that the real world is ever more interconnected and interdependent. We all (face) planetary challenges such as climate change, health epidemics, global poverty, global economic recessions and trade imbalances, assaults on human rights, terrorism, political instability, international trade, and international cooperation. These challenges and opportunities define the contours of our lives, even in their most local dimensions. Yet in spite of growing awareness of the importance of developing global skills, few students around the world have the opportunity today to become globally competent."[582]

ASCD. Our very survival may depend on taking Reimers' challenge seriously. In fact, in a January 2013 *ASCD INFObrief,* devoted to "Teaching in a Global Context," the association suggested ten categories of global and international studies. The ten include: Conflict and Its Control, Economic Systems, Global Belief Systems, Human Rights and Social Justice, Planet Management, Political Systems, Population, Race and Ethnicity, The Technocratic Revolution, and Sustainable Development.[583]

Georgetown University. At Georgetown University, an International Economic Development syllabus included a look at poverty, inequality, health, education, and population. It also focused on international trade, globalization, foreign aid, governance, and the link between investments in human resources and economic growth.[584]

Asia Society and Council of Chief State School Officers. The Asia Society, working with the U.S.-based Council of Chief State School Officers (CCSSO), produced *Putting the World into World-Class Education*. Recommendations of this collaborative effort included: redefining high school graduation requirements to include global knowledge and skills, doing international benchmarking of state standards, making world languages a core part of the curriculum from grades 3-12, increasing the capacity of educators to teach the world, and using technologies to expand global opportunities.[585]

Fairfax County Social Studies: Craig Perrier, who serves as High School Social Studies Specialist for the Fairfax County Public Schools in Virginia, identified *global competency skills* that need attention. Among several others, those skills include:

- Understanding the formation, movement, and function of global systems, processes, institutions, and ideas.
- Foreign language acquisition.
- Understanding of world history, geography, and international institutions.
- Formation of a cosmopolitan mindset open to and informed of global sensitivities, perspectives, diversity, and insight.
- Cross-cultural collaborative work and the communication of ideas beyond their classroom.
- Major concepts and frameworks that include: culture, gender, race, ethnicity, agency, power, structuralism, context, objectivity, and subjectivity.[586]

Diplomatic Qualities and Skills

In a world grown small, countries earn their respect by regularly demonstrating that they are connected to their own citizens and considerate of their neighbors. That simple but stark fact has significant implications for schools and colleges and for international education. For one, it means all students should have at least basic *diplomatic qualities, skills, and knowledge*, such as open minds, natural curiosity, patience, courtesy and good manners, a sense of tolerance, and the ability to empathize with others—to put themselves in someone else's shoes.

Add to that list the need for thinking and reasoning skills, language proficiency, technical and information management skills, the ability to negotiate, and an understanding of the intellectual and social skills that serve as a foundation for diplomatic behavior and protocol. International, global, civic, law-related education, and character education are fast becoming basics in an interdependent world.[587]

> **"The pervasive lack of knowledge about foreign cultures and foreign languages threatens the security of the United States as well as its ability to compete in the global marketplace and produce an informed citizenry."**
> National Research Council
> *International Education and Foreign Languages: Keys to Securing America's Future.*[588]

Soft Power

Walter Russell Mead, as a Henry A. Kissinger senior fellow in U.S. foreign policy at the Council on Foreign Relations, credited "soft power" as being highly influential in reflecting a country's ideals and culture. He recalled that, during its history, the United States used soft power by espousing anti-imperialism, actively encouraging empires to grant independence to their colonies, championing political democracy and human rights, and promoting inclusion in a global community. Mead noted that millions encountered those "values" through business people, government representatives, educators, entertainers, and others traveling, working, and/or living abroad. Humanitarian and developmental assistance from both the public and private sectors and personal contributions directed at people in need secured what became, but wasn't necessarily intended to be, "soft power."[589]

Implications for Education and the Whole of Society
Global/International

• **The urgency of international/global learning will grow exponentially.** The stakes are high and the possible benefits so great that schools and colleges will be expected to strengthen their international/global education programs. As nations and communities, we need to realize that both common threats and common opportunities can bring us together in common purpose. International relations expert the late Frank Method suggested students learn about principles underlying the development of international agreements, the resolution of disputes, setting international standards for acceptable practices, criminality, business law, accounting standards, food safety, environmental stewardship, human rights, and myriad other topics.[590]

• **Curriculum should include the building blocks for international/global education.** The critical and growing demand for international learning has direct implications for world languages and history, international relations, diplomatic skills, and cultural understanding. Students and educators will need knowledge and skills connected to cultural anthropology; geography; communication; culture; civic education; conflict resolution; thinking and reasoning and problem solving skills; political and behavioral science; the arts; appropriate use of technologies; character education; ethics; and an understanding of international issues. To develop even deeper international understanding, students will also need grounding in at least the basics of economics, law, political science, geopolitics, government, human rights, social skills, and the arts. Social studies can help us develop a deeper understanding of issues close to home and an even more expansive view of the world. Extra-curricular and exchange programs and travel by students and educators can help us build on our tradition of using food. festivals, and field trips to introduce other cultures.

The reality. Untold numbers of organizations employ and are run by people from around the globe. Today's students will work in and even lead some of those organizations and need to be ready to collaborate with people whose cultures and languages may not match their own.

• **Attracting, growing, and keeping multinational businesses and other institutions will require an environment of acceptance and support.** As communities, regions, and countries build their futures, they might encourage local or regional entrepreneurs to develop multinational businesses or other institutions. On the other hand, they might want to attract those types of ventures to their community. Any for-profit or not-for-profit international organization will expect an environment of support, including excellent schools and colleges, and an interesting and high quality of life. Community, business, and civic leaders increasingly will need to build relationships with counterparts in other nations.

• **Students could identify characteristics of any country that hopes to become a respected member of the family of nations.** Have students do brainstorming as they consider a measuring stick or a set of criteria that might apply to any nation, not just their own. Taking the process a step further, students and communities could expand their conversation to consider the impact of those criteria on national or even local futures. The students, class, or school might then share those criteria with others across the nation or in other nations to get their views, maybe followed by a visit using Skype, provided time zone differences can be managed, and other social media.

• **Schools, colleges, and communities might stimulate discussions of international/global issues.** Every community should consider convening people to discuss global issues and our local relationship to them. Issues such as war and peace, security, energy, the environment, sustainability, scientific research, becoming a viable member of an increasingly world economy, and a host of other issues and needs can draw people together. Those who get involved discover and discuss the big picture and better understand how each of us can play a part in pursuing the common good. Community education groups, education systems, and other organizations might host or sponsor lecture series or workshops that could be streamed on their web sites or carried on cable. An opening announcement might go something like this, "Leave your biases at the door. We're not here to sell our point of view but to join in a spirited and thoughtful conversation." With online connections and distance learning, educators and students can share ideas and draw on the experiences and inspiration of people continents away.

• **Additional Implications.** Sustainability and possibly survival for a community, a country, or even an individual may depend, at least in part, on the ability and willingness to communicate and collaborate, plus the know-how to do public or corporate business across international borders. Add to that the need for ongoing professional development to provide educators with grounding on international/global education.

Expectations of Nations
Fundamental Questions

What should we expect of nations? Let's give that question some thought.

Unless a nation considers the needs of its citizens, it might not truly represent them. Not listening, not staying in touch, can lead to instability, as conflict overcomes consensus. Lack of appreciation for diversity can also lead to an "us and them" society, which might become divided against itself. A lack of stability internally raises questions about reliability externally, and could cause breakdowns in relationships with other countries. We're not talking about stability that's imposed by an autocratic leader. Instead, we're talking about the stability that comes with inclusive leadership that is directly connected to the common good.

Are citizens captives in their own land or are they free to cross boundaries and explore other parts of their communities, countries, continents, or the world? Can we depend on a country to devote its resources, no matter how abundant or scarce, to improving life for those who live there and to becoming a reliable member of the family of nations? Still other fundamental questions that have a bearing on a country's place as a respected member of the world community might include:

- Does the government operate with the consent of the governed?
- Does the government exist for the people, or do the people exist for the government? Is it totalitarian, democratic, or a hybrid?
- Are crime and corruption under control, or do they divert investments, add cost, undermine democratic decisions, and demonstrate a lack of maturity or self-control?
- Are basic human rights guaranteed?
- Is freedom of the press, speech, and religion encouraged?
- Are citizens generally well informed about issues that affect them?

- Does the country practice equal justice under law?
- Does economic growth benefit a vast array of citizens or just a scarce few?
- Is civic education for students a basic part of the curriculum?
- Are exchange programs with other countries encouraged?
- Does the society cultivate and can it tolerate inclusive, democratic leadership?

These questions are fundamental to an understanding of any nation. The answers will reflect any work yet needed for improvement.

International/Global

Questions and Activities

1. Identify what you consider five important skills and/or bodies of knowledge students (and others) should develop to work effectively with people from nations or cultures different from their own?

2. Prepare a one-page paper on possible concerns that might need to be addressed because *less developed countries* are growing much more quickly than *more developed countries.*

3. Develop a brief, five-minute, PowerPoint presentation explaining why certain worldwide, regional, or even community problems are more likely to be solved through collaboration than strictly competition.

4. Review the Institute for International Education (IIE) slide show or fact sheet devoted to its *2012 Open Doors Report on International Educational Exchange.* The report was produced in partnership with the Bureau of Educational and Cultural Affairs at the U.S. Department of State. http://www. iie.org/en/Research-and-Publications/Open-Doors.[591] Click on "Download Presentation" or "Fast Facts." Identify six specific items that you found to be among the most interesting.

5. If you are a classroom educator, ask students to brainstorm answers to this question, "What are the characteristics of a country that is capable of being a good member of a family of nations?" If you are not in a classroom, discuss possible answers with friends or colleagues, or just think about it yourself.

6. View "The Global Power Shift," a controversial *TED Talk* by Paddy Ashdown, former member of the British Parliament and diplomat with a lifelong commitment to cooperation, recorded Dec. 2011, http://www.ted. com/talks/lang/en/paddy_ashdown_the_global_power_shift.html. (18:29) [592] Identify three things you agree with and three things you might dispute.

7. Study the four factors in the Globalization Index for countries briefly de-

scribed in this chapter. What additional factors would you include, if any? How could discussions of these factors be useful in preparing students for life in a Global Knowledge/Information Age?

Readings and Programs

1. Friedman, T.L. (2007). *The World is Flat*. Picador; Trade Paperback Edition, N.Y.[593]

2. Fuller, R.B. (1969). *Operating manual for spaceship Earth*. New York: Simon and Schuster.[594]

3. Publications from the Center for Civic Education, Calabasas, CA: *We the People: The Citizen and the Constitution (which includes Lesson 29 on Freedom of Expression); Foundations of Democracy; Project Citizen; National Standards for Civics and Government; Civitas: A Framework for Civic Education; Comparative Lessons for Democracy;* Civitas International Exchange Program; and others, http://new.civiced.org/resources/publications/ebooks/new-enhanced-ebook, and www.civiced.org.

4. "Universal Declaration of Human Rights," United Nations, Adopted by the General Assembly on December 10, 1948. Available at http://www.un.org/en/documents/udhr/.[595]

5. Review articles and materials available on and through the web site of the Council for Economic Education, http://www.councilforeconed.org/.

6. Review "The Globalization Index," *Foreign Policy* Magazine, Nov.-Dec 2007, A.T. Kearney, Inc., and *Foreign Policy*, owned by the Carnegie Endowment for International Peace, pp 68-71.[596]

7. View "Wiring the Web for Global Good," *TED Talk*, featuring Gordon Brown, U.N. Special Envoy for Global Education and former U.K. Prime Minister, taped July 2009, (16:46), http://www.ted.com/talks/lang/en/gordon_brown.html.[597]

8. Read "Leading for Global Competency," Reimers, Fernando M., *Educational Leadership*, September 2009, Volume 67, Number 1, http://www.ascd.org/publications/educational-leadership/sept09/vol67/num01/Leading-for-Global-Competency.aspx.[598]

9. Suarez-Orozco, Marcello M. (2007). *Learning in the Global Era: International Perspectives on Globalization and Education*, University of California Press. http://www.amazon.com/Learning-Global-Era-International-Globalization/dp/0520254368.[599]

10. Friedman, George, (2009). *The Next 100 Years, A Forecast for the 21ˢᵗ Century*, Anchor Books, a division of Random House, NY, http://www.amazon.com/The-Next-100-Years-Forecast/dp/0767923057.[600]

Chapter 12

Personalization
Let's get personal

Chapter 12: Personalization. *Let's get personal.*

Trend: In a world of diverse talents and aspirations, we will increasingly discover and accept that one size does not fit all.
Standardization → Personalization

> "Today you are You. That is truer than true.
> There is no one alive who is Youer than You."
> *Dr. Seuss*[601]

Buy clothing lately? There's a good chance a tailor or somebody else who can use needle and thread had to make adjustments. How about your smartphone? You've probably added your own personal collection of apps. Whether we work in the for-profit or nonprofit world, we generally do studies and observe what people might want and need as we create, develop, improve, or market products or services. However, in our education systems, we have too often expected one size to fit all.

That one-size model has been locked-in by imposing standards and high stakes tests. Educators feel they are forced to dig even deeper, hoping to strike a vein of better test scores on a few subjects.

At the same time, the clamor inside and outside those trenches is getting louder. Legions of people realize that our very future will depend on paying attention to the uniqueness of each and every learner. While few are opposed to standards, let's be sure they don't simply freeze the system in the past, as we head into a fast-changing world.

Constructive Restlessness

Restlessness is simmering among students, educators, parents, communities, and nations to personalize learning. Demand is growing for schools and colleges that capture the imaginations of students and secure positive progress as we slip, slide, run, and sometimes fall into an exciting but complex future.

Neuroscientists are clarifying what lights up our individual brains, commands our attention, and motivates our behaviors. Multiplying social media lure and connect people around the planet. Remarkably, we carry a world of information that's just a click or keystroke away. So it's clear that personalization is a force that simply can't be ignored, whether we serve as educators, work for an automaker, or deal with customers in a local bank or grocery store.

What tomorrow brings for education may depend on how willing and able we are to satisfy an ever-growing *constructive restlessness*. Part of the puzzle will be how to meet the demand for a more customized, tailored approach to teaching and learning.

Most of us are good at making observations. Once in awhile, those musings come with a jolt. Consider this from Harvard and Project Zero's David Perkins: "We assume that our life and the lives of our children will be similar. Wouldn't education be more interesting if we connected what students learn to the future?"[602]

OK, what's the take away here? We simply must pursue better alignment with our constituents' and our individual students' interests, abilities, skills, aptitudes, aspirations, and motivations. That's the case whatever their race, ethnicity, first language, national origin, *or* other categories of diversity. Today's students need to be capable of creating a better future, and they are the ones who will have to live in it.

Not a New Concept

Personalizing education isn't new, but it's getting more attention because our stakes for the future are so high. An aim is to provide more options in considering "what is learned, when it is learned, and how it is learned." It is a way to get education even better aligned with a student's individual learning style and multiple intelligences.[603] The U.S. Department of Education defines personalization, at least in part, as "instruction that is paced to learning needs, tailored to learning preferences, and tailored to the specific interests of different learners."[604]

To some degree, personalization as a core idea is supported by: schools-within-a-school; reductions in class size; summer school; alternative schools; magnet schools; and charter schools. Then there are *individualized instruction* (pacing instruction for…the learning needs of different students); *differentiated instruction* (instruction tailored to the learning preferences of students using…methods that work best for them);[605] and *transformative learning* (sometimes connected to deep learning, often among adults, to help them find deeper meaning in their lives).[606]

Academic and personal coaching, tutoring, and mentoring are becoming staples in a torrent of strategies for personalizing education. Another technique is "flipping." In this case, "to flip" might mean doing the traditional classroom lecture online and using actual class time for activities, questions, coaching, and interaction.

Other beams that can support personalized learning might include: teacher advisor systems; after-school programs; in-school and out-of-school tutoring; certain approaches to mastery learning and outcome-based education; Advanced Placement (AP) programs; International Baccalaureate Programs (IBP); individual education plans (IEPs) for special education students; paying attention to learning styles and teaching styles; Response to Intervention (RTI); addressing school climate; and a host of others. Depending on how these things are organized, the commitment of everyone involved, and the resources available, all hold promise for contributing to overall success.

Teamwork: We've all heard something like this. "Sounds like you're saying we need a bunch of Lone Rangers, all trying to have it our own way." Just to be clear, we're assuming that teamwork and getting engaged as a member of a civil society are essential. That's where personal talents, abilities, and strengths blossom and enrich us all.

Personalizing in a Fast-Changing World

In an *Education Week* article on teaching values, writers Katherine Casey and Francesca Kaplan Grossman ask, "Why do so many of our schools still look and feel impersonal, industrial, and disconnected?"[607]

A Market of One Society. Wherever they are, schools are often among the largest organizations in a community. Isn't it ironic that such far-reaching institutions deliver one of the most personal services in any society? That's just one reason why educators need to make every effort to consider individual needs as well as the needs of society when working with their communities and educating their students.

We live in a market-of-one society. Like it or not, people want to feel that they are "exclusive or preferred customers of the firm."[608] On the communication side, growing numbers of for profit and not-for-profit organizations are asking probing questions about the type of information they'd like to receive about clients. On a broader basis, most of us are being sorted by zip codes, demographics, income, and even health concerns. Sometimes we feel more like a target than a constituent.

Educators on Personalization. 2013 National Middle Level Principal of the Year *Laurie Barron* observes that as students rely more for their information on technology rather than teachers, "personalizing the learning environment and building relationships will be increasingly crucial." Even though she recognizes that "test scores will continue to drive public perception," she emphasizes, "the first focus for schools must be on building relationships with students." To personalize, "collaborative leaders need to establish a school culture that is open to and passionate about improving curriculum, instruction, and assessment." Barron believes "relevance and rigor in the classroom" should be focused not only on academic success, but also on "developing students who are well-rounded, productive citizens who give back to their community."

Barron is determined to close achievement gaps. She calls for "multiple sources of data for each individual student, including standardized test scores, universal screeners, classroom pre- and post-tests, and other formative assessments, so that teachers can truly customize instruction for each student." Barron adds "eliminating tracking and ability grouping," moving toward "grouping that is more flexible," and "co-teaching models that allow teachers to work together on content and instruction."

Argosy University Associate Professor and 2012 ASCD President *Debra Hill* calls for a "growth model." She supports "multiple measures to determine where each child actually is on the academic spectrum, rather than assuming that every child ends up in the same place." Hill says she understands this concept is a "direct contradiction to No Child Left Behind." She adds, "The U.S. is not serious about education when preschools and kindergartens continue to be underfunded and continually minimized." *Keith Marty*, superintendent of the Parkway School District in Missouri, sees a movement "away from cohorts and averag-

es to truly being focused on individual student growth and progress."
Joseph Cirasuolo, executive director of the Connecticut Association of
Public School Superintendents (CAPSS) warns, "If public education
systems don't become customized, they will be abandoned in favor of
privately provided programs that are."

Wisconsin Superintendent *Damian LaCroix* forecasts "a highly technological and personalized blended learning environment will emerge
as venture capitalists invest in new programs and tools that foster any
time, any place, and any pace learning." He foresees intensified con-
sumerism as parents shop for the best options to meet their personal
educational preferences for their children."

As a distinguished longtime teacher at Thomas Jefferson High School
for Science and Technology in Fairfax County, Virginia, *Milde Waterfall* raises concern about "requiring that each teacher be on page one on
Tuesday and teach a lesson which has a common assessment." That approach "will be especially damaging in the humanities and sciences but
may flourish in mathematics." She points out that "individual curiosity
and intellectual prowess are crucial in literature, the arts, and the sciences and are badly damaged by the demand for uniform quantifiable
data as evidence of learning." Eugene, Oregon, School Superintendent
Sheldon Berman sees a "transformative breakthrough" as measures of
accountability move from standards for subject matter acquisition toward college and career readiness. That refocusing will lead to greater
attention for "outcomes of education rather than static test scores."

A *Washington Post* article, based on a visit with Potomac, Maryland,
teacher Glenn Whitman, carried some tips for connecting brain research and curriculum. A few examples: Young students perform better
"when they are not stuck in chairs;" "Sleep is an under-rated educational method," especially for students who stay up late and get up for early
classes. Students retain most at the beginning and end of a class; and,
for preschoolers, play may be more effective than memorization.[609]

The Standards and Testing Conundrum

Let's admit it. On the one hand, standards can be helpful. Formative
testing can guide us toward more personalized learning and maybe even

higher achievement. On the other hand, standards and standardized tests can potentially freeze the system into a lockstep that overlooks individual differences and the stark reality of a world that will simply not stand still, not for a minute. Few, if any, tests measure a full range of talents and abilities. Much of what we'll need to know and be able to do in the future may not even show up on our radar, because we'll have to invent it. Apple's Steve Jobs declared, "The most important thing is a person who incites your curiosity."[610]

No Child Left Behind (NCLB), a U.S. federal approach to improving schools during the early 2000s, set requirements, imposed penalties, and called for high-stakes testing as an enforcement tool. Many complained that NCLB caused education systems to narrow the curriculum largely to those things that were easily tested.

As the nation entered the 2010s, Common Core Standards were developed largely by two state-based organizations, the Council of Chief State School Officers (CCSSO) and the National Governors Association (NGA). To begin, internationally benchmarked Common Core standards addressed math and English language arts. The groups went on to explain, "Of course, other subject areas are critical to young people's education and their success in college and careers."[611] That commitment to a broader curriculum has been an incentive for many educators to embrace the Common Core. However, amber caution lights continue to blink among those who have concerns about any over-reach of high-stakes testing.

Futures Council 21 member Ryan Hunter, an American University student leader, cautions that "teachers feel their traditional autonomy and freedom to creatively manage their own classrooms are being undermined" by state-imposed curriculum and standardized test results. He notes the "constant preparation for these exams and preparatory practice exams." Hunter questions whether those tests can measure educator performance. "There are many factors at play in how a student performs on an exam," he observes, "from how much a student studied to sleep deprivation and family/personal issues."

Narrowing the curriculum to match only those things that are most testable could leave many potentially engaged students adrift. Penalties are magnified for those who are disadvantaged or live with inequities or disabilities. If what we are required to learn becomes inflexible and unchanging and no longer seems related to our lives today and tomorrow, we will likely lose interest. Another big question: How will students ever be encouraged to learn what we don't already know?

Standards, after all, can be narrow or expansive, static or flexible. As expectations continue to grow, we'd better realize that the progress we expect as a society depends on how well we cultivate personal strengths, including curiosity and creativity. Any team, any community, any nation is strengthened, often exponentially, by the genius of those who are part of it.

To cultivate that genius, we'll need to cut some slack for students and educators. How can we do that? For one thing, we can try to make sure what is being tested is life worthy,[612] and that test-taking is not keeping future generations from pursuing possibilities that may be beyond our imaginations.

> **"Not everything that counts can be counted,**
> **and not everything that can be counted counts."**
> *Albert Einstein*[613]

Key Questions: Assessment and Personalization

Accountability is important. No doubt about it. Yet, we need to constantly question the assessment side.

• Could some of the time and money we spend on testing be devoted to actually personalizing education?

• Are we conducting and analyzing mind-brain research to figure out how to motivate our students? If not, why not?

• Are we focusing so much attention on testing what we currently teach…that we have little time left to think about what else we should be teaching?

• Is it possible that questions and answers on tests are not always in sync with the realities of a fast-changing world? When those tests are high-stakes, does that stimulate more teaching to the test and pull attention from developing individual strengths?

• Is it possible that traditional testing and assessment have not evolved quickly enough to provide timely feedback that can help us meet individual needs rather than just give us a scorecard? Some technology-based programs do offer instant feedback on certain items that are more easily testable, as well as encouragement and further assignments.

• As the world spins forward, one additional question should be baseline: "Are our curriculum, our instruction, and our tests a reflection of the limits of our imagination?"

Can We Tolerate the Questions? If we were buying a car, we would want to be ready to ask good questions and be sure what we are considering fits our needs. We would probably be concerned about quality, price, service, and whether the model we were test-driving had a record for breakdowns or reliability.

When we're considering something as important as standards for the education of our children, it just makes sense that we would also ask hard questions. Unfortunately, those who ask those questions are often labeled as opposed to standards or "presumably content with lower standards," says high-stakes testing critic Alfie Kohn.[614]

Key Questions from *Sixteen Trends*

In the 2006 book we wrote for the Educational Research Service, these were four key questions we asked everyone to consider in developing standards:

• Do the standards focus on preparing students for the future, or do they freeze the system and its students in the past?
• Will students who don't do well on the tests—who don't measure up— simply give up, drop out, or be pushed out of school?
• Will high-stakes tests narrow the curriculum?
• Will pressure created by imposing standards and high-stakes tests, without adequate resources to personalize education, drive talented teachers and administrators from the field and discourage aspiring educators from pursuing careers in education?[615]

Let's Be Flexible
To freeze is to fail.

In a fast-moving world, any standards and any testing program must be flexible. Neither should, individually or in combination, attempt to freeze the world in place. New programs and mid-course corrections will be essential. To freeze is to fail.

All students, despite their social, economic, or other backgrounds, need to have a chance to succeed. Recognizing and striving to meet the need for personalization is a powerful force for improvement in education, and it could pay huge dividends.

Keith Marty, superintendent of Parkway School District in Missouri believes that "jobs will no longer be defined by hours, nor a place, rather by the outcome and/or product. That means teaching and learning will also no longer be (solely) defined by hours and a place called a school-house. Technology is driving this trend." He adds, "Students will check in at a school, which might also house a day care facility, public library, recreational center, and a center where young people might be cared for while being guided in their learning." Marty also foresees a rise in "learning academies, where students may be learning at medical facilities, financial institutions, retail locations, etc."

"One size does not fit all," Laurie Barron declares. This recognized leader believes educators need to meet an array of education needs, "but how these needs are met should be dependent on each student." She adds, "There should be one goal: high achievement," and "there should be multiple paths to meet that goal: personalization." Barron emphasizes, "Fair does not mean equal," noting that "educators must listen to students and let them have a say in what happens to them in the education system."

Three Crucial Dimensions of Personalization
Technology, Mind-Brain, and Globalization

If a playground is where children *come out to play*, personalization is where every trend highlighted in this book *comes into play*. While we'll have much more for you in other chapters, let's just briefly introduce three dimensions with profound implications for personalization and learning: technology, mind-brain education, and globalization.

Technology: Computer programs can chart individual students' day-to-day progress in discrete areas of the curriculum. Those students can track their own progress, then get online feedback, possibly some tutoring and reinforcement, and follow a prompt to their next assignments. In fact, the state of Connecticut has developed plans to exchange "bubbled-in with pencil-and-paper" mastery tests for "computerized tests, essentially personalized for each student."[616]

Susan Sandler of the Sandler Foundation in San Francisco suggested a "hybrid conception of personalization." In an *Education Week* commentary, she pointed out the importance of making sure students are "personally known at school and have strong relationships." It might be called "Personalization 1.0." The late Theodore Sizer, who headed the Coalition of Essential Schools, suggested that approach for many years. Another approach, which she dubbed "Personalization 2.0," emphasizes "data and customization, which is a good thing…but there is more to high-quality learning than creating the equivalent of a perfect iPod playlist." A hybrid approach, Sandler declared, would combine the two into "Personalization 3.0," which would use technology "to enhance teacher-student relationships, not replace them."[617]

Futurist John Meagher reminds us that calculators have enhanced human computational skills. Google and Wikipedia are replacing memorization of facts. Cursive writing "is falling by the wayside as people communicate less and less via pen and paper in any extensive way." We're seeing expanded use of online, real-time language translation services.

James Harvey, a member of Futures Council 21, suggests that using technology well will allow teachers "to serve as learning coaches, helping students create their own learning…to map and navigate their own learning journey."

Mind-Brain and Education: A fast emerging consideration for personalization is Mind-Brain and Education (MBE). This relatively new field "encompasses educational neuroscience (a branch of neuroscience that deals with educationally relevant capacities of the brain), philosophy, linguistics, pedagogy, developmental psychology, and others," according to Mary Helen Immordino-Yang, an affective neuroscientist and human development psychologist at the University of Southern California at Los Angeles.[618] Not to worry—the issue of nature vs. nurture continues to be an important factor.

Most teachers and principals can readily recount how they have helped individual students succeed in academics and the development of unique talents, skills, interests, and abilities—even in shaping appropriate behaviors. For decades, thoughtful school counselors and psychologists, working in tandem with caring teachers, have been trying to cross the valley between an exponentially more complex world impacting their individual students and a too often slow to change education system.

Globalization: The world is shrinking before our very eyes. Everyone is our neighbor. With massive migration, great throngs of people are moving across the globe. Social and economic, cultural, language, and other differences land at the doorsteps of schools and colleges. Educators, the institutions that employ them, and the people they serve need to fully understand that meeting students where they are and helping them find a way forward is part of the very essence of personalized learning.

What Do We Do If...?
Scans, Pharmaceuticals, and Enhancements
Personalizing has a lot to do with leveling the playing field, but level it up, not down. The *real thing* is not just for those who can afford the extras. A personal approach to learning should be for everyone. While we know that's the way it ought to be, it poses an ongoing but exciting challenge for every education system, from pre-kindergarten through college or university and then beyond through life.

Here are a few "what do we do ifs?" To some extent, they are already facts of life. Apart from their stark reality, some who have avoided them might refer to these possibilities as futuristic and mind-bending. Those wearing blinkers might declare that they are unlikely or impossible. We've heard similar objections about a variety of technologies that, just a few years later, "we couldn't do without."

• **fMRI.** What do we do when those who can afford them start presenting schools with their children's fMRI brain scans, pointing out uniquenesses that need to be addressed. The neuro-psychologist's report might include good bets for instructional strategy that would motivate those particular kids to learn. The value is breathtaking, but the questions people raise will demand answers. How do we maintain a level playing field? What about those who can't afford the scans or whose parents object to the very idea? Who would or should pay for them? How reliable are they? What if the person doing the fMRI had asked different questions? What are the instructional and policy implications? Will this procedure become a basic part of personalization? It's no longer simply the kind of IQ test your grandfather took.

• **Performance-Enhancing or Designer Drugs.** Stories about these pharmaceuticals generally bounce between the sports section and the front page of our newspaper or favorite web site. The topic—the use of anabolic steroids by athletes. Most students contacted for a 2012 *Time* article considered the use of steroids to enhance sport performance as cheating. However, they were less likely to find unprescribed use of drugs such as ADD (Attention Deficit Disorder) medication to be unethical if it helps them improve focus prior to an exam. Based on continuing research, designer drugs to enhance individual learning might become commonplace.[619]

Futures Council 21 member John Meagher spotlights "neurological medical developments and specific biochemical treatments (nutritional supplements brain enhancers, perhaps targeted to individuals) that will enhance learning abilities, potentially reduce neurological disease, impact brain chemistry, and optimize educational and learning opportunities, including behavioral health and/or happiness."

In *Education Week*, writer Sarah Sparks pointed out questions posed during a Society of Neuroscience conference about cognitive-enhancing drugs and commercial drinks. They included "whether it is safe and fair to allow healthy people to boost their brain function chemically, or use drugs to correct environmental factors like poverty or bad instruction."[620]

• **Appliances, Apps, and Augmented Reality.** From smartphones and tablets to creative playgrounds, we've seen virtually dozens of appliances that can enhance our learning. Don't forget the softer side, such as Google and a host of web sites and apps. It seems like every day the curtain goes up on new possibilities. The inventors simply won't stop. Way back in 2005, Ray Kurzweil, an acknowledged and controversial inventor, entrepreneur, and visionary anticipated that, by the 2020s, researchers will have reverse-engineered the human brain,[621] which means they might have a pretty good idea about how this most complex of instruments works. His ideas about virtual and augmented reality are already in play and will continue to be a work in progress.

Personalizing Education
On-the-Ground Examples, Concepts, and Sources

Personalizing education, moving away from the one-size-fits-all model isn't a fad or necessarily even a movement. It is becoming a reality. We now know that we are not dealing with cookie-cutter kids. Each one is unique. Round pegs don't fit into square holes, but that doesn't mean we don't need round pegs.

Expectations for personalization are growing across the whole of society. So naturally we expect something similar from educators. Here are just a few examples, concepts, and possible sources that might be helpful as we consider the challenge. Keep in mind that this is not a list of best or worst. It is only a list of some.

• **Universal Design for Learning.** UDL offers a set of principles for curriculum development that give individuals equal opportunity to learn. According to its web site, UDL "provides a blueprint for creating instructional goals, methods, materials, and assessments that work for everyone—not a single, one-size-fits-all solution but rather flexible approaches that can be customized for individual needs."

David Rose, a developmental neuropsychologist and educator, whose primary focus is on development of new technologies for learning, is founder and head of CAST (Center for Applied Special Technology). That organization has pioneered Universal Design, which is aimed at improving education for all learners. Three primary brain networks come into play: "recognition networks (the 'what' of learning); strategic networks (the 'how' of learning); and affective networks, (the 'why' of learning)."[622]

- **Multiple Intelligences.** Perhaps the most focused contemporary approaches to personalizing education have been driven by Howard Gardner's classic work, *Frames of Mind*, published in 1983. He expanded on that work in *Multiple Intelligences . . . The Theory in Practice*, published in 1993.

Gardner's premise is that each of us has some mix of seven or more "intelligences," which he defines as "the ability to solve problems or fashion products that are of consequence in a particular cultural setting or community." The old definition of intelligence, Gardner says, was largely "the ability to answer items on tests of intelligence." Seven of the intelligences he identifies are: musical, bodily-kinesthetic, logical-mathematical, linguistic, spatial, interpersonal, and intrapersonal. Gardner's theories are based on actual cognitive research.[623]

Some parents and educators complain that schools too often teach to only a few of those intelligences, which leads to student frustration and a loss of a fully developed range of talents society needs to advance.

- **Innovate to Educate: System (Re)Design for Personalized Learning:** The Software & Information Industry Association (SIIA), working in collaboration with ASCD and the Council of Chief State School Officers (CCSSO), released a 2010 report "based on the insights and recommendations of 150 visionary education leaders."

According to ASCD, the report, *Innovate to Educate: System (Re) Design for Personalized Learning*, "provides a road map to accelerate the redesign of the current, mass production education model" to something more "student-centered" and "customized." The hope is that it will "engage, motivate, and prepare our students to be career and college ready." A key element of personalized learning, identified by the group, is "flexible, anytime, everywhere learning."[624]

- **Coalition of Essential Schools (CES).** Coalition of Essential Schools' late founder Ted Sizer told *Educational Leadership* magazine, "Give me time, autonomy, supportive colleagues, and few enough students so that I can understand each one well enough to tailor some of my teaching to him or her—and I will show you students who perform well, today and tomorrow."[625] CES focuses on creating and sustaining "personalized, equitable, intellectually challenging schools."[626]

• **Mass Customized Learning.** Technology & Innovation in Education (TIE), based in Rapid City, South Dakota, has been discussing mass customization as a way to reform education. Julie Mathiesen, TIE's director, notes that the Industrial Age system of education is producing too many students who are unskilled and who fail to graduate from high school. Those students "are no longer told how to think—and don't process and learn through telling. Instead, students learn by doing and by learning anytime, anywhere," she remarked at the 2012 AASA National Conference on Education. She makes the case that appealing to students' learning styles and making learning challenging can lead to students "who want to come back (to school) tomorrow."[627]

Implications for Society
Personalization

Demand for personalization is growing exponentially, and there's little or no evidence that situation will reverse course. No institution is exempt, no matter how new or how venerable. "Beat 'em or join 'em" is a false choice. If we try to maintain business as usual, we'll end up out of touch. Implications of the personalization trend for the whole of society are not only significant but also urgent.

• If businesses, governments, communities, countries, or any other institutions hope for a brighter future, they will need to understand that the ones who are prospering are the ones most able to attract and possibly keep creative, imaginative people, not just those who do well on traditional tests.

• Public officials at all levels need to be sure their education system has the flexibility, know-how, and commitment to discover and build on personal strengths.

• All of us will need to become talent scouts, actively seeking, celebrating, encouraging, and publicly supporting development of the personal and group talents, skills, aptitudes, hopes, and dreams of those around us.

• Accountability measures, often expressed as standards and win-lose tests of what we already think we know, need to allow for and encourage, not inhibit, students and hosts of others from exploring what we don't yet know.

• Generally, we will need to move beyond a "scoreboard mentality," a belief that progress in education can be reduced to the reporting of a simple set of numbers, like box scores for baseball, football, hockey, soccer, and other sports.

• Society should take warning that freezing our formal systems of learning for people of any age could inhibit rather than enable us in our quest to produce the people who will provide leadership in the future.

• In an increasingly interconnected world, every institution, as a matter of survival, needs to define itself as part of a Global Knowledge/ Information Age, even an Age of Knowledge Creation and Breakthrough Thinking.

Implications for Education
Personalization

• **Constructively answering the call for personalized education.** Demand for bringing out the best in every learner can be subtle or direct. Either way, it will require a response. Those who apply pressure will want to know that we understand the need and can clearly describe it. They will also want constructive and effective action.

An expectation for personalization has reached every corner of society and crosses the demographic landscape, from social and economic conditions and racial and ethnic diversity to those who qualify for special education. That demand is likely driven in part by a growing dependence on interactive personal technologies and by a realization that local, national, and world economies as well as civil societies depend on well-educated people. Another fact of life propelling personalization is the stark reality that being competitive for jobs means developing unprecedented levels of knowledge and skill. All of this is nudged along by intense pressure for higher levels of student achievement. Expectations will only intensify. They will not go away.

Virginia teacher Milde Waterfall declares that the concept of one-size-does-not-fit-all is "something that teachers have known intuitively for many, many years." In testing, she leans toward "a creative response" rather than simply collecting quantifiable data for storage and retrieval.

• **Nurturing a climate that includes flexibility, resilience, and adaptability.** When organizations become rigid, anchored in the past, and deeply committed to maintaining the status quo, they can expect people to do an end run around them. Education systems need to be flexible, resilient, and adaptable enough to deal constructively with individual differences. Those they serve also expect the same kind of thing from business, government, and other fields.

Iowa City, Iowa, Superintendent Stephen Murley points out a few specific examples of possible flexibility. One possibility: "modularization of coursework" with classes that are "no longer a year, a semester, or a quarter," and "include smaller subsets that learners can access as part of the current course of study." Another: "The use of competency rather than seat time or hours/minutes of instruction will become the norm. We'll see a rise in badges, certificates, and the like as we shift to demonstrations of mastery." Tokyo International School Vision Navigator Patrick Newell says "personalization equals the opportunity to get involved in something that is interesting, engaging, and relevant." He adds, "Being empowered to create something increases satisfaction. It unlocks self-discovery and the joy of creating." Personalization can take time, he cautions, "trying to figure out what (students) want when they really do not know what they want."

• **Personalizing as a key to more flexible and comprehensive standards.** As pressure intensified for better student achievement, standards and high-stakes tests narrowed the focus of K-12 education. In the U.S., the early 2000s hatched a federal program widely known as No Child Left Behind, which seemed to forcibly limit the focus of education to a few subjects and to lock accomplishment into banks of high-stakes tests. *Salon.com* quotes education historian Diane Ravitch—"When reading and math count and nothing else does, then less time and resources are devoted to non-tested subjects like the arts, science, history, civics, and so on."[628]

Futures Council 21 member Matthew Moen expresses concern that "customized education will fragment common bases of knowledge among the college educated." While shaping education to fit a student's abilities and aspirations is important, he observes customization "as part of a trend that has companies offloading some of their training costs onto universities, which are increasingly seen as places where specific students are given specific skills for particular employers."

• **Eagerly responding to the challenges of technology, mind-brain research, and globalization.** While these persistent forces are disrupting the environment and putting whitecaps on the calm seas of the status quo, we'd better strongly consider their implications for personalizing education.

A growing challenge for our education systems is to make learning as exciting inside the classroom as it is outside. The array of technologies that can help us enhance learning will increase exponentially. While we talk about genetics in the classroom, students can, on their own, buy "biobricks," bits of DNA, online and download recipe books to help them engage in synthetic biology.[629] Brain scans are helping us identify what interests and motivates each of us as individuals. The world has come to our doorsteps and what happens anywhere in the world has profound implications for us.

What are the implications for instruction? For policy? Are we drawing on the richness of *multiple sources of education* to enhance learning?

• **Making education more interesting, exciting, and related to what is important in life.** We know that way too many students drop out of school. We're also concerned about in-school dropouts, those who fail to see the benefit, get lost in the system, or for some other reason, lose interest. Futurist John Meagher suggests schools use something akin to a Myers-Briggs personality test to help identify and adjust to a student's learning style.

Harvard's David Perkins, who has conducted research devoted to thinking, learning, and understanding, has described it as "a relevance gap." During a Harvard University/Project Zero Future of Learning Institute presentation, he lamented that growing numbers of students find school boring. Perkins calls for "lifeworthy" learning and suggests that we consider what content matters in our lives today and will matter in the future—in the lives learners are likely to live.[630]

Some students see school as black and white. On the other hand, they identify the world outside of school as high-definition, full color.

Growing numbers of educators are becoming more creative as they add value to what students are learning though mobile, hand-held, instant, highly interactive communication with the world, using photo and video sharing, GPS, gaming, and a flurry of existing and emerging social media. From *Sim City* to *Math Blaster* and programs devoted to civic education, gaming is getting increased attention. Researcher, professor, and gamification authority James Gee draws a connection between play and learning.

• **Resisting the tyranny of the average.** Try this sad comment that is, in some form, repeated over and over again. "With expectations going up and resources going down, there are two sure ways to increase average test scores—narrow the curriculum to only those things we are required to test and encourage kids who are having problems to drop out of school." It's not a recommendation but a lament. Educators know that the impact of the high-stakes environment has become a tragedy for many students. The fact is that we need to educate everyone, not just a chosen few—not only for employability but also for life as citizens, who might just live next door. Beware—the tyranny of the average. Average scores may go up, but many students are thankfully not *average.*

• **Becoming talent scouts.** On top of all the scientific research, which is vitally important, we need to use our senses to spot interests, talents, and motivations. Like good detectives, we need to always be on the lookout for clues. Those clues (or connections) are all around us if we're looking and listening. Longtime education consultant Larry Lezotte has spoken of "the student who doesn't like calculus but is really turned on to basketball." If we're tuned to sense it, there's a real clue about how to make calculus more interesting.

• **Encouraging educators to prepare students for the future rather than become compliant bureaucrats.** Some educators express concern that they have little time for creating the education system their community needs, because they are too busy managing compliance with mandates. Arnold Fege raises concern about what he sometimes sees as "the transformation of education leaders in our country from intellectual icons of their community to bureaucrats who are increasingly just a part of the infrastructure." He calls for educators to "transform education from the industrial model, characterized by efficiency, to the knowledge model, characterized by effectiveness that depends on the quality of educational, political, private sector, and community leadership." [631]

- **More Implications for Personalization.** While we're personalizing, we need to also be sure we are emphasizing teamwork. Educators and society should be keenly aware that good education is firmly linked to concerns about equity and adequacy of funding. Everyone should feel obligated to understand the benefits and the limits of testing. The purpose should be to provide good information for improving education, not to punish, sort out winners or losers, or give us an oversimplified scoreboard.

Personalization

Questions and Activities

1. Do you believe it will be possible to help students truly achieve without personalizing education in a way that considers their interests, talents, abilities, and motivations? Why or why not?

2. If you are an educator, how are you and your school or college currently attempting to personalize education for each student? Identify at least six additional ways you would like to see your education system personalize its services.

3. View http://www.youtube.com/watch?v=RViuTHBIOq8, *TEDx* Manhattan Beach, Manhattan Beach Education Foundation, 2009, "Embodied Brains, Social Minds," featuring Mary Helen Immordino-Yang, affective neuroscientist and human development psychologist, USC.[632] Then, discuss the significance of what she has to say for inspiring and motivating students, educators, and communities (13:28).

4. In one single-spaced page, briefly explain why you believe mind-brain and education, based on neural research, technology, and globalization are important factors in personalizing education.

5. React to this statement, "If the education system had regularly focused on creating a future rather than defending the status quo, the high stakes environment would not have been imposed in the first place."

6. Review the web site devoted to the Universal Design for Learning (UDL). Develop a brief, approximately three-minute, PowerPoint presentation highlighting four things you found most interesting. You can start at http://www.udlcenter.org.[633]

7. In your opinion, what questions should we be asking about standards and high-stakes testing?

8. What are the "multiple intelligences" identified by Howard Gardner?

Readings and Programs

1. Gardner, H. (1993)., *Multiple intelligences: The theory in practice.* NY: Basic Books, Harper Collins. [634]

2. Immordino-Yang, M.H., & Fischer, K.W. (2009, in press). Neuroscience bases of learning. In V.G. Aukrust (Ed.), *International Encyclopedia of Education, 3rd Edition*, Section on Learning and Cognition, Oxford, England: Elsevier. http://www-bcf.usc.edu/~immordin/papers/Immordino-Yang+Fischer_2009_NeuroscienceBasesofLearning.pdf [635]

3. Council of Chief State School Officers, *Common Core State Standards: Implementation Tools and Resources*, August 2012. http://www.ccsso.org/documents/2012/common_core_resources.pdf [636]

4. Ravitch, D. (2010), *The Death and Life of the Great American School System... How Testing and Choice are Undermining Education*, NY, Basic Books. [637]

Chapter 13

Ingenuity
Flashes of Insight!

Chapter 13: Ingenuity. *Flashes of Insight!*

Trend: Releasing ingenuity and stimulating creativity will become primary responsibilities of education and society. *Information Acquisition → Knowledge Creation and Breakthrough Thinking.*

"Imagination is more important than knowledge. Knowledge is limited. Imagination encircles the world."
Albert Einstein[638]

Ingenuity
Turning it on. Tuning it up. Turning it off.

Quick quiz! Are you ready? **Here's the question:** What is potentially one of the greatest sources of energy on the planet? **Answer:** Human Ingenuity. **Follow-up Question:** What consumes more energy than almost anything else? **Answer:** Our ongoing battle to maintain the status quo.

It's jaw dropping! Every day, each of us wakes up to a stark reality—the challenges we face aren't yielding to business as usual. Vast knowledge is good, and we highly recommend it, but it doesn't substitute for an ample dose of creativity, imagination, mindfulness, and problem solving. If you don't believe it, scan the employment ads. Trying to blow the lid off standardized test scores is a goal worth pursuing, but most of us will only be able to capitalize on the achievement if we have well oiled thinking, reasoning, and contemplation skills and the ability to perform across disciplines and outside the trenches.

Ingenuity, Creativity, Imagination, Innovation...

A working definition of **ingenuity** has to include inventiveness, an aptitude for discovering, skill in devising or combining, originality, resourcefulness, cleverness, an identifying character or spirit, intelligence, aptitude or capacity, and discernment.[639] Ingenuity is not necessarily measured by an IQ test. **Creativity** is, of course, "the state or quality of being creative; the ability to transcend traditional ideas, rules, patterns, and relationships; and to create meaningful new ideas, forms, methods, or interpretations."[640] **Imagination** might be seen as the act or power of forming a mental image of something not present to the senses or never before wholly perceived in reality. It might also be seen as "creative ability, the ability to confront and deal with a problem, resourcefulness, and thinking with an active mind."[641] **Innovation** can be defined as "the introduction of something new or a new idea, method, or device," possibly something novel.[642]

Intellectual Entrepreneurs. We read books, visit exhibitions, go to movies, click on web sites, or immerse ourselves in social media, largely because they pique our imaginations. Most of us are transfixed by stories about those who pioneered an idea or broke new ground. In varying degrees, that pioneering spirit lies deep within each of us. It's an entrepreneur, just waiting to see the light of day—hoping to make it economically or just make a positive difference.

Thankfully, fresh generations of intellectual entrepreneurs are putting the pieces together; seeing certain things in a whole new light; dealing with paradox, controversy, and complexity; developing creative solutions to problems; and conceiving of new knowledge-based industries.[643] Richard Florida, author of a classic, *The Rise of the Creative Class*, calls it "the 'Eureka' step."[644]

It's up to everyone, certainly every educator, to spot, encourage, and help develop the ingenuity that is in each of us and all around us.

Knowledge Creation and Breakthrough Thinking
Making Connections across Disciplines

In our schools and colleges, educators often teach within certain specialties, and we can be thankful that they do. Most organizations are a collection of seemingly disconnected departments. Ideas often emerge from conflict or controversy, but sometimes we feel a frustration similar to what Abraham Lincoln faced when he described "a house divided," with no time for civil discussion or creative solutions.

Big questions. How can we help our students learn and think across disciplines? Who is paying attention to the connective tissue, the white spaces, and the natural links between and among disciplines and across an organization? Who is asking: What are the implications for transportation posed by an aging society? What are the ethical implications of scientific research or interplanetary exploration? What are the implications of generational differences when we make political and economic decisions? When we fire up our smartphones or iPads, do we focus only on those things that interest us most, or do we reach out to discover how what we see and hear might impact others? Are we able to see things *in context*?

The answers to these and dozens of similar questions are urgent as we conceive of and pursue our future. In fact, it is our ability to think, plan, and work across disciplines that has been a driver for our economy and civil society. As specialized as we might be, all of us, whether we like it or not—will live interdisciplinary lives.

Unity of Knowledge. Enter Edward O. Wilson, the noted Harvard professor and biologist. He's made helping us "get it all together" one of his missions. Wilson even wrote a book about it titled, *Consilience: The Unity of Knowledge.* Among his concerns? Legitimate human progress is too often brought to a crawl because people don't think or communicate across disciplines and see things in context. "The ongoing fragmentation of knowledge and resulting chaos in philosophy are not reflections of the real world but artifacts of scholarship," Wilson remarks. This seasoned sage asks us to link facts and theories across disciplines. Rather than insisting that biology, environmental policy, social science, and ethics are solitary disciplines, we need to see where they intersect and how they relate to others.[645]

The Second Law of Thermodynamics. Let's take a minute for the Second Law of Thermodynamics, sometimes called the Law of Entropy. In short, that law concludes that everything disintegrates and generally becomes part of everything else. Think about it. If sunlight just dropped on us like a brick and didn't disintegrate, we probably wouldn't even be here. A few people even see education as that proverbial brick, not necessarily a part of everything else that happens in society. Education has value when it, too, disperses and finds its way into the minds and habits of people.

Becoming a Catalyst. "Schools are fairly good at promoting the acquisition of information," says Avis Glaze, president of Edu-quest International in Canada. "The focus now has to shift quickly to knowledge creation and breakthrough thinking, which support a culture of ingenuity and creativity." Glaze adds, "Many educators value this trend but have never been taught how to nurture creativity and ingenuity in their work. In fact, occasionally, students are punished, or at least not rewarded, for thinking outside the box or for being overly creative."

Contemporary philosopher Ken Wilber observes that what we see as a whole may "simultaneously be a part of some other whole," He reminds us, "The beads of knowledge are already accepted: it is only necessary to string them together into a necklace."[646] Take a look at the ingredients used to make things you use, from shampoo to memory cards. Many are compounds or solutions. They generally got together with the help of a catalyst to make meaning or generate new ideas or knowledge.

All leaders, in any walk of life, should be catalysts. We should expect them to help us think and guide us as we move from data and information toward knowledge creation and breakthrough thinking, even toward wisdom. Eugene, Oregon, Superintendent Sheldon Berman believes that "consciously creating caring and supportive classroom communities" can trigger "success in both the level of conceptual understanding and in students' ability to work collaboratively with others."

Why Ingenuity and Creativity? Why Now?

Most of us have our hands full, whether we're educating students, taking care of business, or making sure we get the kids to and from what might seem like a dozen activities every day. I can hear it now, "Enough already. You're telling me to be more creative and come up with ideas that will disrupt my life even more?"

Releasing ingenuity, stimulating creativity, and encouraging imagination are not just nice things to do. Instead, they are essential to the advancement and very survival of both our economy and our civil society, wherever we live on the planet. Joseph Hairston, former Baltimore County superintendent and a member of Futures Council 21, makes clear that "problem solving skills will be a prerequisite for employment in the new economy. Innovative behavior will be normal practice in society." Call it a new normal.

Finely honed thinking, reasoning, and problem solving skills; the ability to find connections across disciplines; and plenty of sheer audacity will be front and center as we deal with recurring concerns. We need those same qualities and skills to make the most of opportunities that regularly come along, if we are tuned to recognize them.

The Lineup

Consider this. We're at a police station. Six people are standing side-by-side on a riser. They're the suspects. Beyond the footlights are some other people, including a victim and a detective. Now, consider a different kind of lineup. Standing there, shoulder-to-shoulder, are six puzzled people. They are Leonardo da Vinci, Thomas Edison, Grace Hopper, Bill Gates, Steve Jobs, and Mark Zuckerberg. It's a motley crew. Each has been caught red-handed and charged with imagination, creativity, invention, and disruption of conventional thinking. There seems to be no stopping these people.

Your job—Identify the one who has, positively or negatively, maybe both, disrupted your life the most. Why did you choose that person?

Kenneth Robinson, who has U.K. origins, is a widely known creativity expert. He argues, "We are educating people out of their creativity."[647] Sir Ken makes the case that creativity is as important as literacy, but students are conditioned to be frightened of doing anything wrong. "All over the world," he submits, "formal education systematically suppresses creative thinking and flexibility." Robinson points out that "national strategies to raise standards in education are making matters worse. They're rooted in an old model of economic development and a narrow view of intelligence."[648]

John Seely Brown, the former chief scientist for Xerox Corporation and director of its PARC research center during the 1990s, is a phenomenon. The lab, tasked in the 1970s with creating "The Office of the Future," turned out inventions such as laser printing, computer generated graphics, an Ethernet, and the first personal computer with windows, icons, pull down menus, and a mouse.[649] The latter, as widely reported, became a starting point for Steve Jobs' development of Apple.[650]

In their book, *A New Culture of Learning*, Brown and colleague Douglas Thomas, contrast creativity and imagination. They describe *creativity* as "the ability to use resources in new, clever, or unpredictable ways to solve a specific problem in a particular context." On the other hand, they describe *imagination* as "encouraging the question 'What if?'" *Imagination*, they point out, "is inquiry based" and "is the literal building of a world around a new idea." They add, "The goal of education, training, and innovation spaces is to create and structure an environment where imagination can flourish."[651]

Conventional wisdom and "the way we've always done things" may not cut it as we take on worldwide challenges posed by technology; the environment; demographic change; the economy; safety and security; globalization; and advances in learning theory. Even when we deal with issues that involve war and peace, we will increasingly need knowledgeable people who are imaginative enough to forge connections and understanding across political and religious divides, even across oceans.

Words of Wisdom from Howard Gardner. "We live in an era where everything that can be automated will be," Harvard University Professor of Cognition and Education Howard Gardner told a 2011 National Academy of Education Annual Meeting. He added, "Only those of a robust, risk-taking personality and temperament are likely to pursue a creative path."[652]

Gardner includes "the creative mind" in his timeless, acclaimed book, *Five Minds for the Future*, which also features: "the disciplined mind," "the synthesizing mind," "the respectful mind," and the "ethical mind." [653]

During a 2012 Harvard Future of Learning Institute, Gardner noted that many people consider creativity to be "thinking outside the box." He explained, "It's hard to think outside the box unless you have a box." Gardner encouraged educators from around the world to "master a discipline" and "go beyond the known." His clincher— "Don't hesitate to have an iconoclastic temperament."[654]

Watch for Roadblocks, Take Risks, Show Courage

Ellen Winner. What makes creativity hard to teach? Ellen Winner, psychology chair and director of the Arts and Mind Lab at Boston College, has done significant research that gives us clues. One is the difficulty in teaching students (or others for that matter) "the willingness to take risks and learn from failure." Another is "the ability to *transfer* ways of solving problems between *seemingly unrelated situations.*"

Winner found positive risk-taking common among the most creative students. "Risk we tend to think of in negative terms," she said, "but high risk play is endorphin-loading and high-energy, so it is part of what keeps kids engaged in creativity." If students come from disadvantaged backgrounds, Winner noted, they might be even more willing to take risks because they are less comfortable with their current situations. It also helps if those risks "carry meaning for them—such as getting involved in solving local problems."

Robert Sternberg. Oklahoma State University Provost and Senior Vice President Robert Sternberg, an expert on intelligence testing, confirmed "risk is essential to creativity." However, people who "want to get into a good college, a good graduate school, or a good job" sometimes choose *not* to take too big a risk. "Schools," Sternberg remarked, "often encourage you to do the opposite of what you need to be creative." In essence, those who rate entries or applications sometimes reflect "I want you to be creative— and be sure you agree with me." Author and educator Henry van Dyke put it this way, "Genius is talent set on fire by courage." [655]

Stomping on the Dream vs. Searching for the Dream

Let's start with a reality. We've made the case that new ideas are often met with a shower of resistance. Not all ideas are good, but many organizations, including communities and even countries, have developed ways of stomping on the dream before it ever fully reaches the light of day. That's why persistence and courage are important.

Roadblocks and Welcome Signs. Here are a few ways we inhibit new ideas that are deeply important to creating a more promising future. Consider this a how-to guide for *dampening* or even *killing* creativity.

• Immediately play the budget card. "We can't afford it," sometimes can be interchanged with, "You'll never get the money."

- "We tried that before and it didn't work."
- "Look, we've always done it this way, and we're not changing now."
- "It's not our concern. We don't have a dog in this fight. Why should we be concerned? It's outside our field. Besides, you don't have authority to do that."
- "Who in the world came up with that idea? Why doesn't everybody stop daydreaming and get back to work?"

A better way might be to respond with, "Interesting idea! What would you see as the next steps? Let's get that idea on our list. Keep thinking. That's what's going to keep us on the leading edge."

Perils of Circling Wagons and Digging Trenches. The future of careers, industries, for-profit and nonprofit organizations, communities, cities, counties, states, countries, the world, and even individuals like you and me depends on fresh thinking. Yet, when a negative response, a raised eyebrow, or a cold shoulder confronts us, our tendency should be to learn from the encounter but not give up.

Sometimes, we get push back because our idea might seem threatening to someone's self-interest. Often, age-old policies, conceived for another time, lock us in place. It's a form of legalizing our defense of the status quo. Then again, we might specialize to the point of isolation. We circle our wagons and then dig trenches so deep that we lose touch with what's going on in the outside world. That kind of self-imposed isolation is fast becoming an express ticket to obsolescence. We lose sight of the fact that knowledgeable and creative people might just take their ideas somewhere else. As entrepreneurs, they could even start their own *competitive* enterprises.

Even though we are born with an inclination to think, reason, and create, we often have it knocked out of us because some parents, communities, schools, or politicians don't put much value on it. In some cases, those who say they are in charge claim to have all the answers, not realizing that part of their role should be "orchestrator of ideas." Some brilliant people clam up when they find they've been deemed disloyal or precocious—if they *have too many ideas* or *ask too many questions*.

A problem that's endemic in society and alive and well in many executive offices and education systems is simply not recognizing the ingenuity around us. One institution after another falls victim to protecting its flanks from intrusive ideas, only to discover that a competitor takes those ideas seriously. Think of Blockbuster and Netflix. We are constantly confronted by examples of people who win the battle but lose the war.

Innovation…Staying Ahead of the Curve

New ideas are not, by their very nature, necessarily threatening. Some could even put us on the leading edge of the future. Knowing how to deal with ideas in context and understanding a process for innovation are becoming a leadership essential.

Innovation Funnel. One of the more common ways of sorting new ideas is the *innovation funnel*. Here's basically how it works. Pour the ideas in the top. Consider timing, return on investment, benefit for our clients, and consistency with mission. Then it's "go" or "no-go." Some have expressed concern about the *funnel*, since they see it as a filter that may be limited by the creativity of those who judge ideas and the culture of the organization.

Conceiving of Actionable Products. Brian David Johnson, Intel's leading futurist, draws on science, engineering, and visioning to conceive of new "actionable products." He considers the social sciences and works with cultural anthropologists, among others, in conceiving of products and services people might need and want at some point in the future. During a presentation at the 2012 World Future Society Annual Conference in Toronto, Johnson quoted Intel's Chief Technology Officer (CTO) Justin Rattner, who said, "Science and technology have progressed to the point where what we build is only constrained by the limits of our imagination."[656]

Crowdsourcing in Finland.
  Conceiving of actionable ideas? Meet Niko Herlin. Herlin is Foresight Team Leader for Finpro with offices in many countries but headquartered in Turku. Private enterprise and government fund the organization, about half each. In fact, Finpro is the country's trade organization, which follows trends and engages in foresight to develop new products or ser-

vices and improve existing ones. In this case, each employee is asked to submit a few "signals," which they call "observations or signs of change" each month.

These signals are ideas or developments in society that might cause us to say, "Aha!; Huh?; Wow!; No Way!; Taboo; Made me laugh; Unheard of…; Amazes me!; or possibly, They'll get resistance to that." The process uses a "simplified Wiki," and relies to a great extent on the intuition of people, Herlin told assembled futurists.[657] Of course, crowdsourcing can be inclusive of any size crowd you'd like. This particular use might explain, at least in part, why we're all hearing so much about what's going on in Finland.

Time Pressure, Frustration, and Creativity. "If you're like most managers, you've worked with people who swear they do their most creative work under tight deadlines," according to Teresa Amabile, Constance Hadley, and Steven Kramer in *Harvard Business Review on Knowledge Management*. They go on to explain that it's not necessarily so. Saving Apollo 13 after its near disastrous explosion in space is a prime example of creativity-under-pressure in a life-and-death situation.

After observing the realities of "fighting the clock," "energy," and "frustration,' they introduce what they call "The Pressure Trap." The authors write, "Our study indicates that the more time pressure people feel on a given day, the less likely they will be to think creatively."[658]

"Social innovation curricula will spread beyond business schools in American universities into the arts & sciences," observes Futures Council 21 member Matthew Moen. He urges federal government granting agencies to "require more rapid and broad dissemination of the scholarly research they fund," which may cause some consternation "but it will bring good results for society."

What seems like an exponentially growing number of educators are talking about the importance of "reflection" for their students. Consumed by daily demands both inside and outside the system and constantly pursued by an array of communication technologies and stacks of requirements, creativity could become a victim. Talk about something we can't afford! This pressure trap is closing-in at a time when educators need to be thinking about the approaches to teaching and learning we urgently need to get students ready for life in a changing and challenging world.

Got the picture? Look no further than how we do photography. Polaroid and Kodak were powerhouses in that business. Long-term growth seemed certain. In fact, business was so good that they may have had too much success and too little time to bother with the upstart idea of digital photography.

Biomimicry and the Shinkansen. Ever ride the Shinkansen, Japan's bullet train? One of the first things you notice when you are traveling full speed is how fast the clouds are going by. It's a great ride, but it hasn't always been a bed of roses for the rail line. Because the train moves so fast, about 300 kph (186 mph), it was creating the equivalent of a sonic boom when it emerged from tunnels. Two problems were apparent. One, people were upset with the blast. Two, the design of the front end of the train wasn't sleek enough to more quietly slice its way through the compressed air.

What to do? The West Japan Railway Company studied a bird, the Kingfisher, which can plunge from the air into water with hardly a splash." Why not design the lead end of the train like the beak of a Kingfisher? They did it, and as *Smithsonian Magazine* reported, "Japan became a quieter place."

The railroad could have said, "Get used to it. This is the way trains are built." Instead, they took on the challenge, which turned into a prime example of creativity, imagination, invention, and innovation, coupled with biomimicry.[660] The *lesson* may be as big as the *accomplishment*.

> **Iron rusts from disuse; stagnant water loses its purity and in cold weather becomes frozen; even so does inaction sap the vigor of the mind.**
> *Leonardo da Vinci, Italian Engineer, Painter, and Sculptor (1452-1519)[661]*

Letting Go, Giving Permission, and Offering Encouragement

A good first step in releasing ingenuity and stimulating creativity, in ourselves and others, is simply letting go. Some of us inhibit our own creativity and shrink from talking about it. We may not demonstrate it because we don't want to *stand out* or be seen as *off-the-wall*. As we've tried to make clear, being creative can involve a modicum of risk, but that risk is seldom as great as the sheer danger of inhibiting constructive ideas, even if they prove to sometimes be "unacceptably disuptive" or don't work.

Please remember this—You don't need authority to give someone else permission to pursue their talents. That's true in any industry, field, or line-of-work. It's true in education, government, business, and it's true for individuals. In short, each of us needs to become a talent scout—spotting, coaching, encouraging, and marveling at the strengths of others, whomever they are, wherever they are, and whatever their situation in life.[662]

Meet Oleta Grimes, Talent Scout. One of the world's most revered talent scouts is Oleta Grimes. Ms. Grimes was a fifth-grade teacher in East Tupelo, Mississippi. She had a way of seeing the talent, ability, and even the possible aspirations of her students. One was a shy ten-year-old whose family was barely above the poverty line. There was just something about that young man, maybe it was the voice, or maybe she just saw a boy who needed encouragement. In this case, she asked him to enter a talent contest during children's day at the Mississippi-Alabama Fair and Dairy Show. He stood on a chair to reach the microphone, sang *Old Shep*, came in second, won a $5 prize, and got a free ticket to all of the carnival rides. The following January, he bought a guitar at the Tupelo Hardware Store. That little boy became a worldwide phenomenon, a cultural icon, and was known all his life by the name his parents gave him, Elvis Presley. "Elvis Presley," [http://www.history-of-rock.com/elvis_presley.htm] accessed Nov. 21, 2013[659]

The Arts
Helping Us Think, Create, Imagine, and Innovate

We are surrounded and consumed by the arts, and sometimes we don't even notice. Music, dance, musical theater, the visual arts, design, creative writing, and many other art forms stimulate our thinking and fire our imaginations.

Our future might very well depend on our ability to see and think in new ways, across disciplines. The arts help us "maintain our competitive edge by fostering innovation and creativity," said National Endowment for the Arts Chairman Rocco Landesman. They "provide new ways of thinking, new ways to draw connections."[663]

"The future of our nation depends on our ability to create—and to be creative," according to *Performing Together . . . The Arts and Education.* They are a basic means of communication. They help us develop our creativity and creative talents, learn other subjects, lead to a better understanding of human civilization, encourage the development of discipline, prepare students for their adult lives, and develop artistic judgment.[664]

Noted art educator Elliot Eisner has put a spotlight on a bevy of lessons the arts regularly teach us, among them, "multiple perspectives," "complex forms of problem solving," and an understanding that "small differences can have large effects."[665] Many people are convinced that the arts can help us keep kids interested in school, since they stir excitement and creativity. They also strengthen achievement in other subjects.

Boston College Professor Ellen Winner remarks, "The arts are a fundamentally important part of culture, and an education without them is an impoverished education leading to an impoverished society," she insists. "Studying the arts should not have to be justified in terms of anything else. They are time-honored ways of learning, knowing, and expressing."[666]

Consider music and the arts. What was it, in the then-frantic Gilded Age that sparked Irving Berlin to write "Alexander's Ragtime Band?" What stimulated the creative genius of Aaron Copeland to pen "Appalachian Spring," reflecting what he considered "the direct, plain, optimistic, and energetic" American character? Don't forget George and Ira Gershwin and "Rhapsody in Blue," a song that echoes through the ages. Who would have predicted at the turn of the 20th century that it would be remembered for producing a Louis Armstrong, a Glenn Miller, or a Duke Ellington? How was it that, in the 1950s, in a world of Doris Day and Perry Como, Bill Haley and the Comets changed everything with "Shake, Rattle, and Roll"; Ray Charles unveiled his "revolutionary hybrid of blues and gospel"; and Elvis Presley burst on the scene with "Heartbreak Hotel"? What stirred the Beatles to concoct a sound that connected with much of humanity? What laws of nature and sparks of creativity spurred Marvin Hamlisch to write "The Way

We Were," an anthem for our continuous journey from one generation to the next? What nerve did Chuck Brown touch when he invented D.C.'s *go-go*, a sub-genre of funk? What was the social mix that produced hip-hop and rap?[667]

It will take imagination and creativity to conceive of ideas, products, services, performances, and pathways to peace and understanding. The arts can help us find common ground and lift our spirits at the same time. Former U.S. President John F. Kennedy said, "I am certain that after the dust of centuries has passed over our cities, we too will be remembered not for victories or defeats in battle or politics, but for our contributions to the human spirit."[668]

A Few More Thoughts about Thinking

In canvassing various groups about what they think are "the most important things for students to learn," *thinking* usually works its way to the top of the list, according to Project Zero's David Perkins.[669] Perkins and PZ colleague Shari Tishman took on a Patterns of Thinking Project that revealed— "good thinkers have a tendency to identify and investigate problems, to probe assumptions, to seek reasons, and to be reflective."

They also observed that some students have thinking abilities but "aren't disposed to use them." Having that "disposition" depends on "ability, inclination, and sensitivity." When it comes to open-mindedness, the researchers pointed out the need to have a "basic capacity to see a situation from more than one perspective, feel inclined to invest the energy in doing so, and recognize an appropriate occasion to be open to alternative perspectives."[670]

Arthur Costa is an emeritus professor at California State University in Sacramento and co-author of an ASCD series devoted to *Habits of Mind*. He suggests five themes for any "thought-filled curriculum: learning to think, thinking to learn, thinking together, thinking about our own thinking, and thinking big." He recommends frequent questions to enliven thinking.[671]

For example, we might pose ongoing questions such as: Why do you feel that way? What opinions do you think other people might have about it? How could we solve the problem? What should we do first?

What process did you use to form that opinion? How did you make your decision? What more would you like to know about this issue? What are the ethical implications of what you're proposing?

In a world intent on quickly coming up with all the answers, growing numbers of people are concerned about whether we're asking the right questions. Malcolm Gladwell, in his book, *Blink*, warns, "We have, as human beings, a storytelling problem. We are a bit too quick to come up with explanations for things we don't really have any explanation for." All the more reason thinking and reasoning are basic skills.[672]

MacArthur Genius Awards

Each year, the MacArthur Foundation awards five-year grants to individuals who "show exceptional creativity in their work and the prospect for still more in the future." There are no limits on age or area of activity, but all must be citizens or residents of the U.S. According to MacArthur, "the Fellows Program places its emphasis on individual creativity because the discoveries, actions, and ideas that shape our society often result from the path-breaking efforts of individuals."[673]

Implications for Society
Ingenuity, Knowledge Creation and Breakthrough Thinking

Social and economic unrest continue to intensify worldwide. Many feel a sense of urgency. The old solutions don't seem to be working as we try to deal with a flurry of multiplying challenges that demand our attention. That's just part of the situation as we move into what might be called an Age of Knowledge Creation and Breakthrough Thinking. Consider the following possible implications for society. Then think of more.

• Without exception, communities, businesses, governmental and non-governmental organizations, and countries will be faced with re-defining themselves, adjusting their attitudes, and considering pathways to a sustainable future.

• Both elected and selected political and community leaders will likely be expected to have a level of open-mindedness, creativity, and ingenuity as we confront challenges that have massive social and economic impact.

• Remember that cultivating invention, innovation, and entrepreneurial activity should be front and center in discussions about a community's or a country's future.

• Education systems will increasingly be expected to turn out inventive people who have finely developed thinking and reasoning and problem-solving skills and a capacity to use creativity and imagination. The ability to think and learn across disciplines, connect multiple ideas, create new knowledge, and engage in breakthrough thinking could be among basics.

• Expect neuroscience to offer ways to observe what motivates individual students (in fact, all of us) to learn.

• Communities and countries need to become catalysts and enablers rather than roadblocks for emerging ideas. Progress will require flexibility and risk.

• Whole new industries will develop and may find a home in our communities, unless we insist that the past will return. Service, manufacturing, and other jobs and industries will require wall-to-wall education and training.

• Our civil society, economy, and quality of life will depend on our willingness and ability to encourage ingenuity and invention.

• People will become more dismissive of pat answers and arbitrary solutions. More will ask "What if?" questions, consider ideas, develop scenarios, and create new knowledge.

• Those who insist on maintaining the status quo may increasingly feel left out, lose hope, become angry, and strike out at people and nations that are making progress.

• National and multinational organizations as well as communities will embrace a new social context—a mentality that accommodates an age of renewal.

• At all levels, attention must focus on fairness as well as environmental, economic, and ethical sustainability, rather than solely on productivity. Growing numbers of people will realize that progress has different meanings in different cultures.

• Because of life experience—some positive and some horrifying—we hope society will learn from history rather than being "condemned" to repeat it.

Implications for Education
Ingenuity, Knowledge Creation, and Breakthrough Thinking

Social and economic challenges, discoveries in neuroscience, massive developments in technology, globalization, political upheaval, the urgency of invention, the need for new industries and jobs, and a host of other issues and concerns are shaking the world's foundations. Everything that impacts the world impacts schools and colleges. If it doesn't, there's a good chance someone is out of touch. Implications for education?

• **Pursuing active learning, project-based and real-world education, learning through inquiry, differentiated instruction, and learning across disciplines.** "Stand and deliver" has been a staple of Agricultural and Industrial Age education. It helped us learn *what* we might need to know and dispense it in a sort of factory model. Since knowledge and information are expanding exponentially and students can flip on a laptop, tablet, or smartphone to connect with the world, astute educators know instinctively and intellectually that they will be expected to develop teaching techniques that will help us stay ahead of the curve.

While "sit and listen" may still be important, engagement has become essential. Classrooms are turning into hubs for active learning, project-based education, real-world education, learning through inquiry, and learning across disciplines. While some see these techniques as disruptive, others are celebrating the moment, as education gets more exciting for both students and educators. At University Neighborhood Middle School in New York City, *Harry Potter* is stimulating fantasy, while fictional or non-fictional storytelling, as well as essay and editorial writing, are spurring the imagination.[674]

At the Tokyo International School, Patrick Newell, who runs the institution, calls his position, "Vision Navigator." The school teems with excitement generated, at least in part, by learning through inquiry. He says, "I spend much of my time being sensorial, mindful in creating new ideas, creative in igniting movements, and a facilitator to empower others to connect the dots." He adds, we too often "teach students the content we think they need to know, not how to become competent learners who understand how to learn and nurture innate curiosity." Newell suggests "experiential learning."[675]

• **Expecting the focus on education to become more intense.**
During tumultuous times, especially following economic downturns,
the eyes of the world focus on education. When politicians, business people, and the rest of us think about rebuilding our economies, pursuing
innovation, creating new industries, and putting people back to work,
they see connections to what we may need to know and be able to do to
ensure a better future. Both the Agriculture and Industrial Ages had a
profound impact on education. So will our move into a Global Knowl-
edge/Information Age. A challenge for educators will be to understand
that focus as a *growing demand* rather than as *criticism*. Author, columnist, and educator Richard Florida observes, "Crises make us better at
using our most precious and critical economic resource—talent."[676]

• **Ensuring that standards and high-stakes tests don't suck all
the air out of the room.** Standards are helpful, provided they do not
freeze an education system or any other organization in the past. When
they are reinforced by high-stakes tests focused on a few subjects that
are most testable, they can have a tendency to suck all of the air out of
the room. That approach can lead to isolation from emerging realities
as well as frustration and lack of interest by students. Standards need to
be flexible enough to accommodate the massive changes, many driven
by technology, that are taking place by the minute.

• **Nurturing creativity, imagination, ingenuity, mindfulness,
and flexibility.** Sheer necessity, coupled with growing numbers of students who find education decoupled from their lives and hopes for the
future, are driving the demand for creativity, imagination, mindfulness, and ingenuity. Flexibility will be essential for our political and
educational systems around the world to embrace moving from a laser
focus on *what we already know* to *discovering and creating new knowledge*. That will mean paying increased attention to cognitive research,
staying close to forces impacting society, and personalizing our approach to education.

Futures Council 21 member Gary Rowe believes appropriate use
of technology will "offer forms of education that lift students' faces
out of textbooks and provide depictions of the real world, from ballet
rehearsals to biology labs. The opportunity to watch an active world

at work can promote active learning, curiosity, discovery, and achievement." He foresees "a growing cornucopia of learning opportunities, many of them outside the walls of the traditional school." Rowe adds that, "Education will be reformed by disruptive technologies that will transform educational opportunities."

Albert Einstein argued that, "Imagination is more important than knowledge." Courage is needed, especially when budgets are tight, to encourage the dreams and rescue the ingenuity of even those we consider, or who consider themselves, unimaginative.

• **Becoming more knowledgeable about Mind, Brain, and Education (MBE).** Relatively speaking, MBE is a new field and will likely become even more prominent. It encompasses educational neuroscience, philosophy, linguistics, pedagogy, developmental psychology, and other areas. "All human behavior and learning, including feeling, thinking, creating, remembering, and deciding, originate in the brain," according to researchers Mary Helen Immordino-Yang and Kurt Fischer. Various skills "are supported by neural networks…and have important implications for education," they have reported. The researchers ask that educators "understand the logic and constraints of neuroscience research," including neuroimaging techniques.[677]

Questions will arise about how well aligned teachers might be with techniques to use these findings for improving learning. Some will want to know whether the information will become part of students' learning profiles. Ethical consideration might arise if the procedures only benefit those who can afford them. Education systems will likely need to adjust policies to deal with the possible benefits and concerns that will naturally come with MBE.[678]

• **Cultivating Thinking and Reasoning Skills.** Critical and creative thinking are essential. An understanding of logic can help us maneuver inductive and deductive reasoning. Creative or lateral thinking can help us think outside the box and generate flashes of insight. That's the kind of thinking we observe looking at the flying machine that Leonardo da Vinci *drew* and comparing it to the one the Wright Brothers *flew* about 400 years later.

Research tells us that, when we ask a question, we need to allow more than just a few seconds for a student to answer. Those few extra seconds, that "wait time," will send a message that we really care about what they're thinking. Questions, in fact, are essential in our quest to help students become even better thinkers. Examples? "What do I know?" "What do I still need to know?" "How else could I think about this?" "Could I look at it through a different frame?"

Here's another thought. Ask students if what they've learned today has triggered any ideas for them. The question itself lets students know that we aren't finished yet. It encourages reflection. Chances are, if we listen to the answers, the world will be at least a little smarter than it was when class got under way.

• **Additional Implications.** As we embrace ingenuity, creativity, and breakthrough thinking, education systems will need to emphasize the arts as a way to create, express, and think across disciplines. In fact, interdisciplinary learning and leadership will become standard in finding new knowledge in the white spaces or intersections between subjects and issues. Schools and colleges will need to offer gravity-breaking preparation and professional development programs, and open the door for futures studies. Visionary intellectual leadership that goes far beyond organizational mechanics is becoming an expectation. Educators at all levels will be expected to develop qualities we hope to nurture in students, such as creativity, imagination, and enthusiasm about hatching new ideas and connecting the dots. In a world in need of greater collaboration, they'll bring people together in common purpose through community conversations and a host of other opportunities, all aimed at creating a future outside the trenches.

**We will not be able to ride our way into the future.
We will need to invent our way into the future.**

Ingenuity, Knowledge Creation, and Breakthrough Thinking
Questions and Activities

1. List five things you believe must happen to release the ingenuity of educators and students.
2. Watch this Arthur Costa video produced by Big Picture Education in Australia. The program segment focuses on metacognition. http://vimeo.com/5367247 (2:47). Be prepared to discuss TAPS.
3. Produce a five-minute PowerPoint presentation on the basic steps we should take to help students move from information acquisition toward knowledge creation and breakthrough thinking.
4. What do we sometimes do to discourage people from using their creativity or ingenuity?
5. What did biologist and philosopher Edward O. Wilson mean when he wrote, "The ongoing fragmentation of knowledge and resulting chaos in philosophy are not reflections of the real world but artifacts of scholarship?"
6. Consider development of a "futures studies" course or unit for your school or college. List and briefly discuss things students should learn from this type of course.
7. Share five ways the arts unleash genius, promote creativity, and help students learn across disciplines.

Readings and Programs
1. Gardner, H. (2009). *Five Minds for the Future.* (paperback). Cambridge, MA, Harvard Business School Press.[679]
2. Perkins, D. (2010). *Making Learning Whole…How Seven Principles of Teaching Can Transform Education.* San Francisco, CA, Jossey-Bass.[680]
3. Wilson, E. (1998). *Consilience: The Unity of Knowledge.* New York: Borzoi Books, Alfred A. Knopf.[681]
4. View Kenneth Robinson's *TED* presentation about creativity. http://www.ted.com/speakers/sir_ken_robinson.html.[682]
5. Costa, A. (2009). *Habits of Mind Across the Curriculum: Practical and Creative Strategies for Teachers* (paperback). Alexandria, VA, ASCD [683]
6. View *21:21 The Movie, Aligning 21st Century Learning with 21st Century Learners,"* (21:21), 21 Foundation, Patrick Newell, Tokyo International School, http://www.21foundation.com/2121-the-movie/[684]
7. Review ASCD's "Whole Child Initiative" web page, http://www.wholechildeducation.org/.[685]
8. View *Moving Art, TED-X* San Francisco, featuring Louie Schwartzberg, founder, Blacklight Films, (9:56), http://www.youtube.com/watch_pop-up?v=gXDMoiEkyuQ&vq=medium

Chapter 14

Depth, Breadth, and Purposes of Education
What do we need to know?

Chapter 14: Depth, Breadth, and Purposes of Education. *What do we need to know?*

Trend: The breadth, depth, and purposes of education will constantly be clarified to meet the needs of a fast-changing world.
Narrowness → Breadth and Depth

Zooming Out...to the Big Picture

Ask this question, "What are the purposes of education?" In reality, it's a discussion that should never leave the agenda. Of course, purposes often get wrapped up in politics, special interests, biases, economic arguments, and often-firm viewpoints.

Some say education should stick to teaching the 3-Rs. Others submit that the purpose is broader, and should include getting students ready for productive lives as contributors to the economy and civil society.

Thorndike and Dewey. During an era of progressive education, two psychologists, Edward Thorndike (1874-1949) and John Dewey (1859-1952) wrestled with two competing philosophies. In essence, *Thorndike* stood for schools that were "structured around the methods of industrial management" and made use of "techniques of statistical analysis."[686] He saw schools as a delivery vehicle where "experts alone would be able to decide what to teach, how to teach it, and how to evaluate it." On the other hand, *Dewey* believed that "subject matter in schools exists to make the quality of democratic life as good as it can be under given conditions." He asserted that a teacher ought to arouse "a continuing interest in learning throughout a student's life...and develop a love of learning."[687]

While this false choice continues to captivate interest and often dominates the ruminations of educators, politicians, and partisans of many stripes, the world moves on. As it does, one thing is becoming increasingly clear—narrowness will not be enough if we hope to create an even better future.

If mandated education programs such as No Child Left Behind

(NCLB) during the early 2000s were any indication, it would appear that Thorndike prevailed. Requirements have often morphed into narrow standards coupled with high-stakes tests focused on a few things that are more easily testable.

Based on a 2011 study, Lynn Munson, president and executive director of Common Core, a Washington-based research and advocacy group, raised concern that many key subjects have been "abandoned." Specifically, the study pointed out subjects "getting less time than they used to" such as art (51 percent less time), music (48 percent less), foreign language (40 percent less), social studies (36 percent less), physical education (33 percent less), and science (27 percent less).[688]

Add to that a concern that NCLB has memorialized a system of rewards and punishments reminiscent of B.F. Skinner.[689] *Authentic motivation* and maintaining student interest and engagement will depend on our creating a broad, deep, and rigorous approach to education that stimulates an excitement for learning that will last a lifetime.

"This and"…not "Either or." In reality, we may need the ideas of Dewey, Thorndike, and a whole lot more. Somehow, we have wrapped our philosophy and purposes of education around a debate that helped shape schools for an Industrial Age. Now, we're moving into a Global Knowledge/Information Age. To the frustration of growing numbers of thoughtful educators, we are too often faced with the prospect of preparing our students for the future—constrained by a mentality and infrastructure that emerged from another time.

Without seeing a direct connection between what they are asked to learn and how it will be useful to them in their lives, students are likely to tune out or even drop out. Most are already linked to a world of information and ideas. This persistently vexing problem takes its toll across the board, from K-12 schools through colleges and universities. Simplified political rhetoric sometimes combines criticism with demands for even more intense education designed to prepare people for the past. Therein lies a challenge. We simply must rethink the breadth, depth, and purpose of what we do.

Eugene, Oregon, Superintendent Sheldon Berman takes the discussion of breadth and depth of learning to another level. He sees a shift from "I taught it, but they didn't learn it," to "If they didn't learn it, I didn't teach it correctly." It's more of a "diagnostic approach to teaching." Berman adds, "Teachers need to know their subject, but the task of teaching is building conceptual understanding among students."

Many educators are doing what they can to move the system forward, often working against great structural and financial odds. We owe them a thoughtful, open-minded discussion of purposes, hopes, and dreams. Forcing an education system to confine itself to a narrow groove when our economy and civil society need imagination, creativity, invention, and innovation is simply not sustainable.

Purposes of Education

Just to get the discussion started, please let me propose five purposes for education:

- **Citizenship.** Create good citizens of a family, community, country, and world.
- **Employability.** Help students develop the knowledge, skills, talents, and habits of mind they need to be employable, whether they are self-employed or aspire to work inside an organization in the for-profit or non-for-profit sector.
- **Interesting Lives.** Encourage students to see the connection between what they are being asked to learn and real life, noting that, generally, the more we know and the broader our experience, the more interesting life becomes.
- **Releasing Ingenuity that is Already There.** Discover the personal ingenuity of students, their interests, skills, abilities, and aspirations, and build on them.
- **Stimulating Imagination, Creativity, and Inventiveness.** Recognize and cultivate the creative nature of students.

Your assignment: Consider the four broad purposes of education that we've mentioned. As you think about it personally or with large or small groups of colleagues, feel free to make whatever changes you think are appropriate. The challenge is to start an ongoing discussion about the purposes of education. A driving question might be, "In the broadest of terms, why do we educate people?" Of course, *availability of high quality education, coupled with equal opportunity*, should be basic.

Breadth and Depth
Getting Past the Gathering Narrowness

Unfortunately, we've been highly successful at fractionating the purposes of education. *Subjects* have often been forced to compete with each other for attention. That situation has left little or no room in the conversation for the bigger picture. The time has come to do some breadth and depth perception.

Captives of the Cognitive. The pressures of academic standards, testing, college readiness, expansion of knowledge, habits, and demands of several disciplines continue to fuel an unending debate about depth vs. breadth in education. Cognitive knowledge and skills are deeply important, but, alone, they aren't enough. Unease has been rising as we have become captives of the cognitive, since those skills are more easily tested. University of California Vice President for Education Partnerships Russell Rumberger put it this way in an *Education Week* commentary. "Federal and state accountability systems judge the performance of students and schools largely on a very narrow range of cognitive skills, particularly focusing on reading and math."[690]

Just for perspective, ask this question—"Where do math, reading, writing, the sciences, social studies, the arts, and numerous other 'subjects' fit into the purposes we suggested?" The answer—"Everywhere… in *all* of them." The difference is that we are not just teaching math for math's sake or the arts for the arts' sake. We're *also* teaching them because they are important to achieving the larger purposes for education.

Learning across Disciplines. Each of these and many other subjects are critically important. So is the need to learn across disciplines. It's in those multidisciplinary white spaces, in the connective tissue, that we are likely to discover new knowledge. Teaching and learning across disciplines should be considered part of how we operate. The State of Florida, beginning in 2011-12, required that "the reading portion of the language arts curriculum must include civic education content for all grade levels," according to Annette Boyd Pitts, executive director of the Florida Law Related Education Association (FLREA).[691] Thinking, creating, imagining, innovating, communicating, and working across disciplines is and will always be flat-out essential if we hope to build a better future for our economy, our civil society, and ourselves.

"What do you want to do?" vs. "What do you want to become?"
One of the most frequently asked questions for people between the ages of 6 and 25, sometimes lingering throughout life, is, "What do you want to do when you grow up?" It's a legitimate question. However, there are a few questions that should come first— "What do you want to become?" "What kind of legacy do you want to leave?" and "What kind of person do you want to be?" Considering those answers, ponder a possible cluster of career choices, whether in the for-profit or non-for-profit sector, and stay flexible.

During our careers, whole industries will come and go, perhaps yours among them. At some point, we might even want to tap that entrepreneurial spirit deep within us and create our own job, organization, or industry. The depth and breadth of our education will tell the world that we have the curiosity and persistence to tackle big ideas and be part of a successful team.

Immediate Answers or a Learned Person? "The ability of an individual to derive immediate answers on handheld devices undermines the type of person who prospered because of a deep reservoir of knowledge," warns Matthew Moen. However, "education is an intrinsic value for people to become better citizens and lead rich personal lives," he says. While the pursuit of education for vocation will grow, "the breathtaking pace of discovery will make (solely) vocational pursuits less practical and successful, giving new impetus to the concept of a liberal education."

Getting Poised for Innovation and Exponential Change

What we teach and how we teach it has to be aligned with the fast-changing needs of society. We need to consider the emergence of new technologies and a growing restlessness that involves everything from jobs and the economy to systemic innovation and international competition. In *How to Think Like Leonardo da Vinci*, Michael Gelb wrote, "In 500 years, we've moved from a world where everything was certain and nothing changed to a world where nothing seems certain and everything changes."[692]

People make observations about how fast knowledge is accumulating and even speculate on the half-life of education. One estimate put the half-life of knowledge (the point at which half of existing knowledge will be replaced by new knowledge) at twelve to fifteen years in the 1960s and somewhere between six and eighteen months in 2011.[693] Others are simply saying that knowledge is increasing "exponentially."[694] Clearly, if we want to build and maintain a sustainable economy and civil society," our challenge will be to invest in education that gets people ready for the future.

What do we all need to know and be able to do?
Historic Context: Narrowness and Shallowness vs. Depth and Breadth

Right off the bat, you undoubtedly have some firm opinions about what we will all need to know and be able to do if we hope to be poised for the future. Let's hold our horses for a few minutes while we explore how some people and institutions have or likely would have responded to a similar challenge. Trying to include a full range of those opinions would take volumes, so we've cherry-picked a few, just to give us a hint of historic context. Feel free to add others.

We'll start early and work our way forward. Get ready for examples of soaring breadth and depth and remarkable narrowness. Because most of these items don't refer to what we learn each day across disciplines or to the explosion of new knowledge, consider all of that a given.

After this brief review, we'll share a rundown of a fast-growing list developed by the author of this book for a World Future Society D.C. Chapter presentation devoted to what we all need to know and be able to do to be prepared for the future. None of what we share here should be considered the last word. Instead, it is one of several ongoing attempts throughout history to stimulate the conversation. Let's start with Plato.

• **Plato's Academy.** Plato was, of course, an ancient Greek philosopher and educator who founded his Academy in an olive grove around 367 BCE. The lessons centered on what we might know today as art, science, literature, music, astronomy, biology and other sciences, mathematics, political theory, philosophy, and other intellectual pursuits, as well as an array of Socratic Dialogues.[695]

• **Trivium and Quadrivium.** "The medieval western university emphasized a course of study for an educated person that included the *trivium* (grammar, rhetoric, and logic) and the *quadrivium* (geometry, arithmetic, music, astronomy). Those 'seven liberal arts,' as they are sometimes called, are still taught centuries later at colleges and universities because of their critical role in the intellectual development of students," according to the Council of Colleges of Arts & Sciences.[696]

• **Horace Mann (1756-1859).** He was a noted educator, generally credited with being the father of public education and "the great equalizer." Mann noted that education should provide: practical knowledge; moral elevation and a moral compass; an ability to read, write, and spell; an aptitude for mathematics and science; and a discovery of principles and relationships; as well as equal opportunity and a level playing field. He expressed a belief that education should help overcome "the slavery of ignorance." Mann also believed schools should provide a common, unifying experience.[697]

• **John Dewey (1859-1952).** A widely-known philosopher, psychologist, and education reformer, Dewey is often called the father of progressive education. He articulated a relationship between education and democratic life. Dewey believed that investigation, problem-solving, and both personal and community growth were directly connected to sound education. Joined by other thinkers, he saw the need for schools to be more child-centered and to instill a love of learning throughout a student's life.[698]

• **W.E.B. Du Bois (1868-1963).** This noted African American philosopher was widely recognized for his pursuit of social justice, his literary imagination, and his pioneering social-scientific studies.[699] Du Bois declared that training for technical skills had to be enhanced by development of the intellect, "broad sympathy, and the knowledge of the world that was and is, and the relationship of men to it." He noted, "On this foundation, we may build bread winning, skill of hand, and quickness of brain, with never a fear lest the child and man mistake the means of living for the object of life."[700]

• **Seven Cardinal Principles.** In 1918, the National Education

Association, then encompassing what have become several education leadership groups, convened a Commission on the Reorganization of Secondary Education. That Commission suggested the following broad directions for learning, which became known as the "Seven Cardinal Principles." They include: health, a command of fundamental processes, worthy home membership, vocation, civic education, worthy use of leisure, and ethical character. Wisely, the Commission stated that education would likely be driven "by the needs of the society to be served, the character of the individuals to be educated, and the knowledge of educational theory and practice."[701]

• *Paideia Proposal.* Philosopher, theorist, and University of Chicago professor Mortimer Adler, released his *Paideia Proposal* in 1982. In it, he proposed that public school students "acquire three different types of knowledge: organized knowledge, intellectual skills, and understanding of ideas and values." Each would require "a different teaching style." Those "styles" included "lectures for organized or factual knowledge; coaching and supervised practice for intellectual skills, and the Socratic method (questioning) for ideas and values." Adler's curriculum framework included: "language, literature, and the fine arts; mathematics and natural science; and history, geography, and social studies." He also recommended "physical education for twelve years; and manual training, such as cooking, sewing, typing, and machine repair for six years; plus a general introduction to the world of work during the last two years."[702]

• *A Nation at Risk.* This 1983 report, issued by the U.S. Department of Education Commission on Excellence in Education, suggested "a rising tide of mediocrity" that was putting the nation at risk. The report recommended "Five New Basics" for high school students, including: four years of English, three years of mathematics, three years of science, three years of social studies, and one-half year of computer science. For the college-bound, add two years of foreign language.[703]

• **National Summit on Education.** This 1989 meeting, held in Charlottesville, Virginia, involved governors and business leaders but generally no educators. Central to the summit's work were goals such as: readiness for school; performance on international achievement tests, especially math and science; reduction in dropout rates and improvement in academic performance, especially among at-risk students; functional literacy of adult Americans; the level of training needed to guarantee a competitive work force; the supply of qualified teachers and up-to-date technology; and the establishment of safe and drug-free schools.[704] Parent involvement was also mentioned. Some believe this summit became a launching pads for the standards and high-stakes testing era.

• **National Education Goals.** Announced by the President early in 1990, these goals emphasized: demonstrating competence on challenging subject matter; exercising the rights and responsibilities of citizenship; and performing first in the world in math and science achievement.[705]

• *Preparing Students for the 21ˢᵗ Century.* In 1996, as the nation and world approached a new millennium, the American Association of School Administrators (AASA) released a study tapping the ideas of 55 leaders in education, business, and government. The group identified knowledge, skills, and behaviors students would need as the world moved into the 2000s. Here is just a partial list of the dozens of recommendations. *Knowledge*: use of math, logic, thinking and reasoning skills; functional and operational literacy; understanding of statistics; information accessing and processing skills using technology; scientific knowledge; world geography; international languages; multicultural understanding; knowledge of history; and roles and responsibilities as citizens. *Skills*: reading and comprehension; oral and written communication; critical thinking, reasoning, and problem-solving; self-discipline and applying ethical principles; job success skills; conflict resolution/negotiating skills; ability to conduct research/apply data; and knowing languages/being multilingual. *Behaviors*: understanding and practicing honesty, integrity, and the Golden Rule; respect for the value of effort; a work ethic, the need to make individual contributions, and self discipline; understanding and respecting diversity; capability of working as a team member; taking personal responsibility for actions; commitment to family life, personal life, and community; pride in citizenship and knowledge of individual responsibilities in a democracy; willingness to civilly resolve disagreements; recognition and respect for educators; and being excited about life and setting goals for lifelong learning. The author of *Twenty-One Trends* served as project director for that study.[706]

• **Multiple Intelligences.** Howard Gardner is an acclaimed developmental psychologist and professor of cognition and education at the Harvard Graduate School of Education. In his 1993 book, *Multiple Intelligences...The Theory in Practice, A Reader*, Gardner wrote that people have a variety of ways to learn and process information. The list has expanded over time to include the following intelligences: linguistic, logic-mathematical, musical, spatial, bodily/kinesthetic, interpersonal, intrapersonal, and naturalistic. A more recent version is *Multiple Intelligences: New Horizons*, published in 2006.[707] Gardner's *Five Minds for the Future* describes the disciplined mind, the synthesizing mind, the creating mind, the respectful mind, and the ethical mind.[708]

• **No Child Left Behind (NCLB).** The No Child Left Behind Act was signed into law by the President during January 2002. A primary focus was "the creation of standards in each state for what a child should know and learn in reading and math in grades 3 through 8". Student progress in reading and math had to be measured annually in each of those grades and at least once during high school using standardized tests.[709] Schools were expected to show Adequate Yearly Progress (AYP) in their test scores or face a series of highly public labels and corrective measures.[710] One of several concerns about NCLB was that schools felt so hard-pressed to constantly increase scores in reading and math that other parts of the curriculum received less emphasis or support.

• **Whole Child Initiative.** In 2007, ASCD launched the Whole Child Initiative "to change the conversation...from a focus on narrowly defined academic achievement to one that promotes the long-term development and success of children." The tenets declared that "each student: should enter school healthy and learn about and practice a healthy lifestyle; learn in an environment that is physically and emotionally safe for students and adults; be actively engaged in learning and connected to the school and broader community; have access to personalized learning and be supported by qualified, caring adults; and be challenged academically and prepared for success in college or further study and for employment and participation in a global environment."[711]

• **Framework for 21ˢᵗ Century Skills.** This framework was produced in 2009 by the Partnership for 21ˢᵗ Century Skills (P21). Its aim was to help the nation become more able "to compete in a global economy that demands innovation." A recommendation was "to help the U.S. education system keep up by fusing the 3Rs and 4Cs (critical thinking and problem solving, communication, collaboration, and creativity and innovation)."[712]

• **STEM Initiative.** During 2009, the President announced an "Educate to Innovate" campaign to "motivate and inspire young people across the country to excel in science, technology, engineering, and mathematics (STEM)."[713] Many suggested STEM+ to ensure a broad and deep education. STEAM would include the arts.

• **Common Core Standards.** In 2010, the Council of Chief State School Officers (CCSSO) and National Governors Association Center for Best Practices (NGA Center) presented K-12 Common Core State Standards focusing on English language arts and mathematics. Working with participating states, the organizations noted that the standards "represent a set of expectations for student knowledge and skills that high school graduates need to master to succeed in college and careers."[714]

Figure 14.1
Out of the Trenches and into the Future, Twenty Targets (A Work in Progress)
What we need to know and be able to do to be prepared for the future.

We've just reviewed basic recommendations from seventeen individuals, groups, and organizations. Based on those and other suggestions from multiple sources, we've created a draft list of twenty targets for what we all will need to know and be able to do to be prepared for the future. Each is worthy of a book in itself. Here's your assignment. Review the list. Feel free to change it. Most of all, consider it.

- **Communication:** Reading, writing, speaking, listening, concept development, and research.
- **Science:** Foundation in life, physical and materials, earth and space, chemistry, energy, and other sciences, the scientific method, scientific curiosity, relationship between science and technology.
- **Technology:** Technology literacy. Proficient in use to enhance research, productivity, quality of life. Information accessing and processing, problem solving, and creation of new technologies.
- **Mathematics:** Concepts, computation, problem-solving; use of math in everyday life; logic and reasoning; and basic math functions--algebra, geometry, measurement, calculus, statistical methods, and probability.

- **Engineering and Architecture:** Significance of *engineering* in design and integrity of structures, machines, materials, and systems, as well as chemical and other forms of engineering, and of *architecture* in planning and designing the built environment, cities, and communities.
- **Thinking and Reasoning:** Critical and creative thinking. Ability to see relationships; think big; learn through inquiry; deal with complexity, ask clarifying questions; think philosophically, see things in context; and question current thinking.
- **Imagination, Creativity, and Innovation:** Ability and discipline needed to imagine, create, invent, innovate, seek, and discover--to find relationships among ideas and detect ingenuity in oneself and others.
- **Knowledge Creation and Breakthrough Thinking:** Ability to analyze, synthesize, and think across disciplines, to discover new knowledge, new ideas, and possibly new products, services, or solutions.
- **The Arts:** Cultivate creativity, imagination, expression, and an ability to discover and value ingenuity. Develop talents, abilities, and interests in multiple art forms. See connections between the arts and other subjects.
- **Judgment, Ethics, and Character.** Self-discipline, good character, good will toward others, ethical principles, anger management, and self-control. Act civilly, tempering personal gain with empathy and compassion. Good judgment.
- **Civil Discourse and Ability to Overcome Narrowness and Polarization.** Civil discourse, deliberation, and pursuit of the common good. Move from narrowness toward reasoned discussion, evidence, consideration of varying points of view.
- **Employability Skills:** Technology, teamwork, communication, human relations, critical thinking, problem solving, math, budgetary, entrepreneurial, and both general and job specific knowledge, skills, and behaviors. Flexibility and adaptability.

- **Leadership and Management:** Plan, organize, activate, be transparent and accountable. Engage and serve people and clients; be entrepreneurial and ethical. See things in context, mobilize resources, solve problems. Stay in touch, clarify, define, and inspire.
- **Economics and Personal Finance.** Micro- and macro-economics. Approaches that might be needed to create jobs, careers, and industries. Personally, understanding needs vs. wants, affordability, value, interest, return on investment, and applied ethics.
- **Social and Behavioral Sciences.** Role of society, identity, and culture. Grounding in civics, history, law, political science, economics, government, geography, world affairs, behavioral sciences, and other related areas.
- **Civic Knowledge, Skills, Dispositions.** Civic literacy--how laws are made and changed, rights and responsibilities, voting, fairness, justice, equal opportunity, problem-solving, and formulating public policy. Becoming an engaged, contributing member of society.
- **Global/International Knowledge and Skills.** A connected, world-wise person. Understanding of languages and cultures. Ability to develop business, governmental, educational, scientific, and personal relationships across boundaries and cultures.
- **Environmental and Planetary Security.** Balance of economic growth and environmental sustainability. Need for clean energy, air, water, and nutritious food. Wise use of resources. Respect for the natural world. Need to address poverty and maintain security.
- **Health, Well-Being, Life Skills, and Work-Life Balance.** Good physical and mental health habits, diet, physical fitness, parenting skills. Ability to be organized, resilient, resolve conflict, and respect others. Ability to use basic building/repair tools.
- **Futures Processes and Forecasting.** Ability to use processes to develop a sustainable future, such as trend and issue analysis. Understanding importance of futures studies.

That's a Lot to Know
What do we do now?

We hope you've taken a serious look at the recommendations for what we need to know and be able to do to be ready for life in a world of exponential change. Our list of 20 targets provides some idea about how all of these and other recommendations might look when they're brought together. Consider this a work in progress, since each school or school system, college or university, community, state/province, or country will need to make ongoing decisions about what's right for them. The following is a series of logical and likely questions and some food for thought:

What's missing? Great question, but it's one that educators and communities will have to think about. What is not immediately visible in any list of this type is what we can and do learn across disciplines. Also not included is what we learn from daily experience. The world doesn't stop. Neither does our need to be flexible, able to interpret and respond to evolving needs of society, and consider fresh ideas.

Do you really think we can teach all of this? As educators and members of a community, are we only concerned about what happens in our classroom, our education system, our community? That's not enough. We need to back off to the big picture and ask, "What do we all need to know and be able to do to be prepared for the future?" That question should stir a discussion that never stops.

If what is on this list is important, who is going to teach it? How and when will we learn it? Who will make sure that we are learning across disciplines? Those questions should lead us to expectations for parents, communities, schools, community colleges, universities, and other learning institutions. As growing numbers of people in education and other fields have discovered, it's not so wild a dream.

If a student doesn't go to college, do they still need to know these things to stand a chance for success in the economy and civil society? The answer is very likely, "Yes." For example, many students will not end up with degrees in engineering or architecture, but they still need to understand what engineers and architects do and why those professions are important to all of us. All will need to be able to think and reason and have a sense of civic literacy. On top of that, the world needs people who are creative, imaginative, and ethical, whatever job they hold or role they play.

How can we fit all of this into a crowded curriculum? One way to demonstrate the law of supply and demand would be to compare what people want us to teach (demand) with the number of days, hours, sometimes knowledge or experience that we have to teach it (supply). It's a perennial concern, and it won't go away.

After all, every day the earth spins, the story of history gets longer and more complex. Scientific discoveries and technological advances seem to pile one on another. The planet shrinks with instant communication and jet travel, and all of us need to know more about people and cultures half a globe away. "We need to educate about the known, and we need to also educate for the unknown," Project Zero's David Perkins has said. He adds, "We never know when we'll be ambushed into learning something."

Let's take a look at a few possibilities for being more inclusive in what we teach.

• **Lifeworthy.** Among all of the things we are pressed to include in our teaching and the curriculum, we need to give attention to learning that is "lifeworthy," Perkins suggests. At a Harvard Future of Learning Institute, he asked, for example, how much time we should spend on Mitosis if it means there is none left "to teach about communicable diseases and how they spread globally." He asks us to consider how much "richness" there is in what we teach.

• **Smart Sampling.** Perkins points out that literature teachers long ago concluded that they can't fully teach every work that is worthy of deep study. Students may need to be introduced to a number of writings, even though they may not be asked to read all of each one. Those teachers have done sampling. While students will study some pieces of literature in full, they might be asked to read and discuss samples of others.

• **Post Holing (Drilling Deep or Surface Mining?).** For those who teach and learn about history (and that should be everyone) the story keeps getting longer. Considering time available to teach it, starting at the beginning and covering every significant date and episode, could get us halfway through the saga by the end of the term. That's why some educators do "postholing," what Perkins might call "rich cases, works, or artifacts," such as the French Revolution, that can carry lessons about democracy, discontent, diseases, and many other big ideas.[715] Critics sometimes argue that the approach leads to *disconnected learning* instead of *a broad and detailed chronological overview.*

• **Scaffolding.** While having many meanings and applied in numerous ways, scaffolding is a technique for building on what a student might already know, then building a support for even further learning, a process that can lead from possibly narrow understanding to greater depth and breadth.[716] We need to be sure that our scaffolding helps us reach toward the future, not just a carry-over of Industrial Age goals.

If we spend five hours a week on one topic, what else, that may be important in another discipline, is getting little or no attention?

• **Capstone or Finishing School Programs.** Concern has grown that the senior year of high school is often a time to *coast out*, if most of the needed credits are *already in the bank.* One interesting idea is to create a Senior's Academy that "infuses the senior year of high school with engaging and relevant learning for the 21st century," as students prepare to go out into the world, heading for college and/or a career.[717]

Higher education institutions of all types might offer a seminar or series of seminars that attempt to fill gaps between what we know now and the world-wise things we might still need to know and be able to do to be successful in life. The basics of leadership; financial management; ethics; futures studies; and acuity in thinking, reasoning, and problem solving come immediately to mind.

The Liberal Arts & Sciences. Growing numbers of communities and nations are realizing that a broad-based and well-rounded liberal arts education is essential for ensuring a viable future. Breadth and depth, rather than shallowness and narrowness, will be essential if we hope to have any chance of building and maintaining a vibrant society made up of employable people, committed to being contributing members of society.

"A liberal education has nothing to do with liberal or conservative politics," according to the Council of Colleges of Arts & Sciences (CCAS). It is simply "the idea of a broad and well-rounded course of study in the humanities, social sciences, sciences, and the arts. The overarching goal is to liberate the mind from ignorance and superstition." When it comes to employment, the Council highlights "the rapid pace of change in today's world" and its "constant threat to job security of people who have a single vocation." CCAS observes, "education in the liberal arts and sciences provides the skill set for people to learn, adapt, and thrive."[718]

Implications for Society
Depth, Breadth, and Purpose of Education

While the depth, breadth, and purpose of education may not dominate most of our conversations over breakfast cereal, the topic is always there. The newspaper or your favorite online news source calls attention to: a picture of someone who has done the unthinkable; a political decision to do or not to do something that simply doesn't make sense; or a business that's fallen on hard times because some people were twiddling their thumbs rather than looking at potential markets. Our comment is often, "They should have known better." Then, in short order, the conversation comes around to what the education system either should or should not have done to encourage good judgment and get people ready for society.

Thomas Gentzel, executive director of the National School Boards Association (NSBA), is often asked why citizens take on the rigorous task of school board membership. In a 2012 article in *PSBA Bulletin*, he answered, "because they want good schools in their community. They want children to be well educated and their tax dollars spent wisely. They want to be sure their country is stronger in the future than it is today, and they know an excellent public education system is one of the best ways to ensure that happens."[719]

With an explosion in what we need to know and be able to do, we're naturally confronted by the issue of narrowness and shallowness vs. depth and breadth. That brings us to the question, "What are the implications of this trend for society?" Here are a few of those implications for your consideration. You are welcome to think of others.

• If we are more influenced by headlines than we are by the fourth or fifth paragraph, we might end up using bad judgment because we "just didn't know enough about it." We might be blindsided by what we don't know.

- If we want our education system to offer breadth and depth, we may need to get involved in helping educators get the support they need to pursue it. They are likely already concerned about moving beyond the restrictions and narrowness that can come with too many high-stakes tests.

- All of us should realize that we expect schools to provide custodian care (sitting/a safe environment) and constructive socialization while they teach the course of study.

- We should encourage and get involved in discussions about the purposes of education, which need to be expansive, not simply focused on a few subjects.

- Instant gratification will be an ongoing challenge. With limited personal counsel about the length and challenges of life, many young people will choose an hourly job and, for a while, will carry a sense that, "Education can't be that important. Look at those people studying, and I already have a job." Work experience is outstanding preparation in itself. It's also a fact that many young people need to work to support themselves and even their families. Too many students or potential students either can't afford college or don't want to be left with significant debt after they borrow money to help cover tuition and fees. None of that, however, negates the big picture need for a sound education. For most of us, preparation for life is more than landing that first job.

- Elected and appointed officials should understand that narrowing the curriculum could also narrow the short-term and longer-term possibilities for their community, state, or nation. Decisions based on short-term political gain vs. longer-term benefit for society can do damage to the economy and quality of life that could take decades to repair. Chickens, at least those that once had wandering room, generally come home to roost.

Implications for Education
Depth, Breadth, and Purpose of Education

* **Keeping Purpose, Breadth, and Depth Always on the Agenda.** Seriously bring this philosophical, but also very real, set of topics out of the closet and place them squarely on the continuing agenda for discussion, system-wide and community-wide. Make the schools a crossroads and central convening point for the community as people consider substantive issues. At the same time, educators need to know that their keen intellects are respected, their ideas heard, and their guidance appreciated. To the extent possible, give everyone an opportunity and the motivation to say, "We're all in this together."

* **Stimulating Conversations across Disciplines.** Much of what we need to know today and will need to know tomorrow is in the spaces between and among disciplines. Within schools and colleges, educators will need to stimulate intense and energizing conversations exploring sometimes innovative cross-disciplinary approaches to teaching and learning. A short answer to that challenge is often, "We can't do that because…" The response for all of us should be, "Then, what should we do about that?"

* **Considering What is Lifeworthy.** The collective mass of knowledge is expanding exponentially. However, the number of hours in a school day remains about the same as it has been for years. As we continue to face an even more crowded curriculum, discussions within and across disciplines should, as a matter of course, constantly consider what is, "lifeworthy," a term coined by David Perkins of Harvard's Project Zero. Consider the educational payoff when students go in-depth on certain "learning-rich" topics and "sampling" others.[720]

* **Building a Scaffold to the Future.** When we plan and produce our mission, vision, and goals, we need to be sure that the plan is only the beginning. It is a scaffold for reaching out to the future. We often discuss the need to use scaffolds to help students master certain skills or bodies of knowledge. On a more expansive scale, we also need to be sure that our plans are connected to the needs of society, today and tomorrow.

* **Conducting Research Projects.** When appropriate, encourage students to undertake research projects that will help them discover

the breadth and depth of what they are studying. Use the interest students develop as part of those projects to instill further curiosity and persistence in exploring both questions and answers.

• **Creatively Using Technology.** Most schools, colleges, and universities use an array of technologies, including online open-source and purchased learning programs. Those learning enhancements, coupled with targeted experiences ranging from study of the genome to robotics, can help extend the depth and breadth of learning opportunities.

Damian LaCroix, superintendent of the Howard-Suamico Public Schools in Green Bay, Wisconsin, believes, "Learning will become significantly less compartmentalized and lockstep. Instead, learning expectations will be anytime, anywhere, any place, and at any pace. Sophisticated analytics will help to personalize flexible learning experiences based on students' unique interests, needs and abilities." He concludes, "The ability to innovate by harnessing an inherent desire for play, passion, and purpose will fuel the next societal revolution. Technology will serve as both the clutch and the accelerator; hence bringing about transformation on a pace and scale not previously seen."

• **Making Testing and Assessment More Reasonable and Constructive.** Education institutions and communities will need to insist that students receive a well-rounded education that extends student interests and develops talents and abilities beyond what is easily tested. While standards and accountability are essential, those standards should not freeze the system in the past or constantly use high-stakes tests that narrow the curriculum. That approach, when overdone, can be the antithesis of an education that runs broad and deep. Montana Superintendent Laurie Barron, a MetLife/NASSP National Middle School Principal of the Year, recommends mastery based assessment to determine "how much of the content a student has mastered" and then deciding on any further "intervention or acceleration."

• **Focusing on Purpose, Generating Energy.** An ongoing focus on purpose will bring even greater creative energy and a heightened sense of contribution and meaning to all educators. That renewed energy will translate into even more engaged students and a fresh wave of community support.

• **Additional Implications, Breadth, Depth, and Purposes of Education.** Personalization that probes and addresses students' interests and talents and encourages breadth and depth in many ways, including Advanced Placement (AP) and International Baccalaureate (IB) programs can be helpful. *Constructive curiosity* should be met with personal opportunities to explore. Educators need to stay in touch with trends and issues, communicate a bigger picture to their communities, and constantly renew curriculum, instruction methods, and how constituents view the purposes of education. These efforts should be supported by renewed policies, budgets, and programs, as well as preparation and professional development. Holding community conversations, using processes such as trend and gap analysis, can bring each community together in common purpose.

Depth, Breadth, and Purposes of Education
Questions and Activities

1. Consider the five suggested purposes for education included near the beginning of this chapter. Prepare a one-page paper or an approximately five-minute PowerPoint presentation noting why you agree or disagree with any of these items. Suggest what you might add, delete, or change.

2. Review the recommendations included in this chapter from various individuals, groups, and organizations, spelling out or hinting at what students need to know and be able to do. Also review the "Twenty Targets." Develop your own list, no more than two pages. Feel free to draw on anything that is already mentioned.

3. What other historic recommendations would you have included among the individuals, groups, and organizations cited in the "Historic Context" section?

4. View a BBC interview with U.K. Education Secretary Michael Gove (2012) focusing on testing as well as breadth and depth. He addresses his country's General Certificate of Secondary Education (GCSE) exams (2:06). What three to five points does he make that are similar to or different from what you are facing in your community or country? http://www.bbc.co.uk/news/education-18530504.[721]

5. Compose a brief policy statement that you would be comfortable presenting to a school board or governing authority, giving support for greater depth and breadth in education.

Readings and Programs

1. "Most Teachers See the Curriculum Narrowing, Survey Finds," Robelen, Erik, *Education Week*, http://blogs.edweek.org/edweek/curriculum/2011/12/most_teachers_see_the_curricul.html, Dec. 8, 2011.[722]

2. "Centennial Reflections, Intelligence by Design: Thorndike Versus Dewey," Gibboney, R.A., *Phi Delta Kappan*, Oct. 2006, http://www.pdkmembers.org/members_online/publications/Archive/pdf/k0610cen.pdf.[723]

3. "Will Depth Replace Breadth in Schools?" Class Struggle, Mathews, Jay, *Washington Post*, Feb. 27, 2009, http://voices.washingtonpost.com/class-struggle/2009/02/will_depth_replace_breadth_in.html.[724]

4. "Liberal Arts FAQ," University of South Dakota College of Arts and Sciences, http://www.usd.edu/arts-and-sciences/liberal-arts-faqs.cfm.[725]

Chapter 15

Polarization

Don't confuse me with the facts. My mind is already made up.

Chapter 15: Polarization. *Don't confuse me with the facts. My mind is already made up.*

Trend: Polarization and narrowness will, of necessity, bend toward reasoned discussion, evidence, and consideration of varying points of view.
Narrowness ↔ Open Mindedness
Self Interest ↔ Common Good

> "When we understand the other fellow's viewpoint, and he understands ours, then we can sit down and work out our differences."
>
> *Harry S. Truman, Former U.S. President*

Grouping people around two extremes can produce an entertaining television program. However, in the real world, the lack of willingness or ability to reasonably discuss, deliberate, and debate serious issues can lead to constant conflict, gridlock, acts of terror, and even war. Reliability is undermined; evidence is ignored; reasonable alternatives never see the light of day; and progress suffers potentially fatal wounds.

When an organization, community, or country cannot address important concerns because it is chronically fractionated, problems remain unsolved. Then they intensify. Key ideas don't make the agenda and become drowning victims in a sea of narrowness. When feuding factions and unending contentious battles take precedence over progress, communities and nations fall behind. An emergency room physician would likely describe *a self-inflicted wound squarely in the foot.* **Warning:** Look for trouble when we become so polarized that the best we can hope for are watered-down, suboptimal decisions. The common good takes a back seat to gridlock.

The world is moving beyond vertical, top-down, sometimes dogmatic and partisan approaches to problem solving. Those who insist on simply having it their way because that's the way their contributors or clients want it are missing something that is very important. We now live in a collaborative, lateral, networked world, connected 24/7. Those who, out of hand, ignore other points of view may be blindsiding themselves— playing a role that is so defensive (or offensive) that thoughtful and creative people with new ideas simply go over, around, or through them. A few might even become competitors who tout their willingness to collaborate.

Polarization. What is it?

In his *Dictionary of Cultural Literacy*, conservative education philosopher E.D. Hirsch defines polarization this way: "In politics, the grouping of opinions around two extremes: As the debate continued, the...members were polarized into warring factions."[726]

What is Polarization? From the Street...

Let's listen in on the essence of what we hear, see, or read each and every day—in our newspapers, on radio and television, on our choice of web sites or social media, or at the coffee shop. We hear similar things from politicians at every level who dig in their heels, hoping it will help them maintain their bases in bids for election or re-election. The following are not direct quotes from anyone in particular but are attempts to capture the essence of what we're hearing. Even though it's not a laughing matter, a comedian might put it this way: "You know you've got a polarization problem if people keep saying..."

- "It's us versus them."
- "I'm right. You're wrong. It's my way or the highway."
- "Don't worry about doing the *right* thing. Just do whatever it takes to make sure the other party fails."
- "This room isn't big enough for both of us."
- "You're either with us or against us."
- "Are you from a red state...or a blue state?"
- "Take my word for it, it's black or white. There are no shades of gray."
- "I don't care what you or anybody else thinks. Just get the job done."
- "Forget about principle. All that matters is whether we win."
- "Don't bring that up. You know how he feels about that issue. If

you raise a stink, you'll end up paying for it."
- "How much do you think they'll contribute if we sign the pledge to give them 100 percent support?"
- "Deep six this research. It doesn't agree with my point of view. Don't confuse me with the facts. My mind is made up."

A July 2013 *USA Today*/Bipartisan Policy Center poll found that young people had often become disillusioned with polarized politics and were more likely to contribute to society by working through a nonprofit than by running for office.[727]

What's at Stake?
Civility

A major driver of education is a commitment to get students ready to become engaged, contributing members of civil society. That should mean they understand the importance of reasoned discussion; that they know how to gather, consider, and present evidence; and that they have some commitment to comprehending, not necessarily accepting, a variety of points of view.

The late futurist, philosopher, educator, and diplomat Harlan Cleveland, based on his experience, estimated there are, on average, 5.3 sides to most issues.[728] That's between two and three times, maybe even 5.3 times, the number some leaders are willing to acknowledge—despite the fact that any free society should be a crucible for consideration of divergent or even conflicting ideas. We should not rest until every student in our schools, perhaps all of us, are able to withhold judgment long enough to consider at least two conflicting ideas at the same time.

Nonetheless, we have what seem to be a growing number of people whose righteousness has hardened their attitudes and limited their views. Granted, factions are a normal part of any society. Throughout history, they have kept the flame of liberty burning. As much as they might have irritated even those who wrote the U.S. Constitution, factions were seen as stimulating, as extending the conversation about issues and directions. A hope of the nation's founders was that those conversations would help us constantly reach for an elevated purpose or higher ground, not simply wallow in trenches of our own digging.

Even with all of our sophisticated interactive social media, way too many of us are moving from talking *with* each other to talking *past* each other. Listening to the reasoning for another case seems to have gotten lost in the shuffle, replaced by preconceived conclusions, biases, and unbending ideologies. Unfortunately, many have forgotten that by constantly insisting on having everything their own way, without regard for others, they may end up with nothing.

Polarization and narrowness play out inside families, friendships, and communities, and both within and among nations. Talk shows on radio and television are more often shouting matches than civil discussions. Even conversations that seem civil focus on the extremes. Reason is often the victim, dead on arrival. Web sites proliferate to justify a single, narrow point of view. Computer dashboards can be programmed to give us only the opinions we want to hear. Conspiracy theories abound. Hearing about a spine-tingling hoax is, after all, more stimulating than the daily slog of truth.

Political candidates, and many who already hold elected or appointed office, confront the opposition rather than learn from it. Partisanship reigns, too often crushing bipartisanship and even principle, when the principle might possibly be the common good. People declare themselves liberal or conservative and vow against all reason to justify the moniker they've adopted. *The score? Partisanship 10. Common Good 0.*

People come down on "one side or the other" of issues, rather than considering the merits of numerous arguments. Groups *not like us* are too frequently labeled and set apart, causing some of them to demand, protest, demonstrate, shout to gain attention, and revolt. With social media, people can convene almost at a minute's notice to celebrate a birthday or bring down a polarizing tyrant who has stayed in power by pitting one tribe, religious group, or class of people against another. The Arab Spring that began in the early 2010s became a front-and-center example of how quickly people can gather to support or oppose a cause.

Extremists of nearly any stripe have the tools to provoke tension and conflict. A classic is the *straw man*. Convince people of a manufactured danger, crisis, or threat, then rush in to save them. Fundamentalism and extremism can easily tip toward fanaticism. Even labels such as these tend to isolate, stereotype, and discount any legitimacy of concern. Reason disintegrates in the face of bias-driven emotion. What's at stake with this way of thinking? The very future of civil society.

> **"I hear you knockin' but you can't come in."**
> *Billboard R&B Number 2 Single, 1955, written by Dave Bartholomew and Pearl King, Sung by Fats Domino, Gale Storm, Smiley Lewis, Alvin Lee, Dave Edmunds, Bruce Springsteen, Wynonna Judd, and others*[729]

The Missing Ingredient
Civil Discourse

What's missing in this scenario? Absent is the civil discourse that is urgently needed to constantly chart our course for the future. Seemingly discarded is the search for common ground—the quest to turn *my* into *our*. *Multiplication of the common good* is subverted by a *commitment to division*. The purely emotional overcomes reason.

Consider *reason* as a part of damage control. In a May 2012, editorial, "How Political Polarization Hurts the Economy," *USA Today* railed against the inaction of Congress to address fiscal concerns. The piece cited research by *Congressional Quarterly*, pointing out that, in the House of Representatives, "the two parties aligned against each other on 716 of 945 roll call votes, the highest frequency of party unity votes ever for the chamber." Why is this happening? "For one thing," the editorial explained: "To raise campaign funds, candidates pledge allegiance to special interests and ideological groups." *USA Today*'s conclusion? "When lawmakers see themselves as foot soldiers in partisan trench warfare, rather than as public servants, the public is the collateral damage."[730]

In a weekly newsletter, *Rev It Up*, a South Dakota minister, the Rev. Kathryn Timpany, observed: "Some are calling ours 'The Age of Arrogance.' Others wonder how civility and good manners became so old-school and boring. In many places, the newsroom has become merely a stage for theater, the more outrageous the better. On the floors of state legislatures and the federal Congress, both Democrats and Republicans sling slurs across the aisles like unsupervised kids at play in a farmer's orchard, as if being as nasty as you can get away with is an act of political bravado or valor. From the pulpits of churches of all kinds, preachers call each other names and do everything they can to disparage each others' theology and lay sole claim to the most convenient truth they can find."[731]

Former Oklahoma Republican Member of Congress Mickey Edwards wrote "How to Turn Republicans and Democrats into Americans" for the July-August 2011, issue of *The Atlantic*. He observed that elected leaders govern "in a system that makes cooperation almost impossible and incivility nearly inevitable, a system in which the campaign season never ends and the struggle for party advantage trumps all other considerations." Edwards expresses concern that the system is not focused on "collective problem solving" but is too often "divided into warring camps."[732]

Lee Hamilton, who founded the Center on Congress, is known for working across party lines and has raised significant concerns about polarization. For most of our nation's history, he says, we had "a willingness to engage in robust debate over competing ideas, work across ideological divides, negotiate differences, seek consensus, and above all find a way to strike a deal and move forward." Hamilton describes "a motivating philosophy that politicians' ultimate responsibility is to make the country work—not merely to satisfy their own, particular beliefs." That helped make "everything from rural electrification to federal highways" possible.[733]

Hamilton expresses concern that, "We have today a system that rewards politicians for conflict and confrontation, and encourages them to demonize opponents." Some "take their cues from shoutfests and the blogosphere." He remarks, "When goals become moral imperatives, there's no room for compromise. Opponents are not just mistaken; they're immoral. They're cast as evil, ignorant, dangerous, or all three."[734]

The Cold War did not quench our worldwide thirst for common enemies. Perhaps the greatest casualty of an over-polarized society and world is that it keeps us from putting divergent ideas and information together to create new knowledge that could enrich economies and civil societies.

Observations from the Field

With Stops in Canada, South Africa, and the U.S.

Concern about polarization has spread far and wide. It's become pervasive, making decisions even tougher, at a time when social, economic, technological, environmental, and demographic changes are multiplying. Here are a few views from the field:

"The continued polarization of national, state, and local politics is resulting in continued gridlock at all levels of government," observes *Stephen Murley*, superintendent of the Iowa City Community School District in Iowa. Candidates and political parties fan the flames and niche-market media outlets reinforce the belief among true believers of all political persuasions that "they are right and that there is no common ground to be found." Factionalization leads to office holders with mandates from smaller and smaller constituencies, and society is left without the ability to enact new policies or legislation. Moderation and compromise become increasingly difficult. Regretfully, Murley points out, "This inability to work cooperatively has a high probability to destroy all trust in elected officials."

Marcus Newsome, superintendent of the Chesterfield County Public Schools in Chesterfield, Virginia, sees challenges for obtaining needed economic resources, triggered by "individuals refusing to support the common good unless they can see the immediate benefit for themselves."

Jessica Vinod Kuman, a teacher at M.L. Sultan (Pmb) Secondary School in Allandale, Pietermaritzburg, South Africa, notes an increase in "social cocooning" as "individuals retreat to their homes with decreased socializing," but increased visits to social networking sites and more online shopping. She observes "a movement toward extreme religious fanaticism" since the country's transition to democracy and a "re-emergence of nationalism" is bringing with it a greater focus on religion, culture, ethnicity, etc. Kumar senses an "increase or emphasis on traditional leaders in various ethnic groupings in South Africa."

"An emerging narrative of contempt for teachers and public service is extremely dangerous," observes *James Harvey*, executive director of the National Superintendents Roundtable, headquartered in Seattle, Washington. "Educators have been silenced in the debate about how to improve American education" through a process of "finger-pointing, blame, and derogation" that comes from both major political parties. "Educators and teachers at the local level must be brought into this discussion," Harvey recommends.

Avis Glaze, president of Edu-Quest International in Canada, expresses concern about "increasing polarization, as individuals and groups hold on to their dogmas, unwilling to find ways to compromise." She adds, "I am seeing an increase in intransigence. All of this means that we must have leaders with the interpersonal or 'people skills' necessary to bring disparate groups of people together and to work with each other."

Polarization and Suboptimal Decisions

Some opportunities and threats should, by all rights, actually bring us all together and help us find common ground. If we miss on dealing with energy, the environment, health care, our financial situation, and education, every community, every nation, and the world will suffer the consequences. These concerns, among others, need to be addressed through strategies across multiple generations. Each and all of them demand foresight and our very best thinking...across all of the artificial boundaries we build around them. Our future, even our existence, could depend on our working together.

New York Times OP-ED Columnist Thomas L. Friedman is concerned about "the ability of a society's leaders to think long term." In a November 22, 2009, article, "Advice from Grandma," he writes, "What I increasingly fear today is that America is only able to produce 'suboptimal' responses to its biggest problems."

Friedman flags "six things (that have) come together to fracture our public space and paralyze our ability to forge optimal solutions." One is money in politics, the time it takes to raise it, and "the deep pockets that can trump the national interest." Another is gerrymandering that "allows politicians of each party (to) choose their own voters and never have to appeal to the center." He calls out "the cable TV culture," which "encourages shouting and segregating people into their own political echo chambers."

Friedman raises concerns about "a permanent presidential campaign that leaves little time for governing; and the Internet, which can "open the way for new voices," but also "provides a home for every extreme view and spawns digital lynch mobs from across the political spectrum that attack anyone who departs from their specific orthodoxy." Finally, Friedman points to the U.S. business community "that has become so globalized that it only comes to Washington to lobby for its own narrow interests" and "rarely speaks out anymore in defense of national issues like health care, education, and open markets." Rather than truly getting anything big done, "we have to generate so many compromises with so many interest groups," that we end up with "suboptimal solutions that are only the sum of all interest groups."[735]

> **"Politicians are the only people in the world who create problems and then campaign against them."**
> *Charley Reese, Orlando Sentinel*[36]

Scenarios: Choices on a Macro Scale
Contemporary Context

As several dozen people squeezed into a meeting room at a recent conference, a military consultant described the planning process for strategies to meet global threats. He started by noting that everything must rest on a foundation view of the country's role in the world, including its foreign and domestic policy, and offered three scenarios. They included:

• **The country as the sole superpower.** This role involves worldwide engagement and serving as planetary police officer.

• **The country as part of a bipolar world.** The Cold War provided our most dramatic recent example, with the U.S. block facing off against the Soviet block, with some nations either remaining neutral (non-aligned) or falling into one camp or the other. The concept of a Third World emerged from this conflict.

• **The country as a member of the international family of nations.** While maintaining its national sovereignty and security, a country cultivates alliances and partnerships and remains concerned about issues such as trade. The nation works closely with the international community to take on the multitude of challenges and opportunities facing the planet.

The challenge, of course, is to stimulate a reasonable, enlightened, and even heated discussion, not to turn the planning process into warring factions, each running to the nearest camera to get free face time before the next election

Pulling the Problem into Focus

Much of what happens in world affairs and how we're organized in a national and international context has been focused on what divides us. Those divisions might include political boundaries; continents; regional or self-selected economic alliances; economic systems; ideologies; racial, ethnic, social, economic, gender, tribal, and religious differences; or interpretations of environmental science, coupled with personal philosophy.

Political scientist, economist, author, and lecturer Barbara Ward in her 1966 classic, *Spaceship Earth,* observes: "Our planet is not much more than a capsule within which we have to live as human beings if we are to survive the vast space voyage upon which we have been engaged for hundreds of millennia—but without yet noticing our condition."[737] Perhaps we should treat each other like fellow passengers.

If topics are considered taboo, they can become stumbling blocks to civil discourse and barricades on the road to opportunity. This occurs when someone has declared, either directly or through innuendo, that certain things should simply not be discussed. "Don't bring that up! Don't you know he's a conservative, liberal, Democrat, Republican, Libertarian, Social Democrat, Unionist, Independent, Green, you name it?"

It's understood that some people are "polar opposites." The room is eventually filled with a tension no one dares acknowledge, all in the interest of not offending someone's narrow view. A group of politicians asks the opposition for its plan, not as a starting point for discussion but as a first step in building a case against it, even if it includes things they've advocated in the past. The problem for leaders who limit the agenda is that they themselves are eventually blindsided and left out of the loop. Problems that could be easily solved become crises and escalate into catastrophes.

A refusal to consider the views of others and to occasionally compromise, no less collaborate, has led to everything from family break-ups to violence. We'll know, based on our own individual principles, when we've reached the limits of compromise. However, the discussion needs to take place in the interest of the common good. While some are playing overlord for an unbending point of view, others are using interactive technologies to crowdsource, share ideas and information, and find a better way. Collaboration is in. The end run can leave the ideologue isolated, a small island in a vast sea that has washed on by.

Let's go back to the concept of polarity. If you've bought a pair of sunglasses lately, are they "polarized?" That means the shades are designed to block glare on a sunny day while letting other light filter through. In finding perspective in a highly divided world, we need to be careful not to simply polarize or wear only rose-colored glasses.[738]

The Federalist on Factions

Historic Context

When the U.S. Constitution was set for ratification by the states, some of the nation's more eloquent leaders addressed various concerns raised during their discussions and by citizens. Among them was James Madison, who later became President. The following is a segment from a piece on factions that Madison wrote for the November 23, 1787, *New York Packet*. It became part of what we know today as *The Federalist Papers*.

"By a faction, I understand a number of citizens, whether amounting to a majority or minority of the whole, who are united and actuated by some common impulse of passion, or of interest, adverse to the rights of other citizens, or to the permanent and aggregate interests of the community.

"There are two methods of curing the mischiefs of faction: one by removing its causes; the other, by controlling its effects.

"There are again two methods of removing the causes of faction: the one, by destroying the liberty, which is essential to its existence; the other, by giving to every citizen the same opinions, the same passions, and the same interests. It could never be more truly said that of the first remedy, that it was worse than the disease. Liberty is to faction what air is to fire, an ailment without which it instantly expires. But it could not be less folly to abolish liberty, which is essential to political life, because it nourishes faction, than it would be to wish the annihilation of air, which is essential to animal life, because it imparts to fire its destructive agency.

"The second expedient is as impracticable as the first would be unwise. As long as the reason of man continues fallible, and he is at liberty to exercise it, different opinions will be formed."[739]

What is Anyone Attempting to Do?

Can people actually work with colleagues across a political divide? Some are trying. A *National Institute for Civil Discourse (NICD)*, founded in 2011, is chaired by former Presidents George H.W. Bush, a Republican, and Bill Clinton, a Democrat. In its purpose statement, NICD declares, "We hold our First Amendment rights sacred. And yet, the strident tenor of contemporary American discourse is threatening our cherished democracy."[740] A planet-spanning *International AIDS Conference* brings together leaders from around the world.[741] *International Conferences on Climate Change* bring numerous nations together to deal with the impact of climate change on people and the planet.

Their recommendations, such as the Kyoto Protocol, are sometimes accepted but often ignored depending on their more immediate impact on various interests in individual countries.[742] *Bipartisan commissions* are appointed to deal with big issues at several levels and *community conversations* are sprouting up locally.

> **"Once in Washington the system encouraged compromise and governance. Over the past few decades, however, what has changed are the rules for organizing American politics. They now encourage small interest groups—including ideologically charged ones—to capture major political parties as well as Congress itself. Call it 'political narrowcasting.'"**
> *Fareed Zakaria, Host of Global Public Square (GPS), CNN*[743]

Managing our Diversity
Education is Our Hope!

While democracy depends on people who feel strongly and have competing opinions about certain issues, and who will "stand up for what they believe in," it also relies on some degree of acquiescence, or the willingness to compromise in the interest of the common good. At the end of the day, some are generally willing to accept that they haven't achieved total victory in the marketplace of products or ideas.

Taken to extremes, when diversity of any kind is not seen as the norm, conflict is likely to follow. That conflict will probably involve:
- Tribal divisions
- Racial and ethnic divisions
- Religious differences
- Ideological divisions
- Other social and economic divides

Without some level of democracy, people in each of these types of groups will either feel in power or out of power. Samuel Huntington predicted that future clashes will erupt not necessarily along political boundaries, but along cultural and civilizational boundaries.[744] There are those who preach democracy but still want everything on their own narrow terms.

Communicators in a Group Setting

In any group, there are at least four types of communicators:

• **Controlling.** This is a person whose mind is made up, who doesn't want to be confused by the facts. A primary role in life is to impose his or her will on others.

• **Withdrawn.** These are people who have little to say. Their heads move back and forth, as if they were at a tennis match. To their credit, some are good listeners.

• **Relinquishing.** This type of communicator generally spends little time considering facts surrounding the issue at hand. Generally, this person will say something like, "I have a lot of respect for Joe/my political party/etc., and I'll just go along with what they decide."

• **Developmental.** A developmental communicator listens intently to the views of others and constantly tries to move the group toward some level of consensus, combining what he or she considers the best thinking of many, if not all, members of the group.

It's okay to function in any of these categories, depending on the issue, and all of us eventually reach a limit in our willingness to compromise on certain issues. However, the most constructive communicator is the developmental one, the person who listens, considers possibilities, and helps people move away from unbending confrontation toward consensus.

Implications for Education and the Whole of Society
Polarization

• **Identify and teach the basic skills of a civil society.** Education is challenged to bring civility to fractious, sometimes gridlocked or warring communities or whole societies. Civility is, of course, essential to our pursuit of a legitimate, sustainable future.

What are the skills, attitudes, and behaviors we need to be declared "civil?" A few might include: empathy; ethical behavior; respect for others despite our differences; the ability to think, reason, and solve problems; an ability to resolve conflict peacefully and democratically; and a commitment to engage people in discussions and listen to their ideas.

Let's add: the willingness to: collaborate on behalf of the common good; do research, listen, network, thoughtfully consider a variety of ideas, and build a case; understand media literacy skills; and commit to using creativity, imagination, breakthrough thinking, conflict management, negotiating skills, and consensus building—all skills of a developmental communicator. There are many other essentials, so let this list just stimulate the discussion.

Civic education should be seen as essential, since it focuses on civic knowledge, skills, and dispositions; public engagement; policymaking; collaboration; inquiry and research skills; how to evaluate, take, and defend positions; how laws are made and changed; and key principles such as authority, privacy, responsibility, and justice. Social studies and a whole cluster of behavioral sciences are essentials in building a civil society and helping young people and adults develop a *civic temperament*. The motive has remained unclear for why, among the first things a polarized Congress dropped from the federal budget to avoid a 2011 threatened government shutdown, was civic education.

Mastering the art of working within a marketplace of ideas and making change peacefully and democratically may be among our most prominent education goals and survival skills for the future. Working together as problem solving teams in schools, colleges, and communities, we can be challenged to balance our "strong feelings" and "personal interests" with the "common good." We need to work in as much harmony as possible across all diversities. A good place to start is a broad interest in others, including their cultures, clans, tribes, parties, languages, histories, and the thousands of other characteristics of people who enrich any society.

• **Start the discussion of civility with hard questions.** Thoughtfully and honestly involve diverse groups in discussing answers to a few key questions, such as: Are students prepared to become engaged, contributing members of a truly civil society? Do they know how to gather, consider, and present evidence, then build a viable plan? Do they have a commitment to comprehending, not necessarily accepting, a variety of points of view? In a 24/7 interconnected world, do students have media literacy skills? Are they capable of separating truth from fiction? Wheat from chaff? Do they understand that polarizing the citizenry has become a cottage industry for people ranging from radio and television shock jocks to political operatives? Are students, educators, and others on whom they depend for understanding and support also open-minded? Are elected and appointed officials committed to serving the public's interest or their own self-interest? Are we prepared to get past our tribal, racial, ethnic, religious, ideological, social, and economic differences to actually consider the common good?

Veteran superintendent and foundation executive Jane Hammond observes, "Kids in schools need to be having discussions about common values."[745] The big questions: What do we agree on? Where is our common ground?

- **Model civility and engagement in how we teach.** When we thoughtfully engage students, we're sending a message of respect. That's among the reasons why active learning; project-based learning; real-world education; learning across disciplines; learning through inquiry; and thinking, reasoning, and problem solving skills are so important. The *what* of student learning is extremely important, but the *process* of teaching and learning it can have a profound impact on how they act, react, imagine, create, and innovate throughout their lives.

"Inclusion, reason, debate, critical thinking, and commonalities will need to be taught more than ever before." That's a challenge from 2012 National Teacher of the Year Rebecca Mieliwocki. As teachers, "our work knits the fabric of society together like nothing else," she says. "We must model the process we want our future adults to use to solve problems, create jobs, and maintain our democratic ideals."

In an "Eye on Research" article in *Education Week*, Debra Viadero featured a study that focused on students' ability to argue. "That students need to learn how to argue may come as a surprise to parents of strong-willed children." she wrote. "But," Viadero added, "logical arguments differ from the kinds of emotional arguments families experience, and most students possess only weak knowledge of how to recognize, understand, and construct one." Researcher Gerald Graff, a University of Chicago scholar, observed, "Knowing a lot of stuff won't do you much good unless you can do something with what you know," like construct a logical case.

A challenge? Professor of Psychology and Education Deanna Kuhn of Teachers College at Columbia University, who studies student argumentation skills, has observed that students often exhibit a "my-side bias" and "give all the reasons supporting their side, absolutely ignoring the alternative." Some schools use online dialogues and instant messaging software to prepare for whole class "showdown debates" on key questions. In getting ready, students are faced with encountering several sides of an issue.[746]

• **Additional Implications, Polarization.** All of us need to set an example for how to discuss and address issues. Are we reasonable? Are we empathetic? Are we willing to listen? Are we defensive? We need to demonstrate that reasonable compromise to support the *common good* is not necessarily *caving in*. We need to act inclusively, engaging people and their ideas in our conversations and our actions. Each of us can be a *separator* or a *unifier*. Rather than think and act narrowly, we need to consider a global view, especially when we discuss issues such as energy, water, nutrition, and the environment. Students and other citizens need to master the art of making change peacefully and democratically. Civic education is fundamental.

> "Happiness and prosperity are now within our reach, but to attain and preserve them depends on our own wisdom and virtue."
> *George Mason, Framer of the Virginia Declaration of Rights, 1783*

Polarization

Questions and Activities

1. What can educational institutions do to help current and future citizens become more adept at civil discourse in dealing with disputes rather than becoming increasingly polarized? Suggest six ideas in no more than the equivalent of two pages.

2. Read "Advice from Grandma," a *New York Times* column by Thomas L. Friedman, Nov. 22, 2009. Discuss the six things he identifies that "fracture our public space and our ability to forge optimal solutions." http://www.nytimes.com/2009/11/22/opinion/22friedman.html?_r=0. [747]

3. View "How Common Threats Can Make Common (Political) Ground," a *TED Talk*, featuring social psychologist Jonathan Haidt, (20:02), posted Jan. 2013. Haidt is author of *The Righteous Mind: Why Good People are Divided by Politics and Religion*. List three things he says in the video program with which you agree. Also list three things with which you disagree. http://

www.ted.com/talks/jonathan_haidt_how_common_threats_can_make_
common_political_ground.html.[748]

4. What can our school, college or university, community, or country do to
   overcome intense polarization? List three things in each category with a few
   sentences to clarify each item.

5. What are the four types of communicators in a group setting?

Readings and Programs

1. Edwards, Mickey, "How to Turn Republicans and Democrats into Americans,"
   The Atlantic, July-Aug. 2011, http://www.theatlantic.com/magazine/
   archive/2011/07/how-to-turn-republicans-and-democrats-into-ameri-
   cans/308521/.[749]

2. Cleveland, H. (2002), *Nobody in Charge . . . Essays on the Future of Leadership.* San Francisco: Jossey-Bass, A Wiley Co., available for purchase at
   http://www.amazon.com/Nobody-Charge-Essays-Future-Leadership/
   dp/0787961531.[750]

3. Any edition of *The Federalist Papers,* with particular emphasis on James
   Madison's paper on factions. For example, see *The Federalist . . . A Commen-
   tary on the Constitution of the United States,* Bicentennial Edition. (1976).
   Washington, DC: Robert B. Luce, Inc.

4. View an Arizona State University YouTube video featuring John Reich,
   ASU Emeritus Psychology Professor and author of *Radical Distortion...
   How Emotions Warp What We Hear,* (9:53), http://www.youtube.com/
   watch?v=sBSJ4F-p1YE.[751]

Author's Note: The tug between polarization and reasoned discussion, evidence,
and consideration of varying point of view is a very serious matter. Deepening divisions created by an all-or-nothing approach to dealing with every issue or problem, with little regard for the common good, is likely unsustainable.

Chapter 16

Authority
By the power vested in me...

Chapter 16: Authority. *By the power vested in me...*

Trend: A spotlight will fall on how people gain authority and use it.

Absolute Authority ↔ *Collaboration*
Vertical ↔ *Horizontal*
Power to Impose ↔ *Power to Engage*

> **"If your actions inspire others to dream more, learn more, do more, and become more, you are a leader."**
> *Former U.S. President John Quincy Adams*

Nelson Mandela spent 27 years in prison. Dedicated to overcoming apartheid in his country, he was physically deprived of his freedom. That, authorities hoped, would silence his demand that everyone have a voice in their own destinies, whatever their races and ethnicities. It didn't work out that way. While people in positions of elected or appointed leadership cracked down on those who fought to breathe free, Mandela demonstrated what none of them could muster. He was clothed in *moral authority*, so strong that it overcame those who were vested in keeping *themselves* in power and *other people* out. Later, out of prison and serving as elected President of South Africa, he said, "Courage is not the absence of fear—it's inspiring others to move beyond it."[752]

History is replete with examples of moral authority. Martin Luther King, Jr., and Mahatma Gandhi are among 20th century leaders who are revered for their courage and unyielding devotion to justice. Their larger-than-life examples stood taller than any high office or narrow self interest.

King, Gandhi, and Mandela brought enduring change for humankind. What lessons have they taught us about authority? How many of those lessons are we learning and applying in our education, economic, and political systems? In our private and public lives? Will we recognize their type of inclusive leadership in the 21st century, or will we yield to exclusion? The tension continues as the spotlight becomes more intense on how we gain authority and how we use it. Keep in mind that we don't have to hold lofty titles to be leaders. Legitimation takes more than a corner office. For 27 years, Nelson Mandela's office was a prison cell. He led by example.

Authority
Declared? Conferred? Assumed? Earned?

Does education prepare us for the world or just provide an introduction? Do we respect the school of life? Should authority simply be declared or conferred or should it *also* be earned? Should connectedness, creativity, thoughtfulness, productivity, ethical behavior, and a steadfast commitment to the common good trump privilege, title, and position? Is genuine leadership more important than dictum? Does merit count?

World history and current events blaze with examples of declared, conferred, or assumed authority that has led to authoritarianism, anger, suffering, and, in some cases, revolt and overthrow. Despots, when they seize control, often through a coup, false promises, manipulation of the truth, a lack of transparency, inside connections, or even armed conflict, often just announce their authority. Some specialize in creating fear to silence dissent. Over centuries, many have ruled without genuine consent of the governed. Their often flamboyant reigns carry the seeds of their own destruction.

Now, the Good News

Think about it. It's a *good news, bad news* story. The *bad news* is that some people still live with despotism. Too many continue to work in organizations where new ideas are seen as threats and the way forward is *my way or the highway*. The *good news*? As we move more deeply into the 21ˢᵗ century, it's pretty clear that the parade of marching orders is out…and collaboration is in.

We need the vertical but it should be balanced with the horizontal. People in organizations, including countries, are making clear through words and actions that they are intent on having their voices heard and ideas considered. Emerging generations are determined to solve problems that will profoundly impact our future, deal with injustices, and pursue possibilities. They want to be part of the action, not locked in an arbitrary *no-idea-zone*. Liken it to compressed, pent-up energy waiting to explode on society. This human drama has been playing out before our very eyes as dictators have fallen, one by one. We once looked to history for examples. Now we find them in the daily news.

> **"A spotlight on authority may reflect a generational trend of questioning assumptions."**
> *Frank Kwan*

Authority, Authorities, Authoritative, Authoritarian
What's the Difference?

We're generally relieved when we hear that a car accident or home break-in has been reported to the *authorities*. On a massive range of issues, we want those who have been elected or appointed to various positions to act on our behalf. That's true in both the public and private sectors. We delegate *authority* to certain people or groups and expect them to carry out their responsibilities fairly. All should understand that their genuine success will depend on how well they balance competing interests, serve the common good, and avoid simply using their power for personal gain or favoritism.

We refer to some people as *authoritative*, not necessarily because they hold a certain title or position, but because of their depth of knowledge or experience. We might go to them for advice on how to solve a problem or deal with an issue. Generally, we expect their counsel to be clear and reliable. We call them experts, geeks, hotshots, maestros, masters, scholars, proficient, sharp, sharks, virtuosos, whiz-kids, professionals, certified, accredited, or even cognoscente.

Then, we come to *authoritarianism*. That's dicier. One definition: "blind submission to authority." Another: "a concentration of power" in a way that is not responsible to the people. Show me someone who is authoritarian and I'll show you a person who is labeled: autocratic, dictatorial, domineering, imperious, overbearing, tyrannical, arrogant, lordly, presumptuous, pretentious, superior, uppity, controlling, arbitrary, high-handed, narcissistic, pompous, vain, all-powerful, almighty, or stern.[753]

> **The Importance of Authority.** Children depend on their parents, who feel a sense of responsibility to protect them, nourish them, teach them lessons of life, and help them grow and develop. Parents are often a child's first *authorities*.
>
> Throughout our lives, we delegate certain responsibilities and reasonable authority to education systems, law enforcement, the courts, public works, the military, and various levels of government. Many countries have established a separation of powers, checks and balances, between and among executive, legislative, and judicial branches. We declare and limit authority through a Constitution and an array of amendments, laws, ordinances, and public policies.
>
> Most of us hope that those who have been given responsibility and authority to govern and enforce will be able to govern themselves. We trust that they will look after the needs of the whole of society and not simply promote and defend narrow self-interests. We hope for *servant leaders* who exercise that authority responsibly. With some exceptions, people generally agree that an appropriate level of authority is a good thing.

The Evolving Concept of Authority
From Then to Now

Who's in charge here? Who can we trust? Who should make those decisions? Who should be involved? Who are the authorities? Who knows? These are not new questions. In fact, questions about authority have rung throughout history, and they will continue to ring loud and clear well into the future.

Scholars know that the Founding Fathers of what became the U.S. were experimenting with something fairly startling when they conceived of a government of, by, and for the people. Trust the people, **Thomas Jefferson** told us. For centuries if not millennia, people, from tribes to more complex societies, had been convinced that someone, such as an emperor, was needed to protect them from their own human nature.

• **Pericles** had a different idea. He believed citizens of Athens should have their voices heard. His belief was summarized in his funeral oration, delivered by none other than Thucydides. "In a democracy," he said, "the administration is in the hands of the many, not of the few." Some refer to Pericles as the father of democracy. This Greek statesman, orator, and general, lived sometime between 495 B.C. and 429 B.C.[754]

If that seems like ancient history, consider some philosophers who intrigued a young Thomas Jefferson. Central to their ideas was the concept of authority.

• **Niccoló Machiavelli** (1469-1527), a political counsel to Italy's Borgia family, got tired of bickering among Italian cities and urged the Borgias to seize their power. He advised them to forget about ethics, justice, honesty, and kindness, and do whatever might be necessary to preserve their own power, all, of course, in the interest of the state. Machiavelli wrote *The Prince* and gave us the term "Machiavellian," which has come to mean treachery, sneakiness, ambition, and ruthlessness.[755]

• **Thomas Hobbes** (1588-1679) wrote *Leviathan*, defending the absolute power of kings, since he wasn't sure mere mortals could control themselves.

• **John Locke** (1632-1704) favored representative government, such as the English Parliament, but thought it best to limit parliamentarians to men of property and business. Locke spoke of "inalienable" rights—life, liberty, and property. He supported the idea of a king, but with limited sovereignty. Locke's king would support the natural rights of people, not have absolute power. The supreme authority of the government would be an executive, such as a Prime Minister, under the thumb of the Parliament.

• **Charles Montesquieu** (1689-1755), was French, not English like Hobbes and Locke. He was born of a noble family, had no great love for Louis XIV, studied law, and traveled to England, where he studied the political system. Montesquieu thought the primary purpose of government was to maintain law and order, political liberty, and the property of individuals. In his 1748 *The Spirit of the Law*, he expressed a preference for legislative, executive, and judicial branches of government that were separate and kept each other in check. That idea of separation of powers seemed like a good one to framers of the U.S. Constitution. With that kind of checks and balances, a monarchy would not be able to command so much authority that it could engage in despotism.

• **Jean-Jacques Rousseau** (1712-1778), born in Geneva, Switzerland, won an essay contest in 1751. He wrote about the premise that man is naturally good but is corrupted by society. That idea made him a celebrity. In 1762, when he wrote, *The Social Contract*, Rousseau declared, "Man is born free, and everywhere he is in chains." He envisioned people giving up rights to the whole community, not to a king. Rousseau liked the idea of individualism and taking a vote to determine the general will. Note that the first words of the U.S. Constitution are, "We the people…"[756]

> **The Occupy Wall Street movement got under way in 2011 "in the wake of a devastating financial crisis, wars in Iraq and Afghanistan, surging unemployment, the squeezing of the middle class, and a housing mortgage crisis." The movement "expressed frustration and disillusionment with the status quo."**
> *Allison LaFave*

Power and Authority…
War Stories from the Front Line

The room was filled with interested, talented people. I was discussing future-focused leadership. At one point, I said, "Power shared is not power lost…it is power multiplied." A hand shot up. "I've got a problem," the puzzled participant declared, "I know you're right, but I've worked for 30 years to get the power I have today. I'm not about to give it away to anybody." *What counsel would you suggest?*

Let's look at a few real-life examples of how some people have used authority to make decisions. We're not using names.

• **Concertmaster, Cultural Leader, Decision-Maker, Moral Authority.** A school administrator, meeting with a leadership team, said, "Here's what I think we're up against. We need to make a good decision for the community. What do you think we should consider as we draft our recommendations or decide what we should do?" He added, "I know we all agree that, whatever that is, should support the best possible education for our children." This ingenious leader listened and demonstrated his respect for the thinking and ingenuity that constantly flowed from an array of people. He seldom used the word "my." It was generally "our." He never referred to people as "my staff." Instead, he talked about how honored he was to be part of such a distinguished team.

This exceptional leader credited the community and an inspired staff for every success and commended students on their effort and achievement. He made clear to the community that the education system was a product of their high expectations and reinforced what had become a culture of excellence. "We're never as good today as we hope to become tomorrow."

By the time major decisions were made, people knew their voices had been heard. Whether everyone fully agreed or not, they knew those decisions were made in the best interest of the community and its students. What he learned in the process, he shared with educators nationwide and worldwide. His power, which he never discussed, came from his knowledge and experience, his boundless respect for people, his encouragement, his creativity and persistence, his cultural leadership, and his moral authority.

• **Loyalty Detector.** When decisions were discussed, the CEO had already made up her mind. Benevolently, she would discuss an issue with the group but keep everyone guessing about where she stood. A few people always offered their honest opinions. Others held back, knowing from experience that the boss had already decided and that they wanted to be part of the winning team. When they caught a few nonverbal signals that gave them a hint, they emphatically said what they thought she wanted to hear. Those who had other opinions were sometimes considered "disloyal" or "not really part of the team."

Trust groups and factions developed. Fear was an ongoing reality. Heads rolled. Creativity was often blocked. Symbols of respect and encouragement were few and far between: "Who needs that? After all, we're all adults." Incompetence replaced excellence and a driving commitment to the cause. Competition for the boss's favor has its limits. In situations such as this, the most talented are often the first to leave, since they have options.

• **Decision Making Efficiency Expert.** Meeting with a few assistants, a CEO observed, "You know, getting the team involved takes too much time. Why don't the three of us get together and make those big decisions? We can just announce them." Then, verily, as a chorus of concern arose from staff and community, the boss had to pack his bags. He had lost touch, blindsided himself…and had done it unknowingly on purpose, not fully understanding the limits of his authority.

Choosing People for Positions of Authority.

In *Foundations of Democracy*, the Center for Civic Education suggests some of the characteristics we might look for in choosing someone for a position of authority. Included are: "impartiality, integrity, intelligence, diligence, reliability, courage, the ability to work with other people, sensitivity to human needs and rights, the specialized knowledge and skills and physical capacity that might be required, and (their) view on job-related issues."[757]

Responsibility vs.The Limits of Authority and Power

When it comes to how decisions are made, most people worldwide have one thing in common. *They are skeptical about authority.*

Civic Education. Skepticism aside, people "look to authority for resolution of conflict and maintenance of order," the Center for Civic Education (CCE) notes in lessons for *We the People: The Citizen and the Constitution*. This acclaimed education program stirs spirited discussions as it addresses issues of democracy—such as authority, privacy, responsibility, and justice. "The U.S. Constitution provides for authority, but also limits its practice," CCE observes.[758] Of course, assigning authority, sometimes through elections or appointments, means we are asking people to handle a host of issues on our behalf.

Media and Information. Media Consultant and Producer Gary Rowe, president of Rowe, Inc., points out a vast number of people who have gained a version of authority. They include "politicians, professors, celebrities, experts, and those associated with elections, degrees, fame, and academic credentials," to name a few. Yet, Rowe speculates, their authority could "erode before the digital democracy of massively shared information." Information is power, and it is increasingly available to nearly everyone.

"Nothing will be particularly relevant to students just because someone says it is relevant," Rowe observes. For students to be fully interested, he adds, "information must seem personal to them—and they want to participate in the making of meaning." With growing authority in the hands of ubiquitous social media, the ability to make a case in both words and images is getting even more important. "That could be a troubling trend," he concludes, since "information that has value calls for thoughtful writers and editors, talented media producers, and educated recipients." Everywhere, we're experiencing a diffusion of power. Globally, in addition to nation-states, non-state motivated groups fight for their causes on a world stage, whether they're motivated by good or evil.

Authority and Decision Making
How do we use the power?

Being a decision maker is tough duty. Every day, in nearly every language and around the world, they join in a chorus of, "I'm damned if I do and damned if I don't." That poetic outburst usually leads to the rejoinder, "Well, you wanted the job!"

• **Dealing with Demands.** A person of considerable wealth and power, who just happens to run the most watched television station in town, visits a local school administrator. He demands adoption of a curriculum program he believes will specifically benefit his children. The administrator knows from research and personal experience that the program would undercut education for all children in the district. However, it is clear that the person who just left his office has such widespread influence that "he could make my life miserable and maybe even get me fired." The school leader has pledged to use his authority in the best interests of education for all. In this case, what decision should he make? Some have a great deal and want more. Some have very little and want some.

• **Listening and Learning.** Getting and staying physically and emotionally part of the community is basic. Using social media and online or other surveys, plus conferring with advisory groups can help. So can ongoing "listening sessions" or "Futures Councils." *Authority* is generally more *authoritative* when it reflects ideas and opinions of those we serve. In a *Forbes* article, "10 Reasons Your Top Talent Will Leave You," contributor Mike Myatt emphasizes the importance of giving people a voice internally. He remarks, "Talented people have good thoughts, ideas, insights, and observations. If you don't listen to them, I can guarantee you someone else will."[759] The *Achievers* web site suggests eight elements for getting staff engaged: leadership, communication, culture, rewards and recognition, professional and personal growth, accountability and performance, vision and values, and corporate social responsibility.[760]

• **Being inclusive.** "I don't need to modify anything. Everybody I talk to agrees with me!" That sort of declaration begs a response, "Who have you been talking to? Other people also have ideas and concerns we need to hear." We need true advisory councils, not just cheering squads. Inclusiveness is essential. Have we heard voices across generations as well as across racial, ethnic, social, economic groups, and other diversities? Crowdsourcing, using social media, can also help us reach out to build better decisions and to more legitimately use our authority in pursuing the common good. Community Conversations can help enlighten authority and decision making. A concerted effort to build *mutual expectations* can help pave a road to the future.

Legitimate Authority

As we consider how we will use authority that has been entrusted to us, we need to make sure that we consider the pros and cons, those who might gain from our decision and those who might not, laws and policies that govern the situation, and whether our decision would seem logical or reasonable. Democratic decisions imply a need for legitimation, and we need to thoughtfully cover the bases. All of us should try to set a good example as we summon the wisdom and courage to serve the common good.

Joseph Hairston, a former superintendent and a member of Futures Council 21, cautions, "Politics and finance have impacted education in ways that do not focus on what must happen in the classroom for children." He expresses concern that "the focus on power and control has skewed the primary mission of support for education."

Implications of the Authority Trend for Education and Society

- Schools and colleges will be continually pressed to build the leadership knowledge, skills, and experiences of their students, whether those students are advantaged or disadvantaged. Those skills will be seen as basic to success, even survival, in the economy and civil society.

- Students, educators, business and government leaders, and others need to understand the foundations of inclusive decision making and how to legitimately use authority in achieving the common good.

- Authority might be included in discussions across many areas of a school or college curriculum, including: social studies, civic education, government, history, communication, the sciences, technology, leadership, media literacy, and other courses.

- Across disciplines, students and others in the community will need to have opportunities to experience public engagement and teamwork. All of them might learn to identify issues and problems, conduct research, develop plans of action, suggest public policy, and earn public consent. That's what students do who work with *Project Citizen*.

- Students at every level should receive a grounding in foundational principles that are basic to democracy, such as authority, justice, privacy, and responsibility. Each should be able to find a balance between arrogance and empathy, understand human and civil rights, and learn how to identify benefits and consequences of their own actions, for themselves and others.

- All will need to understand the consequences of corruption and authority that has been misdirected for the benefit of the few at the expense of the many. A position of authority should be understood as a position of responsibility and service, not simply a source of personal privilege.

- Preparation programs and ongoing professional development in every field should prepare people to exercise legitimate future-focused leadership in a fast-changing world.

Authority

Questions and Activities

1. What are five characteristics you would look for in a person who is authoritarian? Also, identify five characteristics you would look for in a person who uses authority legitimately.

2. Compare examples of leaders in the section devoted to "Power and Authority...War Stories from the Front Line" with the philosophies of thinkers highlighted in "The Evolving Concept of Authority." Which of the philosophies best matches the approaches used by each of the decision makers?

3. Review and consider use of suggested "Center for Civic Education Lesson Plans," http://new.civiced.org/resources/curriculum/lesson-plans, and http://new.civiced.org/resources/curriculum/constitution-day-and-citizenship-day.[761]

Readings and Programs

1. Consider obtaining and reviewing *We the People: The Citizen and the Constitution*; *We the People: Project Citizen*; and *Foundations of Democracy*, available from the Center for Civic Education, available from http://new.civiced.org/resources/publications/ebooks/new-enhanced-ebook.[762]

2. Review materials available from the Constitutional Rights Foundation, http://www.crf-usa.org/publications/.[763]

3. Rifkin, Jeremy, *The Third Industrial Revolution*, (2012), Palgrave Macmillan, NY, Part 3, Chapters 7, 8, and 9, pages 193 to 270, including "Retiring Adam Smith," "A Classroom Makeover," and "Morphing from the Industrial to the Collaborative Era."[764]

4. "Global Power Shifts," Historian, Diplomat, and former head of Harvard's Kennedy School of Government Joseph Nye, *TED Global*, Oxford, England, U.K., posted Oct. 2010, video (18:16), http://www.ted.com/talks/joseph_nye_on_global_power_shifts.html.[765]

5. "Leads with Moral Authority," James W. Sipe, and Don M. Frick, http://www.leaderswhoserve.com/index_files/Pillar7.htm, excerpt from *Seven Pillars of Servant Leadership*, 2009, Paulist Press, N.J. available from http://www.amazon.com/Seven-Pillars-Servant-Leadership-Practicing/dp/080914560X.[766]

Chapter 17

Ethics
Let's try to do the right thing

Chapter 17: Ethics. *Let's try to do the right thing.*

Trend: Scientific discoveries and societal realities will force widespread ethical choices.
Pragmatic/Expedient → Ethical

> "The first step in the evolution of ethics is a sense of solidarity with other human beings."
>
> *Albert Schweitzer*

What Happened to Ethics?
Who Inflated the Bubble?

Once upon a time, there was "a bubble." Home prices went through the roof. Loans were easy, whether you could afford them or not. In fact, some people became "house rich." Many borrowed a ton of money using their skyrocketing equity as collateral. Then the bubble burst. Suddenly, they were "house poor." Quite a few did not live happily ever after.

The Great Recession that exploded before our very eyes was not a fairy tale. Foreclosures, bankruptcies, stock market losses, retirement savings and earnings depleted, jobs cut, college dreams dashed, and high hopes that turned into deep despair spread like a plague. All the while, some of the people who kept inflating that bubble walked off with fat paychecks and bonuses. A number of their firms cleaned up with major profits until, finally, the worthlessness of the "complex mortgage-backed securities" they were selling even caught up with them.

"How could this happen to us? We were doing so well." It's a fair question. In a 2011 issue of *The Wilson Quarterly*, author and *Washington Post* columnist Robert Samuelson calls attention to "a long list of possible villains: greedy mortgage brokers and investment bankers, inept government regulators, naïve economists, self-serving politicians." Legal and ethical shortcuts piled up.[767] Fingers pointed in many directions.

What actually happened? Was it a failure of the system? Was it greed? Have we accepted predatory behavior as the norm? Was it a lack of oversight? Was there deception? Were there crimes? Were there breaches of ethics? The jury is out.

Ethics! What does it mean?
Try Measuring the Bubble Using This Definition.

While ethics may mean different things to different people, we'll let *Webster's New Collegiate Dictionary* be our arbiter. "Ethics," according to *Webster's*, is "the discipline of dealing with what is good and bad and with moral duty and obligation." Let's read further. It's "a theory or system of moral values" that can involve "principles of conduct governing an individual or group, such as a profession."[768]

Some refer to ethics as a set of guidelines, a code of conduct, a basis for professionalism, and a form of behavior. Many agree that ethics often takes over when we reach the limits of the law, since some things might be technically legal but still don't seem appropriate. One thing is certain: When we're faced with a massive breach of ethics, everything else on our agenda takes second place.

Ethics can involve what we say and do, how we say and do it, and what we decide not to say and do. People run into problems because of *commission* (deliberately committing an act), *omission* (deliberately leaving out critical information, such as a half-truth), *misinformation* (mistakenly providing information that isn't true), and *disinformation* (deliberately spreading information that isn't correct in the hope that people will believe it). Famed Oklahoma comic Will Rogers summed it up when he said, "I'd rather be the one who bought the Brooklyn Bridge than the one who sold it."

Among the best safeguards in guiding our ethical behavior is a commitment to treating people with respect rather than imposing our will without consulting them. Just to provide fair warning: The minute we stray from the Golden Rule, *doing unto others as you would have them do unto you*, we're headed directly toward *an eye for an eye and a tooth for a tooth*.

Pragmatic? Expedient? Ethical? How Important is the Ethical?

Pragmatic: "Whatever works." *Expedient*: "Whatever is easiest." *Ethical*: "What's the right thing to do?" We know that the pragmatic and expedient can be ethical, but we should ask ourselves how any of our actions will impact the lives of others.

How important is an understanding of ethics? The response is generally, "It could not be more important. It's absolutely essential!" Then comes the challenge. How do we develop ethical behavior? In theory, educators hone a sense of ethics through precept and example, but do we really do it? If we do, why? If we don't, why not?

When the Magic Impacts Your Future…
Find Out How It's Done!

Black magic. When we first encounter it, we just might be fascinated. How can that person or organization turn so little into so much so quickly? It's almost too good to be true. Seeing a way to get rich quick or just grow a nest egg for retirement, we invest. Then, one day we wake up to find our money gone and the purveyor of black magic in handcuffs. A benefit might be that we better understand the meaning of a scam, sometimes called a *Ponzi Scheme.*

A powerful purveyor of influence, sometimes called "an influence peddler," using slight of hand, gets wide support from city, state, or national officials for something that ignores the common good in favor of advantage for a few. Then, we discover that votes for or against a line item in a budget or a piece of legislation were just a return for favors or bribes. The common good has a hard time competing with that kind of influence, unless its representatives are committed to an ethical code and we hold them to it. When a breach of trust is discovered, the damage is done. Again, here come the handcuffs.

2001, What Hit Us?

9/11/2001. It's a date that is etched into history. Its very mention sends chills up the spine. Thousands of innocent people lost their lives in a series of terrorist attacks that, like Pearl Harbor in 1941, will live in infamy. Little did we know that destruction of New York's twin towers and a portion of the Pentagon plus a plane crash in Pennsylvania would be followed by the alarming collapse of a spate of marquee corporations.

"Almost beyond belief!" That's how virtually millions reacted to a cascade of alleged ethical breaches that left society aghast. Scandals at Enron, Andersen, WorldCom, Tyco, and other corporations sent shocks through the economy. Retirement funds were lost or threatened. Stocks tumbled. The stories blazed across our newspapers and lit up our television screens. The question loomed, "Can we really believe audited financial statements?"

In describing what happened at Enron, BBC News used terms such as "shady dealings," "concealing debts," "deception," "fraud," "money laundering," "murky finances," and "scandal." Members of the Enron leadership team were portrayed in a 2003 book, *ENRON: The Smartest Guys in the Room*, written by *Fortune* reporters Bethany McLean and Peter Elkind. A movie by the same name hit theaters in 2005. The moral? What good is being smart if you use your intelligence and connections to breach such a wide swath of ethical behavior?

Symbolically, *Time* magazine's 2002 Persons of the Year were "The Whistleblowers," Cynthia Cooper of WorldCom, Coleen Rowley of the FBI, and Sherron Watkins of Enron.[769]

"Without moral courage, our brightest virtues rust from lack of use. With it, we build piece by piece a more ethical world."
Rushworth Kidder, Founder, Institute for Global Ethics[770]

Ethical Challenges and Breaches
They Get Our Attention!

The amygdala, that almond-sized group of neurons placed deep in our brains, is constantly on alert for anything we sense that might cause us harm. It's the same area that triggers fight-or-flight. That warning system is what grabs our attention when we see, read, or hear about ethical violations. "What was that person thinking?" "How will this affect me, my family, my community, my country?" The news media know it. We've all heard the expression from the world of journalism, "If it bleeds, it leads."

Without naming names, here are a few situations that have captured our attention. Feel free to use them as mini-scenarios or as ethics discussion starters.

• A newspaper runs a photo of a man being struck by a subway train. Should the journalist have just recorded the incident or tried to rescue the person in peril?[771]

• A celebrated bicyclist, who won numerous world titles, is investigated for use of performance-enhancing drugs. After repeated denials, he admits the truth. Should he have retained his titles and medals?

- As computers become even faster and technologies even more discretely designed, we will likely see: "neuroprosthetics to augment cognition; nanobots to repair the ravage of disease; bionic hearts to stave off decrepitude." Artificial intelligence (AI) may also aid us in our brain function. Some ask, "When do we stop becoming *us* and start becoming *them*?"[772] Should we welcome these new technologies?

- Educators are accused of altering student answer sheets used with high stakes tests as a way to give the appearance of higher than average achievement scores. An *Education Week* article, "Educators Look for Lessons from Cheating Scandals," explored the need to create "a culture of integrity."[773] How do you react to these incidents?

- A classmate tells you confidentially that he plans to bring a gun to school tomorrow. He adds, "You'll keep this quiet if you know what's good for you." From the time you were in first grade, you've been told, "Don't be a tattletale." "Don't rat on other students." "Don't be a snitch." What will you do? When?

- Someone is suspected of planning what could be an attack, perhaps even an act of terrorism. The person's cyber messages are hacked, phone conversations monitored, and movement tracked using GPS. How do we determine who is a suspect?

- A technology firm's computer scans cyberspace for key words. Within a few minutes, our screens light up with cyber sales pitches for new cars, bikes, sneakers, and migraine medicine, just the things we mentioned in our last tweets to Uncle Jack. Can hacking be ethical, depending on the purpose? What are the ethical limits of our personal privacy? Do those immediate ad alerts constitute identity theft or a breach of privacy?

- A person cleared for access to classified information and sworn to uphold confidentiality releases that information to the media. Is it whistleblowing or espionage?

- The use of drones has raised concerns as it expands from far-off battlefields to the home front to help us manage traffic or follow a bank-robber trying to make a get-away. In the process, we might be seen from the skies as we fire up the grill. Do we consider our loss of privacy minimal compared to the added safety and security?

- You're a psychotherapist…meeting with one of your patients. Late in the session, the patient reveals, "Tomorrow, I'm going to shoot my husband when he gets home from work. I'd do it tonight, but I know he's working on a project that he needs to finish." When she leaves and the door closes, you reach for the phone to call the police. Then you remember your code of ethics, which is quite clear about confidentiality in the psychotherapist/patient relationship. What will you do?[774]

- An elected official takes the oath of office pledging to "bear true faith and allegiance" to his country. Shortly thereafter, he signs a pledge not to support a certain type of legislation in return for a promise of substantial campaign contributions from a number of wealthy donors. Should that be considered a conflict of interest, a violation of trust, an ethical issue, or all three?

- A public official takes bribes from companies that would like to be selected to provide services for the organization. He discretely tells them it is "pay to play," then asks that a premium on all billings be deposited in his private account. Many would call it corruption. Why are some people tempted to take advantage of their positions of authority for unethical, even criminal, acts of this nature?

- A proposal is made to the United Nations to bring countries together to develop a network of sensors to provide advance warning of tsunamis or severe weather events. Multi-thousands if not millions of lives might be saved. A colleague comments that the whole thing is absurd because it would probably do other countries more good than his own. How would you respond?

- A small group of young people taunts a student who is new to the school. The harassment spreads from name-calling and intimidation in hallways to scandalous but false rumors online. What's happening is a classic case of bullying and cyberbullying. The student's parents raise concern, and school officials say they will "look into it." The following weekend, the student takes her life. What should we have done? What do we do now?

These few examples only cover the tip of an ethical iceberg. We could elaborate on tastelessness and a lack of honesty in campaign advertising and a tendency to say anything, true or not, in the pursuit of votes. Each of us may want to spend some time thinking about the need for honor and a moral compass to get past an attitude summed up by, "The problem is not violating a code of ethics. The problem is getting caught."

> **"To be persuasive we must be believable; to be believable we must be credible; to be credible we must be truthful."**
> *Edward R. Murrow, American Broadcast Journalist*[775]

Why do people and organizations use bad judgment?

While the list is by no means complete, here are some of the reasons people walk directly (sometimes indirectly) into an ethical minefield.

- A belief that power means personal entitlement.
- Failure to seek or accept information or feedback that is factual but doesn't support their case.
- Deliberate misinterpretations of information.
- Use of double standards.
- An unwillingness to admit problems.
- Trying to hide information from the media and the public.
- Blaming others or attempting to cover up a problem rather than immediately admitting it and dealing with it.
- *Quid pro quo* (You scratch my back. I'll scratch yours.)
- Easy conclusions about complex problems, without consideration of side-effects.
- Arrogance, over-confidence, a sense of immortality, a superiority complex.

Because education is such a vital service and educators are expected to be among the most trustworthy in society, breaches or even suspected breaches of ethics can cause a high level of concern.

Hot-Button Issues for Education: Problems often surface around issues such as hiring or other personnel decisions, teacher placement, financial concerns, student discipline or placement, class, multicultural sensitivity, the handling of disasters or accusations, attempts to discredit others, privacy and access, and student or staff cheating (sometimes driven by demands that students do well on high-stakes tests).

Markers on the Road...to Ethics Education

Why is ethical behavior so important for people and organizations of all types, especially education institutions? Four possible answers: trust, reputation, credibility, and our ability to get the support we need for educating our students.

We can check out our own ethics in a number of ways. One approach might be to test our decisions or our behaviors against some key markers—*integrity* (Are we who we say we are?), *fairness* (How do we treat other people?), *trust* (Are we honest and dependable?), and *character* (What motivates us?). Together, the answers to these questions help define our *reputation*, and we all know the importance of "a good name."

Widely known as a scholar and civic educator, the late R. Freeman Butts identified key values around the U.S. motto, *E Pluribus Unum.* "Unum" values, he said, include the "obligations of citizenship," such as justice, truth, authority, participation, and patriotism. "Pluribus" values, which he identified as "the rights of citizenship," include diversity, privacy, freedom, due process, human rights, and property.[776]

Michael Josephson, president of the Josephson Foundation, home of *Character Counts*, focuses on the "Six Pillars of Character: trustworthiness, respect, responsibility, fairness, caring, and citizenship."[777] Josephson reminds us that, "Everything you do sends a message about who you are and what you value." He adds, "The ethical person should do more than he is required to do, and less than he is allowed to."[778]

The *marks of a profession* generally include at least three things: a body of knowledge, a language (often with some unique terms), and a code of ethics. In many cases, those codes of ethics are meticulously developed. Among values often mentioned are self control, justice, fairness, courage, and integrity. Charles Kearns, associate professor of applied behavior science at Pepperdine University, also includes *transcendence*, "a recognition that there is something beyond oneself." He warns, "Leaders who are motivated by self-interest and the exercise of personal power have restricted effectiveness and authenticity."[779]

Among principles of personal ethics suggested by Larry Colero of Crossroads Programs, Inc., in his "Framework for Universal Principles of

Ethics," are trustworthiness and honesty, a willingness to comply with the law (with the exception of civil disobedience), refusal to take unfair advantage, benevolence (doing good), and preventing harm.[780] In fact, "Do no harm," is central to the *Hippocratic Oath*, which guides physicians.

In the early 1990s, the Baltimore Public Schools in Maryland developed an expansive list of core values to be considered in driving ethical behavior. Among them were compassion, courtesy, critical inquiry, due process, equality of opportunity, freedom of thought and action, honesty, human worth and dignity, objectivity, rational consent, reasoned argument, respect for other's rights, responsible citizenship, the rule of law, self-respect, tolerance, and truth.[781]

Avis Glaze, president of Edu-quest International in Canada, puts it this way in an article for *ASCD Manitoba*: "If we want a society in which citizens care about one another—in which qualities such as honesty, integrity, fairness, courage, and optimism are pervasive and violence of any kind is discouraged—we have no choice but to nurture these qualities in our homes, in our schools, and in our communities."[782] Glaze urges educators to reach across all diversities, ranging from race and gender to social and economic status and places of origin, to "find common ground on values." That common ground could be enhanced by "character education, building communities of character, and character in the workplace," she concludes.

Allison LaFave of Legal Outreach, a legal education non-profit in New York, urges "the wealthiest quintile of the world to grapple with its social obligations to society." Her concern raises issues of *corporate citizenship* and *social responsibility*, such as "eradicating global hunger and helping improve the quality of life for billions who are poor." While some might attribute their extreme wealth to "a sign of God's favor, dogged determination, and hard work," others acknowledge that their fortune has been built with the help of many who are less fortunate.

In a 1996 study, *Preparing Students for the* 21st *Century*, a council of leaders in business, education, government, and other fields concluded that students needed to develop a sense of civic virtue, grasp the need for a code of ethics, understand and practice the principles of conflict resolution, and exhibit tolerance and acceptance of people who are unlike themselves. Further, the study suggested students must understand the effects of their actions on others as well as the consequences of their own actions.[783]

If we think and care about how our actions will affect our fellow human beings and our environment, we'll be on the road to making better and more sustainable decisions.

> "Power is given to you by others. It is not yours; it is in trust with you and it is a great responsibility. Power is to be used for the benefit of those whose trustee you are."
> *Mahatma Gandhi*

Teaching Ethics

Ethics is often taught through *precept* and *example*. In short, we are generally most successful at delivering lessons in ethics through building an understanding of ethical principles and reinforcing them with real-life stories and role-playing. While individuals and organizations specialize in the teaching of ethics, civic virtue, and character, we wanted to share a few examples directly from the trenches.

The teaching of ethics fits naturally into civic education and character education; however, the scope should be even broader. As students discuss issues in social studies, science, or any number of other classes, the teacher might simply ask, "What are the ethical implications of what we discussed today . . . for us as individuals, for our school, for our community, our country, and for the world?" Then, listen to the discussion.

Getting Students in on the Discussion. In the Tacoma Public Schools in Washington, John Prosser is an instructional facilitator and Ryan Prosser is a math teacher. Both have law degrees, and both teach ethics to 8th grade students. In an *Education Week* Teacher Leaders Network (TLN) article, the Prossers share a few case studies and shed light on underpinnings of how they approach the task.

• One, they note that "ethics is not the teaching of a specific set of beliefs." Instead, it is about "learning to analyze and evaluate beliefs."

• Two, students are already interested and "constantly talk about justice and injustice, right and wrong...fairness of the dress code or whether iPods should be allowed in class." They suggest we "harness this interest for an intentional unit."

• Three, these educators observe that "middle school students are fascinated by morality." Often, ethics is "treated as an off-limits area of expertise 'best left to adults' because 'it's complicated.'" Get students involved in the discussion.

• Four, they observe that "a little controversy can be a good thing." For example, it can "provoke intellectually challenging discussions, present novel ideas, and encourage students to develop and answer meaningful questions." On top of that, it "promotes critical thinking, participation, and self efficacy."[784]

An Ethics Bowl. The Association for Practical and Professional Ethics (APPE) is a catalyst for Ethics Bowls, held in various schools and states as well as nationally. During each of the Bowls, "a moderator poses questions to teams of three to five students. Each team receives a set of ethical issues in advance of the competition, and questions are taken from that set." Topics might include *classroom issues*, such as cheating or plagiarism; *personal relationships*, such as dating or friendship, *professional ethics*, such as engineering, law, or medicine, or *social and political ethics*, such as free speech, gun control, etc."[785] During a Bowl held at Long Island High School, the *New York Times* reported topics ranged from "the morality of using brain-enhancing drugs" to "the consequences of a white lie on a job application."[786]

> "We have a tendency to compare the worst in others with the best in ourselves."
> *Cleric, during Christian-Muslim Summit, Washington National Cathedral, 2010*

What's the right thing to do?
Keeping Ethics Alive in Business Schools

Few decisions are all right or all wrong. Many involve the lesser of two evils. Constantly trying to do the right thing, however, is basic to a civil society.

After fallout from what has seemed like an implosion of ethical behavior, business schools renewed consideration of ethics courses or units. Concerns loomed about "Bernie Madoff-like scandals or insider-trading debacles" and "the irresponsibly risky bets or manipulations of the London interbank offered rate (Libor)." A *Wall Street Journal* article, "Does an 'A' in Ethics Have Any Value?" reported on an Aspen Institute survey indicating that "about 60 percent of new M.B.A. students view maximizing shareholder value as the primary responsibility of a company." Though that might not be a bad goal, writer Melissa Korn notes, "focusing on that at the expense of customer satisfaction, employee well-being, or environmental considerations can be dangerous."[787]

Countering Corporate Arrogance

In the spring of 2002, following the Enron scandal, *The Strategist* magazine ran an article addressing corporate arrogance. Writer Rich Long, professor of communication at Brigham Young University, asked a number of public relations leaders about corporate breaches of ethics. He found them "troubled by what seems like a wholesale disregard for decency and a curious ability to self-destruct." Long concluded, "The common denominator seems to be arrogance, pure and simple."

In the same article, John Paluszek, senior counsel for Ketchum Public Relations, observed, "Arrogance in an organization, not unlike that in a person, is often the product of a culture that proclaims it, alone, has the truth. If that was ever the case, it certainly isn't anymore."[788]

Ethical Issues Facing Today's Students

Can we truly say a student is well educated without a firm grasp of ethical behavior? Probably not. As they move into positions of leadership and take on their role as citizens, students who are in our schools and colleges today will be faced with some of the most monumental ethical dilemmas of all time. A few of the issues they likely will face include:

- A world population expected to grow from approximately 6 billion in 2000 to more than 9 billion by 2050. That's about a 50 percent increase in 50 years. In 2012 about 1.3 billion people were living on less than $1.25 per day.[789]
- Stewardship of the environment and the possible urgency of climate change.
- The need for adequate, accessible, fresh, clean water.

- Use of intellectual or physical enhancements, such as pharmaceuticals.
- The need for greater efficiency in the capture, production, and use of energy.
- The introduction of life forms on other planets...or actual colonization.
- Crime and corruption at every level.
- Governmental gridlock and unwillingness to compromise in meeting the broader needs of society.
- Honesty and transparency in financial and other organizations.
- Attacks by political or religious groups on innocent people to gain dominance, instill fear, and/or capture resources to support their cause.
- Life extension sciences (devoted to extending the length and quality of our lives).
- Genetic research and treatment of genetic disorders.
- Cloning and the growth or building of replacement organs or body parts, in some cases using 3D printing.
- Genetically modified foods (GMFs) and genetically modified organisms (GMOs).
- Ownership of the genome or genetically modified products, such as seed corn.
- Security to protect against clandestine release of manufactured viruses that could impact human and planetary health.
- The use, including possible automation, of drones and other robots for purposes such as combat, surveillance, and traffic monitoring and control.
- Control of chemical, biological, radiological, and other possible weapons that could lead to mass destruction.
- Conceiving of solutions to economic concerns that impact the health and well-being of people worldwide and the condition of the planet.
- Computer and social media ethics, ranging from hacking into private, organizational, or governmental files to cyberbullying, sexting, implanting viruses, and recklessness in the use of

technologies, such as texting while driving.

- Investments that *are* or *are not* made in seeking the prevention and cures of major diseases and avoiding pandemics.
- Clashes among cultures, nations, and sub-national groups as they vie for dominance.
- Violations of basic human rights.

Avis Glaze, observes, "As I grow older, I want to be convinced that the individuals who will be making decisions about my life are doing so from a deep ethical foundation. Education has to focus on educating hearts as well as minds."

Human Rights . . . in Perspective

"The notion of human rights builds on the idea of a shared humanity," according to Mei-Ying Tang of Taipei Municipal University of Education. She adds, "Human rights are not derived from citizenship or nationality but are the entitlement of every human being. In this sense, the concept of human rights is a universal and uniting idea."[790]

Implications for Society
Ethics

- Competition for the world's resources will lead to growing conflict or collaboration in addressing disparities between the *haves* and the *have-nots*.
- Realization will grow that environmental problems cross political boundaries, can threaten all life, and demand solutions—individually, within families and communities, and among nations.
- Governments and other organizations will be expected to be increasingly transparent, openly sharing factual information and their intentions so that citizens can consider them and practice their responsibility as good citizens.
- Despotic or highly authoritarian leaders, especially those who cause harm to people and the environment, will likely be spurned and replaced.
- In general, people will, of necessity, be forced to reflect on the intended and unintended political, economic, social, technological,

and environmental benefits and consequences of any proposal or intended action.

• New and existing technologies will pose increasing threats to personal, corporate, national, and international security. Some technologies that were developed to enhance the quality of life might also be used for destructive, unethical purposes.

• Battle lines will be drawn over intellectual property rights.

• Ethical concern will grow about mass media—their honesty in making a clear distinction between news and commentary, commitment to appropriately defining journalism, and corporate interests balanced against their social responsibility.

• Use of information generated and captured online, including through social media, will raise continuing concerns about identity, privacy, and safety.

• Major scientific discoveries and technological developments will stimulate heated discussion about how they might clash with existing values, beliefs, behaviors, habits, and traditions.

• Violating confidentiality agreements to release classified information.

• Hacking, automated weaponry, and ease in manufacturing viruses that can threaten public health will become even more significant concerns.

• Life-extension sciences, from good health practices to genetic research and pharmaceuticals, will provide benefits and raise concerns, as society ages.

• Expectations will grow that families, schools and colleges, religious organizations, and other institutions will help young people and the whole of society understand and exhibit ethical behavior.

• Industries and professions will be expected to have an established code of ethics that is well communicated and reinforced by periodic training.

• Ethical behavior will be considered a key to justifying the existence of business, government, nongovernmental, education, and other organizations. Investors, consumers, and citizens will expect social responsibility and corporate citizenship.

• In all walks of life, we need to ask a series of questions as we make decisions. Is what I'm proposing: within the law, within policy, ethical, within standards of acceptability in the community, fair, balanced, a step toward consensus, and within our capability? Does it pit one group against another? How will our decision affect others? How will we consider and involve people in our decision making? How will we frame our decision? What kind of example will we be setting? "Is this the right thing to do?"

The High Cost of Corruption

What is corruption? It's often an aberration of ethical behavior, generally involving at least two parties—one who wants to gain undue advantage and another who will provide it for a price. In defining corruption, the State Secretariat for Economic Affairs (SECO) in Switzerland, noted that, "Corrupt practices can range from small favors in anticipation of future advantage to the payment of large sums of money to senior members of governments."[791]

We know that corruption can increase the cost of nearly everything, subvert democratic decisions, deprive people of services that have been guaranteed, suddenly delay scheduled work, divert investments, create massive frustration, intensify disadvantage, and rip away at the moral being of everyone involved.

Implications for Education
Ethics

• **Modeling ethical leadership.** Education institutions, by their very nature, are expected to be paragons of ethical behavior, since they play such a central role in modeling that type of behavior for students. As issues are discussed, administrators, board members, and teachers have an opportunity to make the ethical dimension of their pending decisions a visible part of the process.

For example, school leaders, who are often faced with disparate opinions, are expected to legitimately help their staff and communities find common denominators and reach some level of consensus. A diversity of voices need to be heard, despite social or economic standing or the presence or absence of political influence. Ethically, the bottom line should be drawn at legitimately ensuring equal opportunity for a sound education that will benefit each and every student.

Sometimes, people are reluctant to discuss ethics because of their own perceived shortcomings. It's an unacceptable excuse. No one is perfect. Our imperfections should not stand in the way of helping our students, our organizations, our communities, and even our nations better understand the importance of ethical considerations for our legitimacy, our reputations, and our futures.

• **Including an ethics component in every course.** In any class, daily or weekly, a teacher might present this challenge, "Think about what we've discussed today (or this week). What are the ethical implications for us as individuals, for our school, for our community, for our country, for the world?" Fortunately or unfortunately, the world produces a constant stream of case studies that can trigger discussions about ethics.

• **Expanding programs in thinking and reasoning skills as well as civic and character education.** As students develop their critical and creative thinking skills, they will be better prepared to observe pros and cons as they make choices and solve problems, whether in school, on the job, or as a citizen on the front lines of society. Civic education provides students with a grounding in how the system works and how they might change it, peacefully and democratically. Students are expected to develop appropriate attitudes, behaviors, and skills, and better understand concepts and principles that undergird democracy such as authority, privacy, responsibility, and justice. Character education might focus on areas such as trustworthiness, respect, responsibility, fairness, caring, and citizenship.

• **Dealing with cyber-ethics, as well as physical and intellectual enhancements.** The cyber-world brings its own panoply of ethical challenges that educators will continue to address. Among them are cyber-bullying, identity and privacy concerns, hacking of confidential school district employee or student records, and a host of other issues. Students will need to develop media literacy skills to help them separate truth from fiction as they scan the cyber world and make connections using social media.

Schools and colleges have dealt with the issue of performance-enhancing drugs, primarily in athletics. Future policy issues, as they are recognized, may involve, among other things, special treatment for students whose parents can afford fMRI brain scans for their children and counsel for their teachers on how to motivate them in school. Pharmaceuticals might be used as "intelligence enhancements" or as a way to focus.

• **Getting beyond arrogance and moving toward empathy.** Too often, people willfully do things to benefit themselves at the expense of many others. A lack of empathy and compassion breeds arrogance. Education systems should model concern for others as they make decisions. They should also commit to instilling that type of conscience in others. Doing so is a challenge, but this crucial effort can be reinforced by parents and community, if they are willing to join us in a broader commitment to develop an even more civil society.

• **Additional Implications of Ethics Trend.** We need to answer the question: What is the school's role in teaching *about* ethics? Professional development will be needed to build a capacity to teach about ethics. Not everyone has developed the knowledge and skills to effectively teach an ethics course but many can include an ethics component across all disciplines. Schools and colleges are expected to take the higher ground, modeling ethical behavior, developing codes of ethics, and providing counsel on ethical dilemmas. All need to accept that ethical skills are fast becoming survival skills.

Dilemmas and Moral Questions

James Lukaszewski, a public relations consultant who often addresses the handling of crises, suggests ethical dilemmas will prompt a number of questions. In the February 2002 issue of *Tactics*, he presented several "moral questions you can use to assess appropriate ethical behavior." Among them are:

• What did they know and when did they know it?
• Has all the information been presented honestly and correctly?
• What are the relevant facts of the situation?
• What decisions were made?
• Who was involved/affected?

- What alternative actions were available?
- Are our actions open, honest, and truthful?
- What affirmative action is being taken now to remedy the situation?
- How could this have been avoided?
- Are all the critical questions being asked and answered?[792]

"The Morning Question: What good shall I do this day? The Evening Question: What good have I done today?"

Ben Franklin's Two Daily Questions[793]

Ethics

Questions and Activities

1. Briefly list five reactions to this statement, "Arrogance—the sense they can do no wrong and can get by with anything, coupled with a lack of empathy—gets more people and organizations into trouble than nearly anything else."

2. How can schools and colleges teach about ethics using the premise that the subject is best addressed through precept and example? Develop the equivalent of a one-page summary including at least three of your basic recommendations.

3. Do a 3- to 5-minute PowerPoint presentation that includes up to five "musts" for any organization that hopes to maintain its ethical standing in the broader community.

4. View a television interview with Michael Josephson of the Josephson Institute, aired on Dec. 30, 2012, on KNBC in Los Angeles, (8:10). Be prepared to share at least two of your observations with others. http://josephsoninstitute.org/michael/.[794] If this interview is no longer available, please use your search engine to locate other Josephson possibilities.

5. List 10 reasons you believe people and organizations too often use bad judgment.

6. What are some of the great ethical issues today's students will need to address during their lifetimes?

7. Lead or participate in a discussion of why students need to understand and adhere to ethical codes.

Readings and Programs

1. Read the article, "Character Development: Education at its Best" by Avis Glaze, published in *ASCD Manitoba Journal Reflections*, The Whole Child, Summer 2011, Volume 11, pp 18-23, http://www.mbascd.ca/documents/ Journal2011_Final.pdf.[795]

2. Find information, ideas, and support materials for teaching about ethics at the following web sites: Institute for Global Ethics, http://www.globalethics. org/; Josephson Institute, http://josephsoninstitute.org/ and http://charactercounts.org/sixpillars.html; Santa Clara University Markkula Center for Applied Ethics, http://www.scu.edu/ethics/; Ethics Resource Center, http:// www.ethics.org/.[796]

3. View "Gordon Brown on Global Ethics vs. National Interest," TED Global 2009, interview conducted by TED Curator Chris Anderson, http://www. ted.com/talks/gordon_brown_on_global_ethic_vs_national_interest.html (17:10).[797]

4. As an option, consider *Plato's Ethics: An Overview*, Stanford Encyclopedia of Philosophy, revised May 29, 2009, http://plato.stanford.edu/entries/plato-ethics/.[798]

5. Amundson, K.J. (1991). *Teaching Values and Ethics*. Arlington, VA: American Association of School Administrators.[799]

Chapter 18

Continuous Improvement
You ain't seen nuthin' yet.

Chapter 18: Continuous Improvement.
You ain't seen nuthin' yet.

Trend: The status quo will yield to continuous improvement and reasoned progress.
Quick Fixes/Status Quo → Continuous Improvement

> "A bad system will beat a good person every time."
> *W. Edwards Deming*

Good Today...
Even Better Tomorrow!

Resistance to something new is likely as old as humanity itself. However, in a world of quantum physics and exponential change, simply digging in our heels and passionately defending the status quo won't cut it. Remember the 4-H Club motto: "Good. Better. Best. Never let it rest, 'till your good is better and your better is best."

Let's face it, if we were perfect yesterday, we probably aren't perfect today, because the world changed overnight. Two questions: "Are we committed to continuous improvement?" and "Are we flexible enough to adjust?" The exhortations are often dramatic, such as, "Let's leave well enough alone;" or "We don't have a dog in this fight." That dog may come back to bite us if the world is changing and we're not taking the situation seriously.

Continuous Improvement

People today expect and demand quality, effectiveness, and service. In the past, an organization might get by with defending the status quo or going for the quick fix—the band aid approach. Now, in an impatient world, people want products and services that work, meet their needs, and are delivered on time.

A stark reality. If we continue to unbendingly do what we've always done, we'll likely get to the top of our game, then cycle out into oblivion. If we hope to stay ahead of the curve, we need to constantly reshape what we do and how we do it to meet the needs of a fast-changing world. Continuous improvement might even be considered a *moral imperative*, especially for institutions committed to serving others, such as government, business, and education.

Counsel Rowan Gibson describes leaders we'll need to shape our future in a world that expects constant progress and supreme quality. Those leaders will be looking forward, scanning the landscape, watching the competition, spotting emerging trends and new opportunities, avoiding impending crises. They will be "explorers, adventurers, and trailblazers." He calls for a "hierarchy of imagination" and "a democratizing strategy involving a rich mixture of different people from inside and outside the organization in the process of inventing the future." [800]

"It's not enough to stare up the steps.
We must step up the stairs."
Vaclav Havel

The Quality Scene
PDCA, LEAN, Six Sigma, Kaizen...
• **Deming's Principles.** The principles of quality improvement, unleashed by statistician and consultant W. Edwards Deming, were summarized in his classic book, *Out of Crisis*. Included are: developing a constancy of purpose toward improvement; stopping dependence on inspection by building quality into the product in the first place; improving the system of production and service; ensuring that leadership helps people do a better job; breaking down barriers among departments; substituting leadership for standards; getting rid of barriers that rob people of their pride of workmanship; changing responsibilities to reflect quality rather than sheer numbers; offering on-the-job training and opportunities for education and self-improvement; and driving fear from the organization. [801] Deming's initial claim to fame was the revitalization of Japanese industry following World War II.
• **The PDCA Cycle.** The American Society for Quality (ASQ) offers tools, ideas, and expertise to "make the world work better." Among many ASQ recommendations is to make use of the Deming, Shewhart, PDCA Cycle. PDCA goes like this. **Plan:** Identify an opportunity and plan for change. **Do:** Implement the change on a small scale. **Check:** Use data to analyze the results of the change and determine whether it made a dif-

ference. **Act:** If the change was successful, implement it on a wider scale and continuously assess your results. If the change did not work, begin the cycle again.[802]

• **LEAN.** The LEAN principles of continuous improvement grew from both Deming's work and experiences of the Toyota Production System. Its focus is on eliminating waste and creating value for the customer. Waste, according to LEANING Forward might include: over-production, inventory, transport, processing, idle time, operator motion, and bad quality.[803]

• **Six Sigma Quality.** This process is used fairly widely but has especially been a part of General Electric's continuous improvement efforts. "Today's competitive environment leaves no room for error," GE explains. "We must delight our customers and relentlessly look for new ways to exceed their expectations," with "as close to zero defects as possible." Key concepts of Six Sigma include: *Critical to Quality*, attributes most important to the customer; *Defects*, failing to deliver what the customer wants; *Process Capability*, what the process can deliver; *Variation*, what the customer sees and feels; *Stable Operations*, ensuring consistent, predictable processes to improve what the customer sees and feels; and *Design for Six Sigma*, deliberately designing to meet customer needs and process capability.[804]

• **Kaizen.** Kaizen, used to control waste and improve efficiency, is said to have been created as part of Toyota's production system.[805] Among its principles are: "Throw out all your old fixed ideas on how to do things; no blame—treat others as you would want to be treated; think positive—don't say can't; don't wait for perfection—50 percent improvement now is fine; correct mistakes as soon as they are found; don't substitute money for thinking or creativity; and keep asking *why* until you get to the root cause."[806]

Caution on the Road to Continuous Improvement

We live in a world of big data. In fact, it may take so long to analyze and do something about it that, by the time we produce the *perfect* product or service, our customers or clients have already moved on. It's not easy staying ahead of the curve. In the private sector, the verdict often shows up in sales reports and the success of competitors. In the public sector, it can appear as people who are dissatisfied with the services they're getting and demanding that the system change.

Ron Ashkenas raises the issue in a piece he wrote for *Harvard Business Review*. He quotes thinker Vijay Govindarajan, who observes,

"The more you hardwire a company on total quality management, (the more) it is going to hurt breakthrough innovation." While supporting the idea of continuous improvement, he points out that improving an existing process and "the whole culture that is needed for discontinuous innovation are fundamentally different." In essence, if we continue to even more perfectly do what we've always done, ignoring what's going on outside the trenches, we'll likely become insignificant in a fast-changing world. That is sometimes called *doing the wrong things right*. He adds, "Sure it's important to inject discipline into product and service development, but not so much that it discourages creativity." [807]

Consider this. Offshoring critical components can grow directly from *the efficiency and bang-for-the-buck side* of the house, especially if it can override *the scientific and operational side*. In a leadership article for *Forbes*, contributor Steve Denning passes along some thoughts developed by Gary Pisano and Willy Shih for a *Harvard Business Review* piece titled, "Restoring American Competitiveness." Pisano and Shih observed, "Once manufacturing is outsourced, process-engineering expertise can't be maintained, since it depends on daily interactions with manufacturing. Without process-engineering capabilities, companies find it increasingly difficult to conduct advanced research on next-generation technologies. Without the ability to develop new processes, they find they can no longer develop new products." [808] Cue the requiem.

A Few More Thoughts on the Matter
Collins, Marty, Ouchi, and Peters

A few thoughts from prominent management/leadership consultants, who would likely agree that processes are important but they don't substitute for future-focused leadership. Jim Collins, in an article for *CNN Money*, discusses companies and leaders who do well in navigating a demanding, fast-changing, often chaotic world: "They don't merely react; they create. They don't merely survive; they prevail. They don't merely succeed; they thrive. They build enterprises that can endure." [809]

Keith Marty, superintendent of the Parkway Schools in Missouri, sees further improvement in quality with the collapse of various silos that are "replaced by new partnerships, collaboratives, and/or cooperatives." With limited resources, school districts may need to form legitimate partnerships with other institutions. Coming up short of resources "will force a new way of thinking about organizing teaching and learning." An irony: While education is facing budget concerns, expectations for "student growth and achievement are becoming much more pronounced."

William Ouchi, in his classic work, *Theory Z,* concluded productivity could be enhanced by "managing people in such a way that they can work together more effectively." He described "quality circles" that brought front-line people together to discuss how they could constantly improve both process and product. He contended that people, ranging from managers to manual and clerical workers, should be able to say with conviction, "This is the best place I've ever worked. They know what they're doing here, care about quality, and make me feel part of one big family."[810]

In *A Passion for Excellence . . . The Leadership Difference*, noted leadership consultant Tom Peters stepped into the fray. He declared in this now classic work, "Quality is not a technique, no matter how good." Instead, he said, "Quality comes from people who care and are committed."[811]

Recognition Programs for Quality

Principles of continuous improvement have been impacting policies and decisions in many organizations, including business, government, and education. One of the more prominent recognition programs in this overarching field is the Malcolm Baldrige National Quality Awards, under guidance of the National Institute of Standards and Technology (NIST) and the American Society for Quality (ASQ). These awards recognize a variety of organizations for their ongoing improvement efforts.

• **Baldrige Awards, 2012.** In 2012, recipients included Lockheed Martin Missiles and Fire Control of Grand Prairie, Texas (manufacturing); MESA Products, Inc., Tulsa, Oklahoma (small business); North Mississippi Health Services, Tupelo, Mississippi (health care); and the City of Irving, Irving, Texas (nonprofit).[812]

• **Baldrige Education Awards.** Baldrige criteria for schools or schools systems, colleges, universities, or other education institutions include: student learning outcomes; customer satisfaction and engagement; product and service outcomes, and process efficiency; workforce satisfaction and engagement; budgetary, financial, and market results; and social responsibility.[813] In 2010, a Malcolm Baldrige National Quality Award in Education went to the Montgomery County Public Schools (MCPS) in Maryland for what was called "a comprehensive reform initiative and continuous improvement process that has solidified MCPS as one of the highest-performing school systems in the U.S." Noted were its data-driven decision making; its efforts to close the achievement gap, especially in light of rapidly changing demographics; its approach to distributed leadership in a spirit of collaboration and respect; and its graduation rate, which was "the highest of any large public school system in the nation." [814]

Drivers of Continuous Improvement for Education and Other Institutions

We live in an age of the end run. If we don't think an institution is meeting our expectations, we'll very likely run around it, through it, or over it. Some basic concepts should serve as *drivers* for any organization that wants to be considered as earning its place in the future. Each is firmly connected to continuous improvement.

• **Unmet demands for improvement create a market niche for someone else.** Loyalty is earned every day. If constituents perceive that their expectations are consistently not being met, then we are creating a vacuum for someone else to fill.

• **Improvements might be imposed.** If expectations aren't being met, government might step in to impose *improvements*. In education, for better or worse, they might include more strict standards, high-stakes tests, and funding schemes.

• **Maintaining an initiative for renewal.** As we move into a Global Knowledge/Information Age, all of our institutions, including education systems, will be renewed. We face a choice in this process. On the one hand, educators can initiate the process of renewal. On the other hand, if they don't, someone else will, and simply announce to educators how the new system will work—whether it's workable or not. *Bottom line*: The process of creating a future simply must be ongoing.

Taking the Higher Ground
Identify what you want to become.

How we do things makes a big difference. If we take the whole concept of democracy seriously, we will consider the views of those who'll be affected by our decisions. That's our ticket to higher ground, greater respect, and even more substantial ownership for the decisions we make. If an organization's decisions don't make it easier for people to get an even better product or service on time, they'll either turn to the competition or *become* the competition.

We might get the process under way with Community Conversations, Futures Councils, a series of listening sessions, meetings, and questionnaires. We should constantly keep an ear to the ground, online and offline. Using gap analysis, we might ask people to describe characteristics of the organization we need in a fast-changing world. For an education system, that would mean identifying ideals for schools or colleges that are capable of preparing students for the future. Staying in touch can help us consciously and seriously listen to the wisdom of others and lead us up the road toward continuous improvement. In turn, that can lead to regular enhancements in how we get things done, higher levels of effectiveness, and greater satisfaction. Unless we're willing to make change evolutionary, it might very well become revolutionary.[815]

Implications for Society
Continuous Improvement

Continuous improvement should be central to any organization, community, or nation if it hopes to justify its place in society and in the marketplace. The following are some overarching considerations.

- **Questions will be raised about our progress in a fast-changing world.** Some of those questions might be: Will new technologies wait for our values to catch up? Will some communities, organizations, and countries become more adept at making needed changes than others? What are the consequences (or benefits) for those who reject progress? Will naysayers be treated as negative, wise, or inconsequential? What about the side-effects and unintended consequences of our actions or lack of actions? What are the implications for education if we do or do not address the need for public engagement, civility, and constructive progress?

- **Meeting the competition, whether we like it or not, will depend on quality.** International competition will drive a demand for products and services that are high quality, safe, useful, stylish, and priced so that people in an increasingly global marketplace can afford them. Community development councils will continue to insist on adding value for both the economy and quality of life.

- **A commitment to continuous improvement *should* become basic, but the system *should not* stymie imagination, creativity, invention, and innovation.** Whatever *system* we choose, as we consider systems theory, continuous improvement must be an essential part of it. Process improvements will constantly be needed to maintain reasonable costs for goods and services, boost efficiency, and lead to just-in-time delivery in the supply chain. Efforts to achieve "zero defects" in our processes should not blindside us to evolving wants and needs. Again, to freeze is to fail.

- **Public and private institutions will be expected to constantly create a future.** Communities and most organizations that cling to the status quo will likely lose creative people, who will seek greater civic and job opportunities elsewhere. They will expect a place at the table as decisions are made about process, quality, and usefulness.

- **Citizens, customers, and clients will expect outstanding service.** While all will insist on product quality, they will generally develop zero tolerance for hassles and inefficiency and insist on knowing what value a product or service will add to their lives.

Implications for Education
Continuous Improvement
All education institutions should reflect a spirit of constant renewal, exude positive and constructive energy, get their communities and staffs

on board, seriously listen to the wisdom that surrounds them, and offer outstanding future-focused professional development. They will need to work with staff, governance, and community to turn blocking cultures into enabling cultures, and maintain flexibility. Futures Council 21 Member Gary Rowe looks for "teachers who are always talking about new things, new ideas, and new concepts in the disciplines they are teaching, then actively sharing information with each other."[816] In that spirit, here are a few more implications for schools and colleges.

• **Education institutions will be expected to be vibrant, flexible, and effective.** Leading educators and education systems stay in touch with what their students will need to know and be able to do in the future. Expect a premium on creativity, thinking and reasoning and problem solving skills, and a willingness to consider a constellation of points of view. People are generally better educated and expect more. "Educators cannot let the system stiffen or become atrophied in our fast-changing world."[817] Governing boards, administrators, teachers, and communities need to be key players in helping the system become even better tomorrow than it is today, no matter how great that might be.

• **A commitment to continuous improvement should add intellectual energy and value to the system and everyone in it.** No education system will be able to safely say that it has "arrived." The needs of society and the possibilities for education multiply, meaning satisfaction will come from staying ahead of the curve, not kicking off our shoes and declaring we're the best. Randy Johnson, a Colorado school administrator, warned in a 2001 article carried by *Education Week* and titled "The Zen of Quality," that continuous improvement is a process, not a philosophy. "It can keep the ship in tip-top shape, floating confidently in a turbulent sea of change. It does this by changing the ship to meet the changing sea. But it won't take the ship anywhere. The process is made just to keep the ship floating."[818]

• **Quality and continuous improvement will find their way into the classroom.** To be prepared for a demanding, complex, but invigorating world, students will need to understand and get hands-on experience with principles of quality. In the classroom, as self-motivated learners, they will be able to enthusiastically point out how their lessons and their personal goals are aligned with the aims of the school system. They will be adept at monitoring their progress, conferring with other students who may need help or who might be able to help them, and clearly discussing what they are learning with parents and others in the community. Active learning will pump energy into the classroom, making it a place where "what we learn is connected to the real world."

- **Earning the opportunity to serve.** Organizations take up space. They are there because they've accepted responsibility for providing what they hope will be an essential cluster of services. They create value. With lightning-speed communication, however, competition has grown. Others may also want to occupy that space. That's one more reason why any organization, including an education system, must constantly prove to its constituencies that it is providing value and earning the opportunity to serve.

True Story

"How can we tell people that we're going to improve? Isn't that a direct admission that we aren't as good as I've been telling people we are?"

That concern was voiced in response to a proposed theme, "Getting Better for Kids." It reflected some solid goals a local school system developed to elevate student achievement and start building support for a finance election in less than a year.

After the theme was adopted and rang throughout the community, the response was terrific. One businessperson said, "I've lived here all my life, and this is the first time any public institution has admitted it can do even better. I like it!" We identified mutual expectations and demonstrated how our proposal would help meet them.

On election day, "yes" votes outnumbered the "no."

"Getting Better for Kids" was not just a slogan. Its substance fast became a reality that reflected the culture of the school system and its community.

Continuous Improvement

Questions and Activities

1. Prepare a one-page strategy for how education systems might be able to teach the principles of continuous improvement to students.

2. Review *LLIS.gov*, Lessons Learned by the Federal Emergency Management Agency, in an attempt to constantly improve its emergency services, such as community preparedness, school safety, and mass evacuations. Prepare a one-page report on "Three Things I Observed in Reviewing the Report," http://www.fema.gov/content-offered-llisgov. [819]

3. Develop a five- to ten-minute presentation based on the assumption, "If constituents perceive that their expectations are consistently not being met, then we are creating a vacuum for someone else to fill."

4. What do you think leadership expert and author Rowan Gibson means when he says we should create a "hierarchy of imagination" and "a democratizing strategy involving a rich mixture of different people from inside and outside the organization in the process of inventing the future?"

5. What are W. Edwards Deming's principles that guide the quality movement?

Readings and Programs

1. Gibson, R. (2002). Rethinking Business. *Rethinking the Future.* London: Nicholas Brealey.

2. Deming, W. (1991). *Out of Crisis.* Cambridge, MA: Massachusetts Institute of Technology, Center for Advanced Engineering.

3. Baldrige National Quality Program. Education Criteria for Performance Excellence. Washington, DC: National Institute of Standards and Technology (NIST), American Society for Quality, http://www.quality.nist.gov and www.asq.org

4. Review "Malcolm Baldrige National Quality Award, Montgomery County Public Schools (MCPS), 2010 Award Recipient, In Pursuit of Excellence," http://www.montgomeryschoolsmd.org/uploadedFiles/info/baldrige/homepage/Baldrige-Pursuit-of-Excellence.pdf, and view "Call to Action," an explanatory video program prepared by MCPS, http://www.montgomery-schoolsmd.org/info/baldrige/about/overview.aspx. [820]

Chapter 19

Poverty
Poverty makes us all poor.

Chapter 19: Poverty. *Poverty makes us all poor.*

Trend: Understanding will grow that sustained poverty is expensive, debilitating, and unsettling.
Sustained Poverty ↔ Opportunity and Hope

> **"Mama may have, papa may have, but God bless the child that's got his own."**
> *Billie Holiday and Arthur Herzog, Jr.*[821]

The Many Faces of Poverty

How do we measure poverty? Generally, we compare levels of income. We declare that people who make less than a certain amount of money are poor. In a quantifying society, that seems reasonable to most of us, and it probably is. However, poverty is deeper than numbers on a well-designed chart. It is a condition that impacts the lives of real people.

By 2012, when the Great Recession seemed to be winding down, 22 percent of all children under 18 in the U.S. — approximately 16 million—were living in poverty.[822] Raise your hand if you would like to walk a mile in their shoes. Imagine feeling trapped in the grip of social and economic peril with only rumors of *opportunity.*

One worrisome concern that should command our attention is the persistence of *sustained poverty.* We find it in our individual communities, in our nations, and worldwide. It's a situation that almost seems to be inherited. From generation to generation, it increases exponentially. For too many, except the most robust, poverty can drain the human spirit and weaken the body. It can become devastating as *hope* becomes a long forgotten dream.

While some people are born into poverty, others are limited by discrimination or a lack of educational opportunity. Still others might have been marginalized by a changing or softening economy, downsizing, mistakes, foreclosure, bankruptcy, the loss of a job, illness, an accident, low expectations, or an absence of mentoring and support.

We're not finished yet. There are other concerns. Many can be interrelated and quickly grow exponentially—one on another. Consider dilemmas such as: a lack of adequate and affordable health care; limited access to a predictable, healthy food supply or diet; low birthweight babies and inadequate prenatal care; family stress or breakup; and a vast array of other social and economic conditions. David Berliner at Arizona State University calls some of these conditions "Out-of-School Factors. (OSFs)."[823]

Eugene, Oregon, Superintendent Sheldon Berman believes an ongoing challenge for education and the whole of society will be dealing with a division between wealth on the one hand and poverty and injustice on the other.

> **"Many parents lose the choice of spending more time with their children because they often have to work at more than one job to make ends meet. What used to be a choice is now a necessity. We commonly use the term 'latch key kids.'"**
> *Avis Glaze, president of Edu-quest International, Canada*

What is poverty? The definitions of *poverty* seem too clinical—"the state of one who lacks a usual or socially acceptable amount of money or material possessions" or "debility due to malnutrition." *Destitution*—"the state of a person who has insufficient resources." *Indigence*—"seriously strained circumstances." *Penury*—"a cramping or oppressive lack of money."[824]

The Working Poor: Pulitzer Prize-winning author David Shipler writes in *The Working Poor . . . Invisible in America* about those who labor every day but simply can't break the bonds of poverty. His examples include: "The man who washes cars but doesn't own one. The clerk who files cancelled checks at the bank has $2.02 in her account. The woman who copy-edits medical textbooks who has not been to a dentist in a decade." Shipler notes that the working poor are "shaped by invisible hardships." He describes people who are "climbing out of welfare, drug addiction, or homelessness" or "trapped for life in a perilous zone of low wage work." Some of their children suffer from asthma, made worse by their crumbling housing, or can't see the chalkboard at school because they need eyeglasses that they can't afford, Shipler says.[825]

Global Concerns. In 2010, the World Bank estimated that 1.2 billion people on our planet were living in *extreme poverty*—under $1.25 a day—and nearly 2.5 billion were living on less than $2 a day.[826]

Millions worldwide have been limited by economic and social calamity. Some are born in a place where the land is poor and rainfall is either scarce or comes in massive floods. Hordes of people end up dealing with political repression or warfare, deprived of education that might give them a sense of power to overcome. Some are caught in long-festering or spontaneous revolutions, fomented by religious differences, tribal feuds, or simply the desire of an autocrat to stay in power. One day, they are living a settled life, and the next day their home is destroyed, their lives are threatened, and they join the ranks of refugees. Too many suffer exclusion, forced detention, brutality, and atrocities within their family or community. Whatever the reason, frustrations are universal, and they multiply when the golden ring is always slightly out of reach.

Jim Yong Kim, appointed in 2012 as World Bank president, told *The Guardian* newspaper that he is "passionately committed to ending absolute poverty, which threatens survival and makes progress impossible for the 1.3 billion people (who were, at the time) living on less than $1.25 a day."[827] While poverty and hunger remain highly important issues, "fewer people live in *extreme poverty*."[828] The proportion "of people living on less than $1.25 a day fell from 43.1 percent in 1990 to 22.2 percent in 2008," the Bank reported in *World Development Indicators 2012*.[829]

As the world population expands, poverty will multiply, in some cases, geometrically. Unless it is addressed, poverty can tear away social cohesion. For many, crisis has already become catastrophe. The problems posed by sustained poverty will have profound implications for our individual and collective futures. (Note for Readers: Statistics included reflect several individual years, which generally accounts for variances.)

More Subtle Types of Poverty...
Attitudes and Behaviors, A Poverty of Caring

Other types of poverty? They are not quite so easy to add up on a calculator. However, they demand our attention. They include the absence of a sense of urgency, a paucity of ideas for dealing with the problem, ignorance, a resistance to caring about others who may be less fortunate, a tendency to weigh the relative political strength of those who are caught in poverty, and a lack of courage to stand up for the cause. Some even enjoy a sense of superiority that comes with being "better off" than someone else. While some are born into extreme poverty, others are born into extreme wealth. Occasionally, people isolate themselves from the problem, convinced that it doesn't affect them. A few even complain about "entitlements" that assist those who have fallen on hard times. On the other hand, they refer to benefits they and their industries receive, such as pensions, health insurance, tax-breaks, rebates, or bonuses as simply "incentives."

Too often, we turn a blind eye to the lost potential that surrounds us. Sometimes, we observe people in need and simply say, "They need to pull themselves up by their own bootstraps." A point that often gets lost is that many who are poor do not have bootstraps and can't afford to buy them. In fact, all of us, from time to time, need a hand-up, help, a boost, to get from one rung of a ladder to the next. If we hope to have a future, then we need to be sure those who are coming up behind us can continue, with a sense of dignity, the climb toward even more productive and fulfilling lives. We need to remember that if one person is poor, we are all poorer for it.

Figure 19.1

Children Below Poverty Level in the U.S., 1980-2010, U.S. Census Bureau[825]			Worldwide, Percent of People Living on Under $1.25 a Day[826]	
Year	Number of Children Below Poverty Line	Percent Below Poverty Line	Year	Living on Under $1.25 a Day (%)
1980	11,114,000	17.9	1981	52.2
1990	12,715,000	19.9	1984	55.2
1995	13,999,000	20.2	1990	43.1
2000	11,005,000	15.6	1996	34.8
2005	12,335,000	17.1	2002	30.8
2010	16,286,000	22.0	2008	22.2

Information drawn from the U.S. Census Bureau and the World Bank, 2011-2012.

Poverty and Education
Socioeconomic Gaps Too Often Equal Achievement Gaps

During three decades, from 1980 to 2010, poverty rates in the U.S. have swung in cycles but remained too high, especially for a nation with vast resources. Those rates varied from 15 to 22+ percent. But there is more to the story. According to an Urban Institute (UI) article for *Research of Record*, Caroline Ratcliffe and Signe-Mary McKernan observe that, "49 percent of children who are poor at birth go on to spend at least half their childhoods living in poverty."[832]

The Cost of Neglect. For too many students, poverty has a way of impacting school achievement. It shows up in standardized test scores, student grades, dropout rates, and both college entrance and completion. The Urban Institute adds, "Homelessness, poor health, hunger—poverty's consequences—can be severe. Growing up in poverty can harm children's well-being and development and limit their opportunities and academic success. And poverty imposes huge costs on society through lost productivity and higher spending on health care and incarceration."[833] When a student is lost because of poverty, we all lose. The cost of neglect is expensive.

Poverty among children has not been spread evenly. According to the *2013 Kids Count Data Book*, more than one-fifth of all children (23 percent) were living in poverty during 2011. Here are percentages of children in various racial/ethnic groups who were living in poverty: African American, 39 percent; Native American, 37 percent; Asian or Pacific Islander, 15 percent; Hispanic, 34 percent; non-Hispanic White, 14 percent; and children of two or more races, 24 percent.[834] "In 2010, the U.S. Census Bureau reported that (on the whole) White children made up the majority of total children in poverty (53.6 percent)."[835] The percentages of children living below the poverty line increased between four and five percent when the Great Recession kicked in during 2007-08.[836]

Considering the impact of the Great Recession on children, the *2013 Kids Count Data Book* shared a number of trend lines.

- *Children in poverty* increased from 19 percent in 2005 to 23 percent in 2011.
- *Children without health insurance* declined from 10 percent in

2008 to 7 percent in 2011. *Low-birthweight babies* went from 8.2 percent in 2005 to 8.1 percent in 2011.

- *Children not attending preschool* dropped from 56 percent during the 2005-07 period to 54 percent during 2009-11.

- *Children living in single-parent families* rose from 32 percent in 2005 to 35 percent in 2011, while *children living in high-poverty areas* rose from 9 percent in 2000 to 12 percent during the period of 2007-11.[837]

A soft economy raised the hurdles for educators trying to move the needle on student achievement. In an article titled, "Whose Problem is Poverty?" for ASCD's *Educational Leadership*, Richard Rothstein declared, "Closing or substantially narrowing achievement gaps requires combining school improvement with reforms that narrow the vast socioeconomic inequalities in the United States." Rothstein is research associate for the Economic Policy Institute. He went on to say, "Without such a combination, demands that schools fully close achievement gaps not only will remain unfulfilled, but also will cause us to foolishly and unfairly condemn our schools and teachers."[838]

> **"For every talent that poverty has stimulated, it has blighted a hundred."**
> *John Gardner, former U.S. Secretary of Health, Education and Welfare*[839]

Consequences for Students and Education. A 2012 report from the National Center for Children in Poverty (NCCP) pointed out, "Nearly 16 million children in the U.S. lived in families with incomes below the federal poverty level." NCCP shined a light on some of the impacts: low wages; unstable employment and trying to make ends meet; children's ability to learn; social, emotional, and behavioral problems; and poor health, including mental health.[840]

That child who comes to school after being raised in poverty very likely did not benefit from the quality of health care and nutrition available to other children. The impact can last a lifetime. Other disadvantages, right from the start, likely include fewer learning resources at home, negative stereotyping, placement in lower tracks or ability groups, retention, an anti-school attitude and value system, and test bias, according to the U.S. Department of Education. In some cases, these children might have less qualified teachers in their classrooms, or their highly dedicated teachers might be frustrated by a lack of respect or parental involvement.[841] While students who suffer from poverty may get good grades and have heroically overcome the odds to earn them, some studies have shown that those marks are occasionally based on the tragedy of lower expectations.[842]

We know that socioeconomic factors that trigger poverty can have a significant impact on a child's performance in school. Eric Jensen, in his ASCD book, *Teaching with Poverty in Mind*, notes that many of these children "are faced with overwhelming challenges that affluent children never have to confront, and their brains may be adapted to suboptimal conditions in ways that undermine good school performance."[843]

A Rand Corporation study, "A Matter of Class," conducted in Los Angeles, found that a mother's level of education, often coupled with poverty, is an important factor in children's school achievement in math and reading. The authors especially noted the importance of library visits, of learning some basic skills early, and "understanding the learning process and how to help their kids develop the skills they need."[844] In 2012, the *American Economic Review*, in an article by Gordon Dahl and Lance Lochner, reported that "a $1,000 increase in income raises combined math and reading test scores by 6 percent of a standardized deviation in the short run."[845]

A Poverty, Equal Opportunity Crisis. Ensuring equal educational opportunity for all, despite social, economic, and other factors, is an ongoing quest. Attention to this critical issue heats up and then wanes. At the federal level, the historic Elementary and Secondary Education Act (ESEA) focuses on the needs of disadvantaged students. Offshoots of ESEA, such as No Child Left Behind and Race to the Top, in their own ways, have been intended to be an extension of that effort. In an *Education Week* Commentary, Michael Rebell and Jessica Wolff of the Campaign for Educational Equity at Teachers College, Columbia University, declare, "America does not have a general education crisis, we have a poverty crisis. Results of an international student assessment indicate that

U.S. schools with fewer than 25 percent of their students living in poverty rank first in the world among advanced industrial countries. But when you add in the scores of students from schools with high poverty rates, the U.S. sinks to the middle of the pack."[846] Are we truly ready to address the equal opportunity crisis, even if we are among the advantaged?

NEA Today pinpointed another critical issue—early learning. The publication reported, "When a panel of policy advocates and academics met on Capitol Hill in Washington, D.C., during April of 2012, one spotlight fell on early childhood education." David Sciarra, executive director of the Education Law Center in Newark, N.J., remarked, "We do have a responsibility to build a system of public schools that address poverty needs as soon as the students walk through the door." The group noted a National Institute for Early Education Research (NIEER) report revealing that "funding for Pre-K programs had plummeted by more than $700 per child nationwide over the past decade." That report noted, "Overall, state cuts to pre-K transformed the recession into a depression for many young children." Peter Edelman of the Center on Poverty, Inequality, and Public Policy called for "launch of a full-scale attack on poverty."[847]

"At a minimum, school funding needs to be equalized so that students living in poverty are not shortchanged in school services," remarks James Harvey, executive director of the National Superintendents Roundtable. He hopes to see more "comprehensive wrap-around services, provided through the schools, such as early openings and late closings so that children can be provided with breakfast, lunch, and (if need be) dinner." In addition, in the quest to overcome poverty among children, Harvey calls for "health screenings on school property, adult education programs to attract parents to school, and one-stop social services for families." He adds, "The school building should be a locus for community development."

What do we have to lose?

"Why should we be concerned about sustained poverty?" "After all, the poor will always be with us." As facetious as questions and comments sometimes are, they deserve our thoughtful response.

The answers are glaring: lost talent and productivity; human frustration; increased welfare and other subsidies; crowded shelters and soup kitchens; the need to rely on food stamps; more gang activity, violence, and self-destructive behaviors; fuller jails; expansion of physical and mental health concerns; compromised performance in school; and a demonstrated lack of civility toward those in need.

War Declared…A Perspective

In 1964, U.S. President Lyndon Johnson declared a war on poverty, "because it is right, because it is wise, and because, for the first time in history, it is possible to conquer poverty." That effort was strengthened by the Voting Rights Act of 1965. Also in 1965-66, Project Head Start got under way in answer to a call for "a comprehensive child development program that would help communities meet the needs of disadvantaged preschool children."

In 2011 alone, more than 1.3 million adults volunteered to work in Head Start programs, providing life enriching services to children and their parents.[848] That same year, the National School Lunch Program provided "low-cost or free lunches to more than 31 million (low income) children in public and non-profit private schools and residential child care institutions." That program was enacted in 1946. In 2010, the Healthy, Hunger-Free Kids Act set up guidelines for more nutritious food, beginning with school year 2012-13.[849]

Seeds of the war on poverty had been planted much earlier and started taking root in 1954, when *Brown v. Board of Education* reflected a growing response to unrest. It turned out to be a bold step toward a more just society and a public realization that a lack of education opportunity was a prime factor in reducing poverty.

A "Poor People's March on Washington" stepped off in 1963 as parts of the nation erupted in civil strife. Martin Luther King Jr.'s "I Have a Dream" speech became almost a spoken anthem for those who were committed to solving what had become a sustained/endemic problem for society.

Medicare and Medicaid helped to bolster access to health care for many who were older or vulnerable. Yet, millions were without health insurance. In the second decade of the 21st century, an Affordable Care

Act made that insurance available to even more people, even those with pre-existing conditions. The nation had to deal with the complexity and even the vocabulary of health insurance.

White House Proposal…Giving a Hand Up…

Fifty years after the Poor People's March, in 2013, the White House Office of Management and Budget released proposals for "Giving a Hand Up to Low-Income Families." It declared, "We now face a make-or-break moment for the middle class and those trying to reach it."

Among proposals were: expanding tax cuts for lower-income families, preserving a strong unemployment insurance safety net, supporting growth and job creation, helping states provide paid family leave to workers, reforming child welfare, preventing hunger and improving nutrition, funding a strategic plan to end homelessness, adjusting a program to help struggling families with heating and cooling costs, improving and reforming K-12 education, and expanding access to college.

Other proposals included: equipping American workers with good paying jobs today and in the future, promoting fatherhood and modernizing child support, funding health centers, ensuring that workers receive the pay and benefits to which they are entitled, preserving affordable rental opportunities, preserving funding for HUD block grants, expanding the Promise Neighborhoods to prepare more students for college, revitalizing distressed urban neighborhoods, expanding the child and dependent care tax credit, promoting outcomes for disconnected youth, and expanding low income legal assistance.

The proposals provided a glimpse into the complexity, urgency, and myriad needs that cry for attention.

A Recession and a Depression

Terms such as "the poor" or "those in poverty" are perhaps too generic. There are, after all, the "urban poor" and "rural poor." Some, who by most standards are considered rich, consider themselves "house poor." When it comes to recessions and depressions, the situations are as unique as those who face them, but here are a few.

• **Great Recession.** When the Great Recession hit with force in 2007-08, many who had earned their way into the middle class and beyond saw dreams vanish overnight as home prices plummeted below the value of their loans. Many were caught in a vise. Not only had their equity disappeared, but they also had been borrowing money, using steady increases in home value as collateral. The spectacular increases turned out to be an illusion, a "housing bubble".

When the bubble burst, those who bought high and had adjustable rate mortgages became victims of a system that made some people hyper-rich—at their expense. Many were left with a perfect storm—a mountain of debt, a shortage of cash, a dramatic drop in the stock market, the loss of a job, and commitments to their families to put food on the table and send their kids to college. When they did find jobs, they often paid less and offered less job security. A national debate flared about increasing the minimum wage. As the debate went on, some states and communities went ahead and did it. Consider a minimum wage job at $7.25 per hour. At 40 hours a week, 52 weeks a year, no time off, it comes to $15,080 per year. Feel free to make adjustments as adjustments are made.

According to the Economic Policy Institute (EPI), "nearly half of U.S. households have no savings in retirement accounts." EPI adds, "Households in the 90th percentile of retirement savings distribution have nearly 100 times more retirement savings than the median household."[850] Having enough income to save, coupled with the ability and willingness to save, has implications for later years, especially since we're living longer.

• **Great Depression.** During the Great Depression of the 1930s, people endured similar but in many ways even more intense hardship. Safety nets were yet to be invented. However, even then, innovative programs were developed to put people to work. The Works Progress Administration (WPA), Public Works Administration (PWA), and Civilian Conservation Corps (CCC) employed millions who were victims of an economy that had fallen on hard times. Later, programs such as the Job Corps and summer employment opportunities for youth, attempted to address poverty. Food stamps helped supplement nutrition. In part, those programs also addressed longer-term economic health and concerns about domestic tranquility.

Poverty Background and Statistics

- **Poverty Threshold.** In 1960, the *poverty threshold* for nonfarm families in the United States was $1,490 for *individuals*, $1,894 for a *family of two*, and $3,022 for a *family of four*. In 2013, the figures for the 48 contiguous states and the District of Columbia moved to $11,490 for *individuals*, $15,510 for a *family of two*, $19,530 for a *family of three*, and $23,550 for a *family of four*.[851]

- **Low-Income Threshold.** "On average, families need an income of roughly twice the official poverty threshold to meet their basic needs, including housing, food, transportation, health care, and child care," according to the U.S. Census Bureau.[852] Those living below double the poverty line were under the *low–income threshold*. In 2013, that included *individuals* with income of less than $22,980, *families of two* with incomes of $31,020, *families of three* with incomes of $39,060, and *families of four* with incomes of $47,100.

- **Percentage of Poor in the United States.** In *1959*, 27.2 percent of people of all ages lived in poverty, including 19.5 percent of children under 18, 17.9 percent of people from 18-64, and 35.2 percent of those 65 and older.[853] By *1999-2000*, 11.3 percent of all U.S. residents lived in poverty, including 16.1 percent of children under 18, 9.6 percent of people from 18-54, 9.4 percent of those from 18-64, and 10.2 percent of people 65 and older.[854] With the advent of the Great Recession, in *2010-11*, 15.1 percent of all U.S. residents lived in poverty, including 22 percent of children under 18, 13.8 percent of people from 18-64, and 8.9 percent of people 65 and older.[855]

- **Money Income of Those 65 and Older, 2010.** In 2010, according to the U.S. Census Bureau, among all household units, married or unmarried, over 65: 89.3 percent received income from retirement benefits; 51.9 percent, income from assets; 26.3 percent from earnings; 13.2 percent from public assistance and noncash benefits; 1.9 percent from veterans' benefits; and 0.4 percent from workers' compensation.[856] The Administration on Aging at the U.S. Department of Health and Human Services pointed out that 87 percent received income from Social Security. Median income for those 65 or older in 2010 was $25,757.[857]

- **GDP Per Capita.** In 2012, according to the International Monetary Fund (IMF), as reported in *Global Finance*, the eight richest countries in the world, based on GDP Per Capita/Purchasing Power Parity (PPP), were: Qatar ($106,283); Luxembourg ($79,649); Singapore ($61,046); Norway ($50,716); Hong Kong, SAR ($50,716); Brunei Darussalam ($50,440), the United States ($49,601); and the United Arab Emirates ($48,434). The eight poorest, based on PPP, were: the Democratic Republic of Congo ($364.48); Liberia ($490.41); Zimbabwe ($516.47); Burundi ($639.51); Eritrea ($776.98); Central African Republic ($789.21); Niger ($863.46); and Malawi ($882.67).[858]

Share of the World's Private Consumption:
In 2005, the world's richest 20 percent accounted for 76.6 of all private consumption. The world's middle 60 percent accounted to 21.9 percent of all private consumption. The world's poorest 20 percent accounted for 1.5 percent of all private consumption.

World Bank Development Indicators 2008[859]

Implications for Education and the Whole of Society
Poverty

• **Understanding the history and consequences of sustained poverty.** Schools and communities need a heightened understanding of the role poverty has played throughout history and the challenges it poses for the future. If we choose not to learn from history, then we may, indeed, be forced to relive it.

"The income chasm between the haves and have-nots, a concentration of wealth with a tiny group, and shrinking of the middle class" are issues that should command attention, according to Michael Usdan, senior fellow of the Institute for Educational Leadership (IEL). He calls for a reassessment of "whether equality of opportunity is to be more than empty rhetoric for growing numbers of American youngsters." Usdan points to the clock, noting that "minorities are becoming majorities at breathtaking speed, and they suffer inordinately from the effects of poverty." When it comes to dealing with "these changes that are transforming schools as well as the larger society, we are still asleep at the switch," Usdan observes.

• **Offering education programs that prepare people to avoid or overcome poverty.** If we are *not* in poverty, we need to understand how to avoid it. If we are *in* poverty, we need to understand how we might possibly get and stay out of it. From dwindling retirement savings to job losses and home foreclosures, the Great Recession of the early 2000s pushed thousands of people into financial uncertainty. Their future and the futures of their children were threatened.

An obvious but important response to this challenge is that we desperately need better economic education. Consider the importance of personal financial knowledge and skills, from responsible credit card use and balancing a checkbook to understanding profit and loss and return on investment. *Caveat emptor* or *buyer beware* should be burned into our habits of mind, especially following financial crises spurred on by cheap loans that turned out to be so expensive that they crashed individual and family budgets and stanched hopes and dreams. Ethical behavior might have given those who took huge liberties with our financial futures to line their own pockets at least a modicum of conscience. We certainly need resilience and an education that is broad and deep enough for us to apply what we know and are able to do in a variety of future jobs and careers.

"The school must be the great equalizer while it is assuring high achievement for its students," submits Arnold Fege, president of Public Advocacy for Kids (PAK), noting that "this has never before been accomplished." He points to the importance of "an investment in education, from birth to, in some cases, adulthood." Programs offered before and after school, tutoring, Head Start and other opportunities for early education, free and reduced-price lunches and other meals, counseling, and many other programs, at least in part, attempt to address many of the challenges that accompany poverty.

• **Motivating students by personalizing education, not just aiming for higher average test scores.** Granted, we aim to top off average test scores in our schools. That's commendable, but we need to be sure that we don't lose our focus on those who end up on the lower end of the curve. The tyranny of the average can be brutal. All students need to get an education that is broad and deep, not just a scaled down curriculum that has too often sliced away nearly everything that isn't easily tested.

Many children and their parents endure often unimaginable difficulty. For them, disadvantage is not just a word, it is life. If, in our attempts to raise average test scores, we shortchange or strip away programs that might provide motivation for a broad range of students, valuable incentives to stay in school might disappear with them. If students who are having problems drop out, our average scores might go up, but every thoughtful educator knows that leaves whole societies unprepared for the future. All students need rigor that is sensitive to their interests. All students need to see their own strengths and aspire to reach for higher ground. All students need a full range of thinking, reasoning, and problem solving skills. All students need good nutrition to strengthen their bodies and outstanding counseling to help them find direction, to explore possible careers and opportunities for service, lift their spirits, and strengthen their confidence.

> "Shameful child poverty levels call for urgent
> and persistent action. It's way past time to eliminate
> epidemic child poverty and the child suffering, stress,
> homelessness, and miseducation it spawns."
> *Marian Wright Edelman, Children's Defense Fund, Huffington Post*[860]

Household Expenditures for Food

Sometimes, food prices go through the roof. As they do, people often reduce the quantity or quality of the food they consume. Those prices can be impacted by drought, poor land, lack of delivery systems, use of food products for other purposes, corruption, and a host of other things. The lack of nutritious food and clean water can, of course, lead to disease and conditions ranging from obesity to starvation. A shortfall can also invite political unrest and magnified impatience between the haves and have-nots.

To illustrate the stark difference between rich and poor, let's look at the average percentage of total household expenditures that are spent on food in various parts of the world. During 2009, in the *U.S.,* with average annual household expenditures of $32,051, six percent went for food. In the U.K., it was nine percent and in France 14 percent.

However, in India, where 2009 average household expenditures were $620, more than a third, 35 percent, was spent on food. In Kenya, where average household expenditures were $541, a massive but, at the same time, meager, 45 percent was spent on food. Food security is and will remain a prime concern, along with education, in lifting people, whose talents the world needs, out of poverty.[861]

Poverty

Questions and Activities

1. Identify five to ten ways poverty interferes with student achievement.
2. If the "war on poverty" is an *unfinished agenda*, what do you believe should be added to that agenda to further reduce poverty in the nation? In the world?
3. Prepare a paper or PowerPoint presentation on the topic, "What education needs to do now to help students overcome sustained poverty."
4. Review *"World Development Indicators 2012"* or a more recent update from the World Bank, a progress report in working toward a series of Millennium Development Goals. In each major category, such as hunger, universal

primary education, etc., find at least one thing you believe everyone should understand. http://data.worldbank.org/sites/default/files/wdi-2012-ebook. pdf.[862]

5. View a mini-documentary, done by *ABC News*, "Martin Luther King, Jr.'s Speech, I Have a Dream." (Includes some video and text from his 1963 "I Have a Dream" speech at the Lincoln Memorial in Washington, D.C., http://abcnews.go.com/Politics/martin-luther-kings-speech-dream-full-text/story?id=14358231 (1:50).[863] **Activity:** Why did King deliver that address? What are two of his "dreams" mentioned in the speech? In your opinion, on a scale of 1 to 5, with 5 representing the greatest progress, how far have we come in fulfilling each of those dreams? Why those rankings?

Readings and Programs

1. *Kids Count Data Book*, Annie E. Casey Foundation, Baltimore, MD, *Kids Count Data Book*, 2013, Annie E. Casey Foundation, Baltimore, MD, http://datacenter.kidscount.org/files/2013KIDSCOUNTDataBook.pdf , for download or purchase.[864] Search for annual updates.

2. "God Bless The Child," sung by Billie Holiday, lyrics to song written by Billie Holiday and Arthur Herzog, Jr., Carlin America, Inc., Warner/Chappell Music, Inc., EMI Music Publishing, written 1939, recorded in 1941, http://www.youtube.com/watch?v=Z_1LfT1MvzI.[865]

3. Shipler, D.K. (2004). *The Working Poor . . . Invisible in America*. New York: Alfred A. Knopf., ordering information: http://www.amazon.com/Working-Poor-Invisible-America/dp/0375708219/.[866]

4. "History of Head Start," video, (7:08), testimony in both English and Spanish, http://www.acf.hhs.gov/programs/ohs/about/history-of-head-start.[867]

5. "LBJ, The War on Poverty," *American Experience*, PBS, 2 hours, six minutes, aired 2013, http://www.pbs.org/wgbh/americanexperience/films/lbj/player/.[868]

6. Worldwatch Institute. See annual *State of the World* and *Vital Signs* reports, available for purchase from www.worldwatch.org.

7. View "Two American Families," *Frontline*, PBS, http://billmoyers.com/2013/07/10/two-american-families/, July 9, 2013.

8. Current world almanacs.

A Discussion Starter

As with other trends, a brief section in this book cannot contain more than a very small part of what we need to know about poverty. This chapter is only a discussion-starter.

Chapter 20

Scarcity vs. Abundance
What's enough?

Chapter 20: Scarcity vs. Abundance. *What's enough?*

Trend: Scarcity will help us rethink our view of abundance.
Less ↔ More   What's Missing? ↔ What's Possible?

> "He knows we're out here without supplies.
> How does he think we're going to live 'till spring?"
> *Laura Ingalls Wilder, The Long Winter. The train can't make it through the snow.*[869]

The Overwhelming Power…
Of Needs and Wants

What is it about us? We're captivated by the struggle. We lean forward and soak up stories about people who have overcome adversity, beaten the odds, made it through a storm, found a legitimate way to go from rags to riches, or simply just survived.

Times were tough out on the Dakota plains during *The Long Winter* of 1880-81. So challenging, in fact, that Laura Ingalls Wilder has put generations of us on the edges of our seats as coal runs out, food supplies dwindle, and snow drifts get too deep for mere mortals to penetrate. In that sea of white, the Ingalls family wondered how they would survive until the trains could break through. The Ingalls *wanted* their Christmas barrel and the gifts that were inside, but first they *needed* to survive. Thankfully, Pa Ingalls and a whole lot of other people in the small community of De Smet looked at scarcity as a problem to solve. To take the chill off, Laura and her sisters twisted hay to burn. Pa hitched his team to a sled and nearly froze finding a farmer, miles away, who was willing to part with some of his prized seed wheat. The Ingalls ground it into flour to make bread. When everything is scarce, even a short supply seems like abundance.

The Ingalls family faced two overwhelming questions: "What do we need?" and "What do we want?" The wants were strong as ever but the needs had to be satisfied just to survive. We know one thing for sure—that family's *scarcity* fired their imaginations and left all of us with an *abundance* of inspiration.

Scarcity
Badge of Honor or Kick Start for Progress

Some cultures thrive on scarcity. They take pride in it. You might hear a story something like this, "Times were tough, but we made do with what we had. We just stuck with it." Others might add, "You don't need as much stuff as you think you do." A lot of us wear our scarcity as a badge of honor, as an expression of our beliefs, or as a way to reduce our footprint on the ecosystem.

Intolerable Scarcity

Genuine scarcity is not a myth. In fact, it is a stark reality for millions of people every day. That could mean a scarcity of nourishing food, clean water, adequate health care, safety and security, arable land, or energy to power transportation and industry.

Some communities and parts of the world are left with a scarcity of know-how and an unstable civil society because of inequity in education, injustice…or a brain drain as people move to greener pastures. Some folks just pick up and leave to take advantage of an opportunity or escape the ravages of war.

When some level of scarcity is tolerable, that's one thing. When it isn't, that's something else. Let's take a look at the lack of just two of the many things that are among building blocks of abundance in the 21st century.

- **Electricity.** A quarter of the world's population doesn't have access to electricity.[870] Millions rely on gathering twigs, brush, or dung to burn for heat and light. Huge portions of the day are devoted to finding something to burn. The smoke leads to respiratory diseases, incapacity, and shorter lives. Despite that scarcity, the same people often live in parts of the world with an abundance of sunshine and wind. Solar panels or wind turbines could bring them plenty of electricity to improve their quality of life, jump-start their economies, and put them squarely on the road to abundance.

- **Fresh Water.** The U.N. estimates that "783 million people lack access to clean and safe water, and 37 percent of the world's population doesn't have access to sanitation facilities."[871] People spend much of their lives fetching water from a river, stream, waterhole, or a village tap. The lack of clean water for hydration, hygiene, and cooking keeps these millions on the borderline of survival.

Physicist Dean Kamen, an inventor of the Segway personal transporter, has developed a relatively small water purifying system called *The Slingshot*. Using less power than a hand-held blow dryer, it can turn out about ten gallons of purified water an hour, enough each day for a village of 100 people. In some places, that might mean a few solar panels to power it. Another compact device is known as *Portapure*. These are ways of turning scarcity into abundance.

In Senegal, my longtime friend Boubacar Tall, a leader at his country's Ministry of Education, took me to communities large and small. We visited schools in Dakar, Kaolak, Louga, and Ross Bethio to check out how students were doing as they identified and addressed community problems. They were using a Center for Civic Education (CCE) program called *Project Citizen*. In Ross Bethio, 10 to 15-year-olds took on the challenge of getting fresh water for their community. For millennia, the river was the prime source of water and persistent water borne diseases. Thanks to their genius and persistence, coupled with support from community leaders, Ross Bethio now has taps for fresh water. Those kids proved for all of us that even the most historic and persistent problems can be solved. They turned scarcity into abundance.[872]

Respect for the Natural World.

Realize it or not, we are surrounded by abundance that flows from the natural world—air to breathe; energy from the sun; plant life that sustains us; even bees that help pollinate our crops. "For the first time in history, a majority of human beings live (their) lives in artificial environments, virtually cut off from the rest of nature," observes social thinker Jeremy Rifkin.[873] A challenge we face is to live in harmony with the natural world. If we don't, it could cost us big time, even our survival.

> **"Necessity is the mother of invention."**
> *Source Unknown. Often attributed to Plato*

Scarcities of Our Own Creation

Often, potential abundance is turned to scarcity because we've made bad choices in our financial or in our personal lives. "We were doing well. Owned a big house and fancy car. Had quite a lot of money in the bank. Some good investments. Kids doing well in college. Then, this guy came along who said he could double our assets within a year. He showed me examples of how he'd done it for some fairly famous people. I went for it, hook, line, and sinker. The guy is now in jail. I'm back to square one. I guess I made some bad choices. Trying to avoid bankruptcy. Next step for me is a personal and financial turnaround."

Transference of Abundance or Transference of Scarcity?

Many Baby Boomers and members of the Silent Generation had done well. Not all, but a significant number, had saved, invested, and were feeling safe as they moved into their retirement years. Some economists were suggesting that, as they passed on, we would see the greatest transference of wealth in history. With the Great Recession, except for the wealthy, nest eggs were depleted. People who hoped for a comfortable retirement continued to work or looked for part time jobs. On top of that reality, working families were taking a flurry of hits, such as foreclosures and escalating energy costs.

With fits and starts, the price of fossil fuels will continue to go up. Along with manipulation in the marketplace, supply and demand will naturally lead to higher prices at the pump. How does that impact real people?

- A single mom in Los Angeles couldn't afford the gas to visit her family in Sacramento.
- Parents had to pull their child out of pre-school one day a week to make ends meet. They were living on the margin, working full time, paying off college loans, and trying to get ahead. Then gas prices went up.
- A part-time student attempting to get the education to qualify

for a better job, faced with an increase in the cost of fuel, was concerned that she might have to drop the class she was taking. She told CNN, "If it costs $200 for me to drive back and forth from Albany, there's just no way."[874]

While honest-to-goodness real live people sacrifice, an abundance of renewable energy is all around us, if we have the political will and sense of common purpose that we need to pursue it. At some point, the transference of scarcity to those who can least afford it has to stop. It's unsustainable.

Sometimes, things are scarce because we just can't pay the price. However, investing in research and development and rolling out new ubiquitous technologies can turn scarcity into abundance. In a 2012 issue of *The Futurist*, the flagship magazine of the World Future Society, Peter Diamandis and Steven Kotler wrote an insightful piece titled, "The Abundance Builders." In it, they pointed out the racks of equipment of a few years ago…now tucked into a smartphone. Diamandis is chairman and CEO of the X PRIZE Foundation. Kotler is a best-selling author and journalist.[875]

Remember what Archimedes said, "Give me a lever and a place to stand, and I will move the world." Let's call this Archimedes-2: *Give me a computer, a connection to the Internet, an unlimited selection of apps, and a 3D printer and I will change the world.* What should we do about that? How about making computers, smartphones, and Internet connections available to everyone. Add microfinancing if and when it's needed. Then, stand back and watch the entrepreneurial magic.

Scarcity vs. Abundance…In the News

While few refer to scarcity and abundance directly, the airwaves, cyberspace, newspapers, and other media have been buzzing with stories focusing directly on the contrast. A challenge: When we read factual information, let's consider the implications for our future and the futures of our fellow human beings. Let's try to avoid immediately seeing every fact through a partisan political lens—simply as information that confirms or disputes our preconceived notions. That's the kind of objectivity we need if we ever hope to address the issues we face as a society.

• In 2010, the top one percent of households in the U.S. owned 35.4 percent of all privately held wealth. The next 19 percent had 53.5 percent, and the bottom 80 percent had 11 percent, according to G. William Domhoff, research professor in psychology and sociology at the University of California, Santa Cruz.[876]

• In 2011, "Donations to charity (in the U.S.) rose to $298.42 billion, but were still $11 billion below a 2007 record as nonprofits battled through the sector's second-slowest recovery from recession in 40 years," according to a study, *Giving by Americans*. The study's findings were reported by Reuters.[877]

• According to author Jonathan Foley in a 2011 article in *Scientific American*, "Right now about one billion people suffer from chronic hunger. The world's farmers grow enough food to feed them, but it is not properly distributed, and, even if it were, many cannot afford it, because prices are escalating."[878]

• If everyone in the world lived like the people of *Costa Rica*, it would take 1.4 earths to meet their needs. Like the people of the *United Sates* 4.1 earths. Like *China* 1.1, *France* 2.5, and the *United Arab Emirates* 5.4 earths, according to Tim De Chant in a 2012 *popsci.com, Popular Science* article, based on information compiled by the Global Footprint Network.[879]

Pathways to Abundance

Consider this a brainstorming session. First, we'll share a few of the many ways our friends, colleagues, and fellow citizens are trying to achieve abundance, even if they have to redefine it. Then, you can add to the list.

• **Education.** Even though jobs were tight during and following the Great Recession, young people, in fact, people of all ages, were flocking to post-secondary education. While graduates often searched for months to find jobs, facts from the U.S. Bureau of Labor Statistics (BLS) echoed a stark reality known as the *education dividend*. While the March 2013 unemployment rate for those without a high school diploma was 11.1 percent, it was 3.8 percent for those with a Bachelor's

Degree or higher.[880]

• **Downsizing, Less is More.** Facing the prospect of limited income, people moving toward retirement are downsizing. Often, they are looking for smaller living space and a compact, more fuel efficient car. Ironically, so are more college graduates faced with paying off student loans that had accumulated to $1 trillion by 2011. The average for the Class of 2011 was $26,000, according to NBC News.[881] In some cases, that meant moving back in with the folks or micro living space and no car. In downsizing, some have found a kind of abundance in accepting that, for them, less is more. Waste not. Want not. *By the way, we can also expect more do-it-yourselfers (DIY), who will try to save a bit of money as they get back to basics and learn or relearn some valuable skills. So many people are making a variety of their own things that the whole phenomenon is being called the makers movement. Even the multi-owner share economy is getting traction.*

• **Entrepreneurship.** As a society, we will need to invent our way into a more sustainable future. For the well-educated, idea prone, risk-taking, creative, and courageous, invention may be the ticket to abundance. All the more reason for making sure everyone is equipped with the know-how to manage a budget, an organization, and themselves. Among the most well-known of the entrepreneurs have been people like Henry Ford, Thomas Edison, Warren Buffett, Bill Gates, Steve Jobs, and Mark Zuckerberg. At age 28, Zuckerberg's net worth was estimated at $14 billion, and his annual salary set at $1 per year.[882]

• **Become Rich and Famous.** Television and the Internet carpet bomb us with examples of people who *have it all*. A Pew study released in 2007, just before the Great Recession hit like a hammer, found that the top goals for Millennials were—to get rich, 81 percent, and to be famous, 51 percent. One respondent qualified her hopes, "It's *famous*, but it's much less than stardom. I want to affect society." Their top concerns on the road to abundance included: money/finances/debt, getting a college education, getting a job and into a career, plus family and relationships.[883]

• **Saving and Investing.** Often, when some people are in their peak earning years and awash in money, they simply spend it. Later, they might be faced with a gigantic yard sale when the outflow exceeds the inflow. No matter how hard it might seem, saving and making wise investments is generally the pathway to abundance for the greatest number of people. Making good choices is important, and enduring some scarcity, the level depending on income, can pay dividends later on. *Caution:* Avoid being stingy and selfish. What you give and contribute to others can create greater abundance for your community, your family, and the world, and can, itself, help achieve an abundance of satisfaction as well as a legacy.

• **Invest in Infrastructure and People.** If we hope for personal abundance, then we need to invest in physical and social infrastructure. Our hope to move from scarcity to abundance rests on the foundation of our environment, our society, and our economy. Unless we contribute through our taxes and other investments, improve education, build or rebuild roads and bridges, develop renewable sources of energy and energy distribution systems, conserve a sound environment, and maintain a system to safeguard our safety and security, we may discover that what we thought would be wealth is actually worthless. If we hope for abundance for ourselves, our community, our nation, and the world, we simply must invest in education and the well-being of all people.

• **Appreciating What We Have.** For most of us, in our quest to overcome scarcity, we don't notice the abundance that is all around us. Despite what we're seeking, do we consider the gifts we already have in our lives?

Education: Four Insights on Scarcity and Abundance

Futures Council 21 member Sheldon Berman believes that "one of the greatest challenges we'll face over the next decades is the growing division between wealth and poverty." That situation can lead to injustice, such as "the degree of control that can be exerted over media, politics, and culture."

A second, highlighted by Peter Diamandis and Steven Kotler in their book, *Abundance: The Future is Better Than You Think*, is the Industrial Age model of education where "conformity is the desired outcome." They add, "an emphasis on facts is no longer necessary. Facts are what Google does best. But creativity, collaboration, critical thinking, and problem solving—that's a different story." They are "fundamentals of today's jobs" and "have become a new version of the three Rs."[884]

A third insight is expressed by veteran teacher Milde Waterfall, who observes that too much information can sometimes distract students from playing with ideas. She believes students' impressive "try it" spirit is a force to build on. Yet, she cautions, that spirit could be paralyzed by fear. "Teachers are afraid of not having students pass tests. Students are afraid of not being successful in the current culture." If we hope to deal with scarcity, then we need to "allow our classrooms to be sanctuaries of invention."

Growing Tension, Intensifying Urgency

Tension between wants and needs is, in itself, a powerful force. Throughout history, wars have been fought over land or control of people. Periods of mass migration have been driven by climate change, drought, and shortages of food and water or drastic changes in temperature. Uprisings have boiled when people have been deprived of their liberty. Some have migrated or staged revolutions when they faced what they considered an unacceptable scarcity of opportunity to provide for their families, develop their talents, or pursue their dreams. The sense of urgency is growing to stimulate the conversation about scarcity and abundance. Make having that discussion a goal.

"How many times has someone told you, 'Necessity is the mother of invention.' It's true. In fact, it is so true that, in the future, the new disadvantaged may be those who have never known disadvantage. Blindsided by abundance, we might never have a chance to learn the most basic survival skills."

Implications for Education and Society
Scarcity vs. Abundance

Implications of this trend are becoming more glaring, magnified by the realities of food that people *can't* afford to buy; unproductive land; the inability to obtain rare minerals that are basic to manufacturing or energy storage; or high quality education that is only available to those who can afford it. A few of the many implications include:

• Most of us are likely to complain that we have a scarcity of time…to get a job done, to enjoy a vacation, or to live out our hopes and dreams.

• All should understand the basics of living with scarcity. While some may only know material and financial abundance, they should be prepared to cope, in case that abundance turns to scarcity. Examples might range from financial crises leading to home foreclosures to a business leaving town, devastating a local economy.

• Students will need a grounding in financial literacy, including the ability to separate needs from wants, the importance of saving and investing, and the ethical imperative of making sure as many people as possible are within reach of the golden ring.

• We should develop in ourselves and in our children the qualities that lead to generosity, volunteerism, giving, charity, contribution, support, and philanthropy. We often have an abundance of gifts to share with others who are dealing with scarcity.

• People will need help in maintaining physical and mental health during times of scarcity. Schools, colleges, and communities will need to offer opportunities to learn problem solving and entrepreneurial skills, then encourage those who pursue new ideas.

• Opportunities should be provided to learn from people whose hopes, dreams, and persistence have led them to accomplishment. All should understand that progress often requires learning from others and enduring a sometimes long bumpy ride.

• Communities and whole societies should fully grasp the lost potential of talented people who do not have opportunities for affordable, high quality education.

• Inventiveness and innovation skills should be seen as basic, if we hope for a constantly improving economy and quality of life.

- As a society, all must be able to face facts and identify problems and solutions free of pre-determined political dogma or posturing. The urgency of opening the door for everyone to move beyond scarcity, if they wish, is growing exponentially in a fast-changing world.

- All will be faced with the need to understand that simply accepting scarcity as a badge of honor can also limit possibilities for a viable education. That attitude could lead to stagnation and even conflict. Scarcity may help *build character*, but carried to an extreme, it can turn a blind eye to possibility and deprive society of the ingenuity of those whose voices might not be heard and whose ideas may not be considered.

Scarcity vs. Abundance
Questions and Activities

1. As parents, educators, or members of a community, identify five first steps we might take in helping people move from scarcity toward abundance.

2. As a school or college class, discussion group, or family, individually identify five things that you would most want to have in abundance. Then, identify five things you would least want to be scarce. As members of a group, discuss the main questions, making it possible for people to introduce some of their items for conversation.

3. Review and consider lesson plans, such as: "Scarcity, Choice, and Decisions," *The Mint*, http://themint.org/teachers/scarcity-choice-and-decisions.html; "Socialize" (scarcity and making choices), *Council for Economic Education*; and "Books for Teaching Economic Concepts, and Books with Examples of Scarcity," *Scholastic*, http://www.scholastic.com/teachers/lesson-plan/books-teaching-economic-concepts?pImages=n&x=67&y=20, (as well as others).[885]

Readings and Programs

1. Diamandis, Peter H. & Kotler, Steven. (2012). *Abundance...The Future is Better Than You Think*, Free Press, A Division of Simon & Schuster, Inc., NY.[886]

2. Gore, Al. (2013). *The Future: Six Drivers of Global Change*," Random House, NY.[887]

3. "Water for Ross Bethio, Senegal Project at the World We Want Foundation," posted Dec. 4, 2012, video at http://www.youtube.com/watch?v=8qjljBx-wJ-E (3:06).[888]

Chapter 21

Personal Meaning and Work-Life Balance
Honey, Let's Get a Life

Chapter 21: Personal Meaning and Work-Life Balance
Honey, Let's Get a Life!

Trend: More of us will seek personal meaning in our lives in response to an intense, high tech, always on, fast-moving society.
Personal Accomplishment ↔ Personal Meaning

> "There's always more to life than money, Will.
> Money is just a means to an end.
> It shouldn't be the goal."
> *Novelist David Baldacci advising a character in The Innocent* [889]

The Thrill of Accomplishment...
The Agony of Losing the "Life" in Work-Life Balance [890]

Mayo Clinic Oncologist Dr. Edward T. Creagan tells the story this way in his *Stress Blog*: "Once upon a time, when we left the office, the factory, or the firm at the end of the day, we left our work behind. Our homes didn't contain computers, high-speed Internet connections, or fax machines. We had leisure time to connect with our family and friends. We had the luxury of time to reflect." Writing for *mayoclinic. com*, Creagan warns, "The walls of our homes no longer protect us. We're always connected; we never truly leave the work behind. We're now impaled by 'weisure,' —being on the clock even on leisure time." He adds that nothing closes the virtual office, not even a snow day, and calls "the expectation that we be available 24/7...soul-crushing." [891]

Wired
And Without Life Support

All around us, growing numbers of people are discovering that being wired, accessible day and night, always multitasking, can have devastating consequences for families and their own personal interests and well-being. For some of us, the relentless pursuit of money and thrill of accomplishment have turned us into workaholics. During tough economic times, when the job goes away, we too often discover that we've destroyed our life support system in the quest to get ahead of the Joneses. Even our greatest accomplishments can seem hollow when we have no one left in our personal lives who really cares.

Studies show that, for many people, life, liberty, and the pursuit of happiness are becoming more than making a bundle and accumulating a few more luxuries than anyone else at the office or in the neighborhood. When that reality sets in, we often look for remedies. Some, for the first time, discover the difference between needs and wants. That means we're likely to run squarely into the law of diminishing returns—the stuff we buy doesn't seem to make us any happier. Others get into stress-busting habits, from eating or drinking too much to an intensified pursuit of personal health and well-being. We try to exercise away the stress, which isn't a bad idea. Many turn to the spiritual, finding themselves in music and the arts, walks in the woods, travel to interesting places, reading, meditation, service to others, reconnecting with family, and visits with friends. Some find peace and meaning in religious beliefs.

Too often, we fail to realize that there is much more to each of us than our job, whatever it happens to be. When people ask us, "What do you do?" maybe we need answers that go beyond our work. Work-life balance is essential for everyone. Even children and young adults need to understand how important it will be to their futures. For many of us, it could just be a matter of *getting a life*.

So, You Want to Attract and Keep Talent!

"Organizations across the globe continue to ask their employees to 'do more with less,' leading to increasing dissatisfaction with work-life balance," reports Mark Royal, senior principal at Hay Group Insight, a management consulting firm. Hay research found that "employees who perceived work-life balance support from their organizations" had greater confidence that their companies would be able to recruit and keep top talent. [892]

Juggling Demands
Living Our Lives

"It's not just the destination. It's the journey. Life's too short to spend all of my time worrying." We know how to explain the need for spending at least part of our time living in the moment, being present, and the importance of mindfulness. Sometimes, we think about enjoying the journey...then the smartphone rings and we're immersed.

In the beginning, we were amazed at how productive we could be with a string of technologies that have developed over the centuries. The technologies became appliances, then morphed into extensions of our bodies and minds. At some point, we reach a tipping point: Do we command the technologies or do they command us? Maybe we can reconcile the situation by declaring ourselves technology-enhanced humans.

Educators, Students, and Emotional Well-Being. Brad Kuntz teaches at Gladstone High School in Oregon. He was 2011 winner of ASCD's Outstanding Educator Award. In the association's *Education Update*, he urged fellow educators, "The better we look after our students' emotional well-being, the more trust we will earn from them." He advises "Do whatever it takes to keep yourself at the top of your game. You owe it not only to your students, but also to yourself."[893]

Work-Family Balance

Erin Callahan, former CFO of Lehman Brothers, told her story in a *New York Times* op-ed: "First, I spent a half-hour on Sunday organizing my email, to-do list, and calendar to make Monday morning easier. Then I was working a few hours on Sunday, then all day. My boundaries slipped away until work was all that was left. Inevitably, when I left my job, it devastated me. I couldn't just rally and move on. I did not know how to value who I was versus what I did. What I did *was* also who I was."[894]

University of Massachusetts Professor of Psychology Susan Krauss Whitbourne, writing for *Psychology Today*, observed, "The emotional boost we get from our loved ones can help us feel better about the work that we do." How work-family balance is treated in the workplace impacts our families. "Restricting employees from telecommuting, at least in jobs that are amenable to off-site work, can make them *less, rather than more*, productive," she points out. Inevitably, things don't go the way we plan, Whitbourne says. "Babysitters become unavailable, children get sick, school beckons you to parent-teacher meetings, work deadlines get shortened, you need to put in extra hours, etc., etc." Given a supportive situation on both sides, we can draw strength to cope, get both jobs done, and "enjoy a fulfilling an enriching life in both spheres."[895]

Help is on the way

Some wisecracks don't help, such as, "He doesn't have to worry about burnout because he's never been on fire." Easily said. In reality, most of us are under fire to be at our best, whether we're nurturing our children, dealing with a disability, hunkering down on a battlefield, convincing people to accept our new idea, dealing with a crisis, or facing the loss of a job or foreclosure on our home. Sometimes, the pressures are over the top. Whether it's post-traumatic stress disorder (PTSD) or bouts of anxiety about almost anything, it's time we 'fessed up to the importance of emotional health.

"If you're finding it more challenging than ever to juggle the demands of your job and the rest of your life, you're not alone," says Jen Uscher in a *WebMD* Feature, "Beat Burnout by Making More Time for the Activities and People That Matter Most to You." Uscher suggests five ways to bring a little more balance to your daily routine:

- **Build downtime into your schedule.** "Make it a point to schedule time with your family and friends."
- **Drop activities that sap your time and energy.** Spend less time on web or media sites, and you might finish your work early. Limit the time you spend with people who are "constantly venting and gossiping."
- **Rethink your errands.** Think about "whether you can outsource any of your time-consuming household chores or errands."
- **Get moving.** "Make time for exercise."
- **Remember that a little relaxation goes a long way.** Plan a holi-

day. Get stuff done so that you can go home. Enjoy nature. Work out. Exercise. Jog. Create. Play. Take up a hobby. Do some gardening or lawn work. Get interested in genealogy. Do carpentry. Volunteer. Help someone build a house. Get a pet. Clean out the garage. Do some cooking. Take or sort through your pictures. Do scrapbooking. Go to a concert or a ballgame. Sing. Learn to play an instrument. Encourage others. Take some time for friends and family. Collect something. Tinker with your car. Read a novel. Listen to music.[896] A friend told me that he took up golf to forget his problems. Now, he says, he has another problem—his putting.

Mindfulness
Are We Missing Out on the Moment?

What is mindfulness? *Psychology Today* puts it this way: "Mindfulness is a state of active open attention on the present. When you're mindful, you observe your thoughts and feelings from a distance, without judging them good or bad. Instead of letting your life pass you by, mindfulness means living in the moment and awakening to experience."[897]

Sometimes, we have a hard time focusing on what we see around us or the nuance of what someone is trying to explain. The flowers are there to enjoy and the ideas are there to enlighten us, but our minds are miles or months away. Occasionally, we become overly judgmental, and see everything and everybody through our predetermined lens.

Harvard Professor of Psychology Ellen Langer, tells us we need to be mindful of our environment and actually make a conscious effort to see, admit, and even reinforce the qualities we observe in others. Stop and smell the roses. Each moment is a gift. We can deny it, ignore it, or let it fill us with a sense of appreciation, even awe.[898]

Emotional Intelligence. Daniel Goleman, in his classic book, *Emotional Intelligence: Why It Can Matter More Than IQ,* ponders why some people, despite their IQ, do well, and others don't. Some of the difference can be attributed to emotional intelligence. He observes, "The ability to control impulse is the base of will and character. By the same token, the root of altruism lies in empathy, the ability to read the emotions of others. Lacking a sense of another's needs or despair, there is no caring. And if there are two moral stances that our times call for, they are precisely these, self-restraint and compassion."[899] Of course, we need to keep in mind that some conditions impair or limit control over emotions and may call for various therapies.

What can we do to teach this skill? Based on research, *GoodThera-py.org* suggests: practice active listening; teach empathy; teach impulse control; talk about social skills; and encourage discussion." Emotional intelligence "enables people to have positive interactions with others, to predict others' thoughts and feelings, and to engage in appropriate levels of empathy. The article notes "a strong correlation with career and academic success."[900]

Dark Cloud or Silver Lining

Our attitudes are contagious. As much as we hate to admit it, some people seem to get their kicks by depriving colleagues of positive reinforcement, even putting them down and playing the heavy. Generally, these are people whose egos are much greater than their talent. They seem to claim their *silver lining* by putting a *dark cloud* over everyone else. As those who are empathetic are dispatched, only the dour, riskless, and talent-deprived remain. The company logo could very well be a dark cloud.

Even in their personal lives, some people thrive on being destructively judgmental. Some even act out through violence, harassment, or bullying. Imagine for just a moment that a 70-year-old has fulfilled a lifetime dream of pulling some musicians together and singing "At Last." We have a choice for what we might say after she plays the recording for us, looks our way, and says, "Well?"

Dark Cloud: "You know, I've never liked that song. You were a little off key. Who ever suggested you could sing? I'll stick with Beyoncé."

Silver Lining: "Congratulations. You did it. I love it when people sing. You've brightened my day and lifted my spirits. What are you recording next?"

"The slogan of the hour shifted from the cheery
'Have a nice day'
to the testiness of 'Make my day.'"
Daniel Goleman[901]

Observations about Work-Life Balance From the Field

Members of Futures Council 21 were asked to share views on work-life balance, including health and wellness, especially for students and educators.

• **Avis Glaze,** a leading counsel on character education and leadership, calls for greater attention to "growing incidence of depression, despair, and disenchantment among students." She adds, "We need to pay more attention to the emotional health needs of both students and educators."

• **Damian LaCroix,** superintendent of the Howard-Suamico School District in Green Bay, Wisconsin, observes that "a fast, hyper world sometimes leaves students, staff, and families feeling exhausted, out of control, and disillusioned." LaCroix expresses concern that many students may not get the support they need at home, and their parents might not be involved. Some of those families are rocked by divorce, poverty, transiency, and a lack of connection to a faith. He adds, "All of this is playing out against a backdrop of increasing demand from Common Core standards."

• **Two Millennials.** Allison LaFave is a marketing, communication, and development associate for Legal Outreach in New York, a Harvard graduate, and a Millennial. LaFave observes that "unemployment, the squeezing of the middle class, and the mortgage crisis" have left "people across the political spectrum eager to express their views." She also sees many people becoming increasingly secular, as religious organizations seem to become captives of political groups, exacerbating a "culture war."

Another Millennial, Ryan Hunter, echoes concerns about whether the political system is poised to offer solutions to pressing political, economic, social, educational, and environmental problems. Hunter also notes that some Millennials "are less likely than their parents… to adhere to the tenets and beliefs of a specific religion." However, he believes a majority unconsciously hold "a common unifying set of principles," which he hears expressed in interviews. Some sociologists have "identified this underlying phenomenon among youth as egovonism." The *Urban Dictionary*[902] explains that egovonism, in part, involves borrowing ideas from many established religions, often personally interpreting our own professed religion in a way that better suits us.

> ## "He who keeps his cool best wins." and "Laughter is inner jogging."[903]
> *Norman Cousins,* longtime Saturday Review editor and author of Anatomy of an Illness. Cousins believed that laughter is the best medicine. Research indicated that laughter reduced stress levels and increased HDL (good cholesterol). Dr. Lee Berk, a psychoneuroimmunologist, noted that "the best clinicians understand that there is an intrinsic physiological intervention brought about by positive emotions such as mirthful laughter, optimism, and hope."[904]

How Important is Work-Life Balance?
From Humor and Health to the Pursuit of Happiness

In an always-on world, driven by demands for our attention, tons of twitters, emails as far as the eye can see, and schedules that press us from all sides, we depend on adrenalin to help us stay ahead of the curve. Sleep can sometimes be scarce. We often spend our weekends catching up on work or wearing ourselves out running errands or trying to reclaim our personal lives. Rather than continue to burn the candle at both ends, the time has come to admit one of the limits of our lifestyles and technologies—they can actually work us into a frenzy or even work us to death. Before we immerse ourselves in further implications of this trend that is so important to the fullness of our lives, let's consider the thoughtful insights and wisdom of a few more observers:

• **Michael Miller, M.D., on Laughter.** "Half of Americans are searching for meaning and purpose in life," according to Barna Research, based in Ventura, Calif. The American Heart Association's *Scientific Study 2000* was among the first to document that "laughter and an active sense of humor may influence heart and artery disease." "The old axiom that 'laughter is the best medicine' appears to hold true when it comes to protecting your heart," says Michael Miller, M.D., director of the Center for Preventive Cardiology at the University of Maryland Medical Center in Baltimore.[905]

• **The Late George Carlin, Comedian.** Speaking of laughter, comedian George Carlin, who was known for pressing the limits, is quoted as saying, "The paradox of our time in history is that we have taller buildings, but shorter tempers; wider freeways, but narrower viewpoints. We spend more, but have less. We buy more, but enjoy less."[906]

• **Gregg Easterbrook, Author and Social Observer.** "Most men and women in the Western nations have attained the conditions of which previous generations dreamed," says Easterbrook in his book, *The Progress Paradox.* "Although this is excellent news, the attainment makes it possible for society to verify, beyond doubt, that personal liberty and material security do *not* in themselves bring contentment." He adds, "That must come from elsewhere, making it time to awaken from the American dream." Easterbrook observes, "Society is undergoing a fundamental shift from 'material want' to 'meaning want,' with ever larger numbers of people reasonably secure in terms of living standards, but feeling they lack significance in their lives."[907]

• *Career Builder* **on EQ and IQ.** Writing for *Associations Now,* writer Daniel Ford reported on a 2011 *CareerBuilder.com* study. The study, Ford said, "found that emotional intelligence, defined by the nonprofit research firm Six Seconds as 'a set of skills for understanding and using emotions effectively,' accounted for almost half of an individual manager's performance level." It also found that emotional intelligence increased organizational engagement and achieved higher bottom-line results. On top of that, employee turnover dropped. The study indicated that people with high EQ are "more likely to stay calm under pressure, know how to resolve conflict effectively, lead by example, and make more thoughtful business decisions."[908]

• **The Arts and Sports.** To maintain our equilibrium, some of us pick up a brush or musical instrument and create our own version of a work of art. Others find satisfaction in being a consumer of the arts, buying a music download, CD, DVD, photo, sculpture, or a painting. No less than the 2012 U.S. Census Bureau *Statistical Abstract of the U.S.* reported that, in 2008, country-wide, 224.8 million people participated in arts and leisure activities or were visitors or audiences for: movies, sports events or playing sports, other outdoor activities, exercise, gardening, volunteering/charity work, community activities, jazz or classical music concerts, musicals, non-musical plays, art museums/ galleries, craft and visual arts festivals, parks/historical buildings, and reading of literature.[909] *Denver Post* Theater Critic John Moore reported that *local arts events* produced $1.4 billion in revenue and attracted 14.1 million people during 2005, compared to 4.5 million fans who attended *home games of professional sports teams.*[910]

• **Health and Wellness.** Personal, organizational, and community health and wellness are like a foundation that supports our ability to enjoy and function. Exercise; eat healthy; quit smoking (or don't start); rethink drinking habits; hydrate; avoid drugs that aren't prescribed for your health; choose supplements wisely; try not to get hooked on video games; avoid excessive noise; take care of your eyes; watch your weight; get checkups and follow your physician's advice; take care of your heart, your circulatory system, and your bones.[911] Try to avoid burnout or being too much of a *couch potato* or a *mouse potato.* While you're at it, promote healthy buildings as well as both indoor and outdoor air quality. If you're getting a bit older, get some advice on maintaining geriatric health, which might depend on how well you've taken care of yourself earlier in life.

The Centers for Disease Control (CDC) has developed a *Worksite Health ScoreCard*, which includes items such as: organizational supports, tobacco control, nutrition, physical activity, weight management, stress management, depression, high blood pressure, high cholesterol, diabetes, and signs and symptoms as well as emergency response to heart attack or stroke.[912] Moving to the school and college front: The Santa Barbara[913] and San Francisco[914] Unified School Districts and the University of California at Santa Barbara[915] have developed different wellness programs that are an idea fest for anyone building or improving that kind of health program in their communities.

Director of the National Institutes of Health (NIH) Francis Collins, a doctor of both medicine and philosophy, counsels us to choose the right foods and exercise programs. He calls it, "one of the most important investments you can make in your future." Collins adds, "America, it's time to change your lifestyle—it just might save your life."[916] Add this: *USA Today* reports on a survey by Principal Financial

Group showing that a wellness investment pays off—improvement in
medical costs by $3.27 and *absenteeism costs* by an average of $2.73 *for
every dollar spent on wellness.*"[917]

> **"Don't say you don't have enough time. You have exactly
> the same number of hours per day that were given to
> Helen Keller, Pasteur, Michelangelo, Mother Teresa,
> Leonardo da Vinci, Thomas Jefferson, and
> Albert Einstein."**
> *H. Jackson Brown, Jr.*

Implications for Schools and the Whole of Society
Personal Meaning and Work-Life Balance

Implications are endless because this trend is so deeply ingrained in our
lives. Hands down, we need to pay greater attention to emotional health,
emotional intelligence, and the vast reaches of the affective domain. We
need to recognize students and educators—give them some credit—for
their current and future contributions to society. In difficult economic
times, schools, colleges and other institutions simply must be there to
help individuals and families as they deal with stresses and fatigue, often
driven by unemployment, downsizing, and delayed retirement. Baseline,
we should rally a commitment to preparing students, educators, and our
communities to pursue even greater work-life balance. We should encourage "seekers," who want to serve others, and make sure they have
opportunities to do just that. Here are a few related implications.

• **Considering how business, government, education, and other
institutions can contribute to work-life balance.** Many institutions
can play a role. Some will find perspective in religion or their community of faith. Some might seek renewal of the spirit by spending more
time with their families and friends. Others will want to create a work
of art or commune with nature—watching birds, walking through a
park or forest, listening to the wind in the trees, or tracking Saturn as
it drifts across an evening sky.

Two challenges. First, we need to understand that it's OK to do
those things, to unwire long enough to unwind and refresh. Second,
we need to know how. Some people have been so busy for so long that

they either feel guilty or have no idea what to do when they are presented with leisure time. "A sustainable world cannot be built without full engagement of the human spirit," says Gary Gardner, director of research for the Worldwatch Institute.[918] The quest is on for purposeful lives and meaningful legacies.

• **Making way for more people of all ages who are driven to move into education, public service careers, or just volunteer.** Among the many beliefs and attitudes we need to develop is an ingrained appreciation for people who work or volunteer in the public service. Too often, people, including politicians, judge our success by the size of a bank account rather than on our contributions to society. Through our schools and colleges, parenting, and in our own lives, we can discover our worth by contributing to those who need our help and to a society that can be enriched by our generosity and spirit. Financial collapse and foreclosures remind us that what has happened to others can happen to us. We need to treat others as we would want to be treated. Carol Peck, a National Superintendent of the Year, observes, "after experiencing the increased intensity, competition, and depersonalization of the job market, people are coming to place a higher value on loyalty and engagement in work environments—like teaching—that value relationships and the opportunity to change lives."[919]

• **Paying more attention to both physical and emotional health and well-being.** With limited life experience, students need skills to cope with situations that may seem overwhelming—a bad grade, the loss of a boyfriend or girlfriend, a move, being bullied, or an economic downturn that puts college funds at risk. Gangs are identity groups, often providing a homebase for people who feel they have been excluded or marginalized. Combined emotional upset and a scarcity of experience in dealing with conflict resolution too often leads to acts of violence. A thoughtful array of essential life and leadership skills could help students find perspective and give them a framework for dealing with emotional trauma that might otherwise lead to self-destructive behavior.

Patrick Newell, who leads the Tokyo International School, believes, "the time has come to reconnect with nature." Too often, he says, "children are connected to their digital tools and are not connecting with their five senses." Newell notes that "perception and reality are sometimes digitally blurred," as people overly depend on technologies for their relationships. "Many are in search of personal meaning and purpose to help them find and fuel their passions." All of us need to understand that "some type of magic happens when humans are connected with the harmony of nature."

• **Additional Implications.** We've spelled out some of the benefits: heightened personal effectiveness; better relationships; the confidence to communicate positively and effectively; enhanced ability to manage personal emotions and help others deal with theirs; and increased flexibility to deal with conflict, change, and growth. Even greater levels of creativity and productivity flow from empowered people, who feel more confident working in teams, and who have greater respect for their organization or community. Obviously, emotional health has significant implications for personal relationships that range from being a good citizen of the world to being a balanced and positive force in a family.[920]

Personal Meaning and Work-Life Balance
Questions and Activities

1. How can organizations give people a greater opportunity to pursue a balance between their work and their personal lives? Develop a five-minute PowerPoint presentation to explain.

2. Develop a list of five reasons why giving short shrift to our personal lives in favor of making ourselves personally more productive can eventually become damaging to productivity.

3. In no more than one page, suggest the basics of what we should be teaching students about "emotional intelligence" and "work-life balance."

4. View "Daniel Goleman Explains Emotional Intelligence," interview, *You Tube* (26.37). Goleman discusses his book, *Working with Emotional Intelligence*, http://www.youtube.com/watch?v=NeJ3FF1yFyc.[921] What three ideas stand out for you? Discuss them with a group.

5. Read "Be Good to Yourself," an article by teacher Brad Kuntz, *Education Update*, ASCD, Oct. 2012, http://www.ascd.org/publications/newsletters/education-update/oct12/vol54/num10/In-the-Classroom-with-Brad-Kuntz.aspx.[922] Identify up to five suggestions he makes for teachers to take good care of their students and themselves. Discuss the ideas with a group.

Readings and Programs

1. Goleman, D. (1997). *Emotional Intelligence . . . Why It Can Matter More Than IQ*. New York: Bantam Books.[923]

2. Easterbrook, G. (2004). *The Progress Paradox, How Life Gets Better While People Feel Worse*. (Trade Paperbacks Edition.) New York: Random House.[924]

3. View "How to Make Work-Life Balance Work," Nigel Marsh, *TED Talk*, taped May 2010, in Sydney, Austrialia, http://www.ted.com/talks/nigel_marsh_how_to_make_work_life_balance_work.html,[925] (10:05), posted Feb. 2011.

4. View "Wellness Policy Video," Santa Barbara Independent School District, California, (6:42) http://www.youtube.com/watch?v=3VZ2Zq6NrTQ; and San Francisco Unified School District (SFUSD)-High School Wellness Program as Public Policy, Video (10:02), http://www.youtube.com/watch?v=tWtT_vF5a-0.[926]

Conclusion

Dealing with the Trends
Creating a Future is the Essence of Leadership

Conclusion

Dealing with the Trends
Creating a Future is the Essence of Leadership

> **"Courage is not the absence of fear—it's inspiring others to move beyond it."**
> *Nelson Mandela*

Bold Steps
To a Brighter Future

Congratulations! You've just completed a tour de force of massive trends that affect everyone, everywhere. We've touched base with demographics, science and technology, energy, the economy and environment, globalization, education, work-life balance, and a host of other forces impacting society. All have implications for our future, in some cases, even our survival. A door to the future is wide open. What do we do now?

Everyone knows how it goes. We're pumped. The ideas are swirling. An hour later, our good intentions are still flaming. We rush to a meeting about an issue that's bubbling inside the trenches. A few conversations and text messages, plus a slight revision in policy, should handle it. Then, we respond to emails, return calls, and have four appointments in a row. Countless stuff keeps drifting into our lives, and before we know it, those dreams for creating a future are lost in the shuffle.

I had just done a presentation on trends and future-focused leadership. One thoughtful person raised his hand, stood, and declared, "You're absolutely right. We need to get connected to these trends and get people involved in shaping a future. Give us 20 years, and we'll be there." I thanked him for his eloquence and commitment to the cause but had to add, "In a world that's changing exponentially, we don't have 20 years. We need to constantly deal with the complexity, find it energizing, and make sure we're connected with the escalating needs of a fast-changing world."

We all know that just maintaining the status quo can take more hours than there are in a day or days in a year. Therein lies the challenge. The world is moving at warp speed into a Global Knowledge/ Information Age, and we're often trying to do even more perfectly what it takes to function in an Industrial Age. Then we remember the wisdom of leadership counsel Warren Bennis, who told us eloquently that "managers administer, but leaders innovate. Managers maintain, but leaders develop. Managers imitate, while leaders originate. Managers accept the status quo, while leaders challenge it." The leader, Bennis says, always has "an eye on the horizon."[927]

Aspirational Leadership: Think for a minute about *aspirational leadership*. Most of us admit that, without aspirations, circumstances will drive our future. The desk will always be full. We need to see a complex, fast-moving, sometimes chaotic world as part of a new normal. Some think of these massive, ongoing forces as attacks. Others see them as opportunities. Our challenge is to step atop the fray, join hands with everyone around us, declare our successes, and reshape what we do and how we do it to meet the needs of a world that is being transformed before our very eyes.

Every day, we give hundreds of answers, but we have little time to think about whether we, ourselves, are asking the right questions. Things that worked two years ago may not be cutting it today. Some of us defend the old ways, because we haven't taken time to consider that accelerating trends are creating new expectations.

How can we get a handle on the future? How can we demonstrate our intellectual leadership and breakthrough thinking? How can we energize our education system and our community? How can we take those exciting steps toward creating a new agenda for a world that is resetting for a whole new era?

> "I have traveled the length and breadth of this country and talked with the best people. I can assure you that data processing is a fad and won't last out the year."[928]
> *Editor in charge of business books for a major publisher, 1957*

Getting a Handle on the Future
A Plan of Attack

What can we do now? Based on what we've learned from *Twenty-One Trends*, what's a logical plan of action to help us shape the future? We'll take a look at a number of possibilities, such as planning; turning our plan into a living strategy; generative thinking; using social media; agenda time; professional development; conferences and seminars; college, university, and school courses; futures studies; book clubs; and probing a few trends a week or month with the team. This book, and the processes it suggests, can serve as an environmental scan that is essential to planning, goal-setting, and accreditation.[929] For scholars, teachers, professors, and those with inquiring minds, study *Questions and Activities* and *Readings and Programs* at the end of each chapter and get connected with our more than 900 references.

Whatever we do, we should keep in mind that the action we take (or don't take) is *not* just a one-shot-deal. In a world moving at what seems like hypersonic speed, the quest to redefine an organization and refresh our thinking is an ongoing proposition.

Of course, state associations, local chapters, and national associations, such as ASCD, the American Association of School Administrators (AASA), National Association of Elementary School Principals (NAESP), National Association of Secondary School Principals (NASSP), National School Boards Association (NSBA), Council of Chief State School Officers (CCSSO), National Education Association (NEA), American Federation of Teachers (AFT), National Governors Association (NGA), and National Council of State Legislatures (NCSL), are only a few of the organizations that can encourage use of these ideas to create a more sustainable future.

Futures Processes
Getting and Staying Connected

One of the most basic of futures processes, or futures tools, is identifying, monitoring, and considering the implications of trends—but we won't stop there. We'll take a brief look at several of those tools, ranging from trend and issue analysis to gap analysis, and explore the idea of convening Community Conversations and Futures Councils. As we get the journey under way, let's also not forget we can become trendsetters.

• **Trend Analysis.** *Twenty-One Trends* zeroes in on revealing and dealing with trends. We've spotted those 21 trends from the hundreds sweeping across the landscape. Each of us should try to become a *trend spotter*. How can we do that? Intuition can help. Things we see, hear, and read start adding up. We might, in fact, identify trends through content analysis—by systematically considering the frequency a topic is mentioned in the media—or through a number of other systematic processes or futures tools we'll be introducing. We need to pay attention to every field, not just our own.

Then, do follow-up research to confirm that what you've identified could become a trend, based on hard evidence (strong signals) and glimmers of a tendency (soft signals). Also consider whether the current trajectory of a product, issue, or mentality is sustainable. Just to illustrate: The level of polarization we have experienced in society is likely *unsustainable*, if we hope to maintain domestic tranquility. Business as usual probably won't be possible in an economy headed into a massive reset.

Our challenge now is to identify implications of the trends we've highlighted in this book. Consider an example:

Trend: Identity and privacy issues will lead to an array of new and often urgent concerns and a demand that they be resolved.

Question: What are the implications of this trend for how we operate our (school, university, or other organization)?

Question: What are the implications of this trend for what students need to know and be able to do, their academic knowledge, skills, behaviors, and attitudes?

• **Issue Management.** This process kicks off by engaging a group in identifying some of the dozens of issues facing an organization, industry, profession, government, or nation. An issue can be a trend, condition, or problem, internal or external, that could, will, or does impact the successful achievement of our objectives. Together, you identify what those issues are. Next, working as a group, judge each one based on the *probability* or *likelihood* it will become *major issue*. Of course, 10 percent would be low probability; 90 percent would be high, and 50 percent would be 50-50. Then, consider the potential *impact* of each issue—high, medium, or low. We'd better pay attention to issues that rank high in probability and high in impact. If we don't manage certain of those issues, they may eventually manage us. Even issues that rank low in probability and high in impact could become significant wildcards or game-changers. The process can also spur a discussion of possible side effects, including unintended consequences.

Figure C.1 provides a hint of how to sort issues using a *probability/ impact matrix*. Using this matrix, we can classify each of the issues by type. Some are *critical*, some *ongoing*, and some just *emerging*. Spotting issues as they're emerging gives us a better chance of dealing with them before they turn into problems. Based on that analysis, we can then decide what priority to assign each of them. We might follow up by assigning people or groups to track and make recommendations on certain issues. In some cases, our analysis will drive policy or action.

Figure C.1
Probability/Impact Matrix

Issue Type				Issues Identification and Sorting	Impact			
Critical	Ongoing	Emerging	Priority	**Issue Statement**	Probability (%)	High Impact	Medium Impact	Low Impact
X			1	High-stakes testing has led to a narrowing of our curriculum to those subjects most easily tested.	90	X		
	X		2	Parents are expecting teachers to immediately respond to email and voicemail messages.	40		X	

• **Gap Analysis.** Among the most exhilarating of these futures processes is *gap analysis*. Like trend and issue analysis, it gets people involved. In this case, it helps us envision characteristics or ideals for the organization we want or need to become, if we hope to prepare our students for life in a Global Knowledge/Information Age. We can break into small groups to come up with several series of six to eight single-sentence characteristics or ideals for the organization of the future.

Then, ask an inclusive group to use a scale—perhaps 1 to 10—to rate us on how well each statement currently describes how well we're doing. Don't be concerned if we come up with an occasional *zero* or *one*. That number just gives us an idea how far we have to go if it's a direction we want to take in pursuing certain ideas. Some might be brand new. The process helps us discover gaps between where we are and where we'd like to be. Of course, the next step is to figure out how to fill those gaps.

Here are a few examples of statements developed in a couple of school systems using this process as they responded to trends: ◊ "Learning in the classroom will be interconnected with experiences and careers outside the classroom." ◊ "We will prepare students in process skills, including creative thinking, diplomacy, teamwork, ethics, and individual curiosity, that they will need to function effectively in the future." As we all know, sometimes we are more animated about what we *oppose* than *what we hope to achieve*. This process can clarify the dream and get even more people on the dream team.

• **Flexibility/Innovation Analysis.**[930] Involve a diverse group in identifying five to 10 statements describing a flexible, innovative organization. Then, again on a scale of perhaps 1 to 10, ask knowledgeable groups of people to rank how the organization currently compares to each of those descriptions. An example of one of those statements might be, "The organization can readily introduce new programs and services." The insights give us a place to start in identifying and filling any gaps.

• **Historical/Defining Moments Analysis.** In this case, a group identifies examples of programs that have been successful or have faced persistent problems. Members of the group discuss various decisions, budgetary events, staff changes, or other defining moments that impacted the program one way or another. It's a way to learn from history and from both our actions and inactions, as we plan for the future.

• **Thinking across Trends.** We've made clear in *Twenty-One Trends* that new ideas are often hatched in the white spaces or connective tissue between and among trends. We might ask, for example, "What are the implications of *aging* for *technology* or for *jobs and careers*?" That's prime ground for knowledge creation and breakthrough thinking.

- **Scenarios.** Based on our discussion of trends and issues, our history, how flexible we are, the characteristics of the organization we want to become, and numerous other types of data, we can develop sets of *assumptions*. (For example, they might address *increased, decreased,* or *stable* enrollments or budgets.) Based on a variety of assumptions, we can compose a series of three or four scenarios describing *alternative futures.* Scenarios have been described as "a coherent picture of a plausible future,"[931] also as a way to demonstrate foresight.
- **PEST, STEEPV, STEEPED.** For the PEST process, consider political, economic, social, and technological forces that will likely affect the organization and the whole of society. For STEEPV,[932] add environmental forces and values. For STEEPED, further add educational and demographic forces.
- **Demographics and Psychographics.** If we hope to know who we are as an organization, community, or country, we need to be familiar with demographic research. Among common demographic factors are race and ethnicity, average age, and socioeconomic data. Obviously, all people within certain demographic groups don't share the same opinions or sets of values. Psychographics can help us identify people by their interests, their views on certain issues, and what motivates them.

Our companion book, *Future-Focused Leadership: Preparing Schools, Students, and Communities for Tomorrow's Realities*, published by ASCD, focuses on these processes and provides greater detail on how they work.[933]

Benefits of Staying in Touch with Trends. Studying and considering the implications of these and other trends enables us to, among other things:

- Keep our organization fresh, energized, and open to new ideas.
- Encourage creativity and imagination.
- Get connected to forces affecting the whole of society.
- Use futures tools to identify problems far enough in advance that they don't become crises or even catastrophes.
- Identify and stay in tune with possible tipping points.[934]
- Spot opportunities we otherwise might not have considered.

- Discover that far-reaching trends go beyond today's issues, such as class size, standards, accountability, and testing, but have a direct impact on all of them.
- Overcome the isolation of our disciplines, disagreements, and other differences to find the connective tissue that unifies us.[935]
- Forecast possible futures and even become trendsetters.
- Turn our institution into an even more indispensable, relevant force.
- Stimulate teamwork, public engagement, and a sense of ownership.

Engagement and Collaboration

In our chapter devoted to authority, we shared an example of a person in a position of leadership who decided he could save a lot of time by just involving a few close colleagues in making big decisions. He suffered from not getting the insights of people who worked in the trenches and stakeholders in the community, ended up with a lack of ownership for decisions, and eventually had to find work elsewhere.

We've tried to be clear that we are moving from vertical (top-down) toward horizontal (collaborative or lateral) decision making. That doesn't mean vertical decisions will disappear, but they will be informed by engaging people in the conversation and clearly communicating what we're doing and why it will benefit our bottom line, which could be the education of students, service to the community, or a more viable product or service.

How can we connect people? How can we bring them together in common purpose? How can we tap the diversity and richness of thinking in an organization? How can we create a rallying point and ultimately a sense of ownership for what we want to accomplish? How can we stir a sense that "We're all in this together"? The answer is not that complicated. We simply need to *engage people and let their ideas inform our decisions*. Not everyone will get what they want every time, but they'll know their voices have been heard and their ideas considered.

The possibilities? *Community Conversations* engage diverse group of people in the processes we've just explained. *Futures Councils* and other *advisory bodies* can be an ongoing ear to the ground for us as we ask staff and community to share what they believe might be emerging issues or trends and to think with us about the implications of trends we've addressed in *Twenty-One Trends*. We can expand the process with *crowd-sourcing, systematic or informal surveys,* and *interviews.* Reserve *agenda time* and do *listening sessions* at every type of internal and external meeting or get-together. Use every reasonable occasion to share ideas and harvest questions, concerns, and suggestions. All are among the many on-the-ground opportunities for connected and collaborative leadership.

Putting the future squarely and constantly on the agenda is invigorating. It encourages *generative thinking.*[936] Put simply, we're ensuring that significant groups of people are identifying, studying, and considering the implications of trends, as well as critical, ongoing, and emerging issues that will likely affect our organization and its mission. These groups can provide ideas that will help us keep our decisions connected to the realities of today and possibilities for the future.

Rather than complaining that "we don't have time to think," let's make *thinking* a part of our everyday business and engage a network of groups in the process. The discussion moves from excuses for yesterday to the excitement of creating a future. Everyone is enriched by a free and open, ongoing discussion of trends.

Starting with the big picture and moving toward specifics generally makes more sense than trying to build a future on steamy arguments about current issues that suck all the air out of the room and leave little or no time for consideration of the future. If we want everyone in the same boat—all hands on deck—we need to let them on board.

Consider Multiple Types of Councils.

Think about holding these types of sessions with teachers, professors, administrators, board members, and various professional associations and groups. For example, math, reading, science, language arts, social studies, civics, early childhood, vocational and career, art, and other teachers and departments will likely want to explore implications for their particular disciplines. Some councils could be formed across disciplines. College presidents; superintendents; principals; deans; academic directors; business officials; department heads or chairs; planners; elementary, secondary, and postsecondary educators; curriculum and instruction directors; professional developers; human resource professionals; those who prepare teachers, administrators, and other professionals in many fields;

accountability directors and specialists; technology experts; communication professionals; special educators; counselors; and transportation, facilities, and other groups may want to consider the specific implications of these trends for them and what they do. Broadening the education base, P-16 and P-Life groups might consider how these trends could or should influence education, pre-school through college or life. Add Governors, State Legislators, Members of Congress, Presidents, Mayors, departments and agencies, associations, industries, and professions. The possibilities don't stop with this list.

Developing a Plan, Living Strategy, Strategic Vision
Making It a Continuous Process

Linear planning. It's been around for years, likely because it perpetuates itself. Planning is always hard work. However, it's easier when we put most of our energy into just moving the same things up, down, or sideways. How does anything new ever get on the agenda or into the plan? Linear plans might even be dangerous to our success or survival in what is fast becoming a non-linear world.

Flexibility, nimbleness, and vision should be built into our plan. That way, we can turn on a dime to pursue an opportunity or deal with a concern in a 24-7 world that just won't wait. To some extent, *we've lost the between* as we move from here to there. That makes most of us uncomfortable, but it's a reality we face. Bottom line: We need to turn our *strategic plans* into *strategic visions* and *living and evolving strategies* that can guide us as we build for a new tomorrow.

Daily, even hourly, we are confronted with problems and opportunities that may have, even yesterday, been off the charts—beyond our imaginations. One of many reasons we pay attention to trends and issues? It helps us get a glimpse of what's just over the horizon or simply headed our way. We need to be true to our values and our heritage, but also nimble enough to quickly adjust our plan.

Freeze the plan, and the world will surge past us, leaving us as an island, disconnected from those we are pledged to serve. If that sounds disconcerting, it isn't. It's just a new way of thinking, a fresh mentality, as we pursue a constant process of creating the future we need, rather than simply defending what we currently have. That approach can breathe life into an organization. Our mentality shifts to becoming even better tomorrow, no matter how good we are today.

A laser focus on defending what we have always done is not a substitute for going on the offensive. Taking initiative has become basic. In flying language, that means we need to constantly make in-flight adjustments as we hit stormy weather or get a boost from a tailwind. The analogy applies to our education system and many other institutions. Too often, we get instructions from *our own control tower* to jettison our fuel and climb to a higher altitude—tantamount to cutting budgets and improving achievement, all, supposedly, without breaking a sweat. Those types of commands are evidence that we may have a long way to go in teaching thinking and reasoning skills.

New trends will emerge. A new mix of generations will populate our planet. Astounding new technologies will burst into our lives and become commonplace. Industries and careers will change. Cultivating human ingenuity will become an increasingly important part of education. New ethical dilemmas will confront us. Peace will need to be won and diseases conquered. All of us, and certainly students in our schools and colleges, must be prepared to coherently deal with problems and opportunities that are beyond our imaginations.

Make *Twenty-One Trends for the 21ˢᵗ Century* an essential part of the planning process. It is designed to provoke discussion, raise important questions, focus on far-reaching trends, provide essential data and ideas, and stimulate thinking about what our education institutions and other organizations will need to *be like* and *look like*. Think of this book as an *intelligence report*, designed to connect us with forces that should inform our decisions, since they will profoundly impact our future. The reverberations of many of these trends are already going global.

While we surely need to tackle today, we simply can't take our eye off tomorrow. We have an opportunity to build an even better world; create what truly are leadership organizations; demonstrate our ability to engage in knowledge creation and breakthrough thinking; and, perhaps selfishly, build our own legacies. It's time to get the process under way and keep it going.

Enjoy the journey as you constantly create an even better future.

12 Guiding Principles for Leaders Capable of Creating a Future

Twenty-One Trends for the 21ˢᵗ Century published by Education Week Press, and *Future-Focused Leadership,* published by ASCD, can be considered companion publications. In this book, we directly address massive trends and consider their implications for education and the whole of society. In *Future-Focused Leadership,* we take an in-depth look at the many ways we can scan the environment and develop strategic communication and a vision for the future. Included is a chapter devoted to "Twelve Guiding Principles for Leaders Capable of Creating a Future." Here is a brief listing of those principles.

1. Curiosity, persistence, and genuine interest are the main power sources for futures thinking.
2. Breadth and depth are both important.
3. Leaders connect the dots and seek common ground.
4. There are more than two sides to most issues.
5. The future is not necessarily a straight-line projection of the present.
6. Enlightenment and isolation are becoming opposites.
7. Peripheral vision can help us avoid being blind-sided.
8. A belief in synergy can spark knowledge creation and breakthrough thinking.
9. Collateral opportunity and collateral damage both deserve our attention.
10. Bringing out the best in others is basic.
11. Courage and personal responsibility need to overcome fear and self-pity.
12. The role of strategic futurist is part of everybody's job.

Future-Focused Leadership: Preparing Schools, Students, and Communities for Tomorrow's Realities, Gary Marx, published by ASCD, 2006.

"Insanity is doing the same thing over and over again and expecting different results."
Albert Einstein

Deaing with the Trends
Questions and Activities

1. Using what you've learned about trends from this publication as well as other readings, discussions, and activities, develop a plan that will further connect you and your organization to forces impacting the whole of society. You might call it "a constant process for creating a future."

2. Convene a diverse but knowledgeable group. Together, identify five issues your organization faces. Judge each one based on the probability/impact matrix explained in this chapter.

3. Engage a diverse and knowledgeable group in studying each of the trends discussed in this book. Then, ask that they work as a group to identify the implications of those trends for the education system or other institution.

Extend the effort to a Community Conversation by including issue analysis and gap analysis. The processes are explained in this chapter.

4. What steps would you recommend for turning your strategic plan into a living strategy or strategic vision?

Appendix
Conclusion-Plus

Getting People Engaged: Context, Considerations,
Pointers, Follow-Up

Invariably, we are looking for fairly concrete steps or at least idea-starters to bring people together in common purpose and help us shape the future. In some cases, an education system or other organization might start with a *Community Conversation*, an event that could involve a diverse group of people. An ongoing process to do follow-up and stay in touch with a fast-changing environment could also involve a network of *Futures Councils*. The next few pages include notes on how we can convene these types of "brain trusts" or "talent pools" and tap their genius.[937]

What's in a name?
We'll be discussing Community Conversations and Futures Councils. While you will want to develop an approach that is best for you, you might consider starting with a community-wide meeting of 150 to 250 people (or another number) who represent the diversity of those you serve. Then, to stay in regular contact, form Futures Councils, perhaps a network of them, with regularly rotating memberships, who can help you spot trends and issues, stay in touch, and build an ongoing sense of ownership and community. It's a way to tap the good thinking that is all around us.

Develop a Context for Consideration of Trends

Whether we're holding *Community Conversations*, preparing for a *Futures Council* gathering, or getting connected in a host of other ways, we need to have our ducks in order and make good use of everyone's time. Here are some steps we might consider.

• **Appoint a council or councils.** Consider who might serve on a system-wide Futures Council, a diverse group from a span of interests and backgrounds—and representative of the community. Staff should also be involved in the process. In fact, you might want to appoint a network of these councils, with rotating memberships, that meet once or twice a year.

• **Develop an Agenda.** Review background materials you've shared, possibly in advance. Arrange for a 60-90 minute presentation on trends and issues. As part of the agenda, include small-group (table group) brainstorming sessions (six to eight people to a group) focused on identifying possible implications of trends. Then ask table facilitators to share their reports with the group. The implications activity, including group sharing, might take up to 45 minutes, depending on how many are participating.

Trend analysis is often followed by sessions devoted to issue analysis and gap analysis. During gap analysis, members of the group brainstorm from six to eight full-sentence descriptions (characteristics or ideals) of the education system we need to get students ready for the future. Then, they rate how we are currently doing in relationship to each of those statements...on a scale of 1 to 10. Note the gaps, since they will likely drive later discussion about plans of action to get us from where we are to where we'd like to be. Following the small group sessions, have everyone come together to briefly share their ideas. In each small group, all ideas should be captured on recorder sheets, which are then collected and summarized for consideration in decision making.

• **Inform the discussion.** Ask members of the council(s) to study trends and issues, using resources such as *Twenty-One Trends*. If possible, provide copies of this book and other materials. Distribute those same materials to all board members and administrators, as well as staff, community leaders, and possibly members of the media. We should demonstrate that we are all in this together.

• **Clarify the advisory role.** Be clear that the council is not making decisions. Instead, it is advising on the possible implications of these trends and even on ideas for dealing with them. No votes. No consensus taking. Just ideas. However, emphasize that the council's thinking will be seriously considered but not always accepted as we plan for the future. The group is sure to turn out some ideas whose time has not yet come. Decisions, after all, are often a hybrid of several ideas and points of view.

Our goal is to enrich decision making as board and staff constantly work toward a better education for students and the community.

• **Broaden the scope.** In considering the scope of the councils' advisory efforts, keep in mind that the effects of these trends are important to every institution in the community. They have direct implications not only for education but also for quality of life and economic growth and development. You'll find businesses, industries, professions, governmental and nongovernmental organizations, foundations, even states and nations, discovering the possible impact of these trends for them. *By broadening the context of the discussion, the education system can become the crossroads and central convening point for addressing needs and possibilities for the entire community.* [938]

Hold Community Conversations/Futures Council Meetings

• **Leadership, funding, and frequency of meetings.** Futures Councils or Community Conversations can take many forms. These groups might: come together once or twice a year; hold productive meetings that range in length from two hours to a full day; consider meeting at a venue that has special significance for the community; ask a respected, capable member of the community to serve as honorary chair; and think about symbolism. If you're hoping to stay in touch with the fast-changing world of the 21st century, for example, you might call these groups Futures 21 Councils.[939]

Of course, there are many other possibilities. Consider asking local sponsors to cover the costs of the meetings and follow-up activities, making clear that their sponsorship should not be seen as a way to directly promote a product or service or influence results of the effort. Certainly, a variety of online possibilities might be on the agenda before, during, or after; however, if at all possible, the primary meeting should draw actual participants together in the same place at the same time. Capturing and building energy, tapping a wealth of ideas, and reinforcing a sense of community around building for the future are essential ingredients.

• **Aim for productive thinking.** Use a highly facilitated process to ensure productive thinking and efficient use of Council members' time. Appoint a facilitator who is adept at keeping the meeting moving while bringing out the best, most productive thinking of everyone involved. Facilitated small-group brainstorming sessions followed by sharing with the entire group generally stimulates productive thinking, releases ingenuity, and creates a great deal of energy and excitement.

• **Use generative thinking and brainstorming.** As part of the agenda, explain the idea of generative thinking and the process you'll use for brainstorming. Ask participants to avoid using "killer" phrases that shut down communication and resist the temptation to limit discussions to a few things some people want to get off their chests or to seize a captive audience for a "pet" issue. Make sure everyone is engaged in the process; exchange long-winded discussions for idea-rich productive thinking.

• **Identify and address operative questions.** If the group is considering education, Council members might be asked to respond by focusing on driving questions, such as: "What are implications of these trends for our education system?" and "What are implications of these trends for what students need to know and be able to do to be prepared for the future—their academic knowledge, skills, behaviors, and attitudes?"

If we have time, we can broaden the discussion to the community at-large, asking small groups to brainstorm answers to questions such as: "What are implications of the trends for economic growth and development and quality of life in our community?" "What concerns are we seeing in our community that are directly or indirectly related to each of these trends?" "What roles should educators, parents, nonparent taxpayers, public officials, and others play to get us from where we are to where we need to be as a community?"

Do the Follow-up

• **Organize results of the meeting(s).** Collate ideas that emerge from the process. After editing for clarity, consider summarizing and sharing the ideas with a broader cross-section of the community. That can be done through the news media, newsletters, web pages, speaking engagements, public discussions, and other means. likely a combination.

• **Consider a follow-up questionnaire.** Moving beyond the actual Community Conversation or Futures Council(s), compose a brief questionnaire that invites an expanded group of people to consider and comment on some of the trends and implications that emerge from the meetings. It's an opportunity to increase public engagement and gain further insights about the local impact of these trends from both staff and community. This follow-up questionnaire might be distributed through email or regular mail, and serve as a discussion starter at parent, community, or service club meetings. Questions could be placed on the agendas for faculty or school family meetings and used to stimulate thinking among people in many walks of life. The purpose of the follow-up activity is to elicit the thinking of an expanded group of people in the community, to benefit from the ingenuity that is all around us, and to help us shape the organization for the future. The purpose is *not* to determine the popularity of one issue over another or promote a project or program.

- **Summarize the expanded comments.** Put together a brief summary of comments that includes the gist of ongoing thinking of Futures Councils and responses to questionnaires.

- **Develop suggestions for possible action.** Think about the wisdom of appointing a representative Futures Action Team. That group could be made up of staff, along with a few representatives of Futures Councils. Together, they might develop a coherent set of suggestions for addressing the trends. At each step, participants should be reminded to focus their thinking on what the education system *may need to consider* as it prepares students for a Global Knowledge/Information Age. The groups are not intended to make policy or program decisions.

- **Keep the leadership team informed.** Prepare ongoing updates and reports for the superintendent, CEO, president, or other appropriate organization leaders.

- **Update board and community.** The executive leadership team will want to provide regular reports on the activity for board and community. An ongoing flow of ideas further demonstrates collaboration and will be helpful in making even more connected management and policy decisions. Again, for clarity, the purpose is not necessarily to influence immediate decisions but to stimulate expanded thinking as we shape our future.

- **Make adjustments in plans and activities systemwide.** Consider how the ideas might affect the systemwide plan. Use whatever approval process is required to move forward, with adjustments that will help address concerns and make the most of opportunities. What we do will help us turn our strategic plan into a living strategy.

• **Consider departmental and individual adjustments.** In departments, across departments, and in classrooms, teachers and administrators will be able to consider longer-term possibilities. In some cases, ingenious ideas might be put to use right away.

• **Put the plan in motion.** After adopting or adjusting a plan, develop a theme; set targets; put the plan into motion; and make clear to the community that its support will be crucial. In some cases, an important part of this effort might be establishment of a sense of urgency. Expectations should be developed that are mutually shared and reasonable. If additional resources are needed, the community-wide involvement should increase chances of getting them. Board members will also be involved in parts of the process and may be called upon for policy decisions when they are required.

• **Be prepared to address concerns and to energize everyone.** Be ready for objections. Some will try to shut down idea generation by declaring that we'll "never get the money." *If we take the objections of habitual naysayers seriously, then very little will ever move forward.* We'll be stopped in our tracks. While some of the ideas generated through this process might come with a healthy price tag, many others can simply help us focus or redirect our current energies. Others will provide logic for making further investments that will pay off down the line. If we don't make some reasonable investments, we may suffer the cost of neglect. The dividend is a sense of being connected to an organization that is truly making ongoing contributions to people's lives.

• **Set an example.** Remember that this gravity-breaking effort will be seen as an act of leadership and will set an example for others in your community, your state, and possibly even the nation.

References

1. Martin, Antoinette, "Defining the Buyer of the Future,"*New York Times*, Feb. 6, 2009. http://www.nytimes.com/2009/02/08/realestate/08njzo.html.

2. Gillis, Justin, "Heat-Trapping Gas Passes Milestone, Raises Fears," *New York Times*, May 10, 2013. http://www.nytimes.com/2013/05/11/science/earth/carbon-dioxide-level-passes-long-feared-milestone.html?pagewanted=all&_r=0.

3. Gallman, Robert E., "Trends in the Location of Population, Industry, and Employment," Ohio State University, Table 1. http://ageconsearch.umn.edu/bitstream/17629/1/ar610017.pdf, based on information from Historical Statistics of the United States, Bureau of the Census, 1960, p. 74.

4. U.S. Bureau of Labor Statistics, "Employment Projections. Employment by Major Industry Sector, Table 2.1." http://data.bls.gov/cgi-bin/print.pl/emp/ep_table_201.htm.

5. Chase, Howard, *Issue Management: Origin of the Future*. Issue Action Publications, Stamford, CT, 1984, p. 38.

6. *Webster's New Collegiate Dictionary*. Merriam-Webster Inc., Springfield, MA, 2003, p. 1,334.

7. Marx, Gary, *Sixteen Trends...Their Profound Impact on our Future*, Educational Research Service, Alexandria, VA, 2006-2008, Editorial Projects in Education, Bethesda, MD, 2012, p. 1.

8. Marx, Gary, *Sixteen Trends...Their Profound Impact on our Future*, Educational Research Service, Alexandria, VA, 2006-2012, Editorial Projects in Education, 2012, Bethesda, MD, 2012.

9. Marx, Gary, *Ten Trends . . . Educating Children for a Profoundly Different Future*. Arlington, VA, Educational Research Service, 2000.

10. Marx, Gary, *Future-Focused Leadership: Preparing Schools, Students, and Communities for Tomorrow's Realities*. ASCD, Alexandria, VA, 2006.

11. Petersen, John L. *Out of the Blue...How to Anticipate Future Surprises*, Madison Books, Lanham, MD, 1999.

12. Feynman, Richard, other Feynman quotes. *The Feynman Webring*, 2004. Retrieved from the *Bill Beaty Science Hobbyist online*. http://www.amasci.com/feynman.html.

13. Bennis, Warren, & Goldsmith, Joan, *Learning to Lead*. Cambridge, MA: Basic Books, Cambridge, MA, 2003, pp.8-9.

14. Barry, Dan, "Boomers Hit New Self-Absorption Milestone: Age 65," *New York Times*, Dec. 31, 2010. http://www.nytimes.com/2011/01/01/us/01boomers.html?_r=0&pagewanted=print.

15. "Turning 30," a series of articles about Millennials turning 30 year of age, *Huffington Post*. http://www.huffingtonpost.com/news/turning-30.

16. "As the World Turns." http://www.astheworldturns.net/.

17. Strauss, William and Howe, Neil, "Cycles in U.S. History, Remembering the Future," based on research by generational experts such as, *The Time Page*. http://www.timepage.org/time.html#cycles.

18. Strauss, William, and Howe, Neil, "Global Generations and Global Aging: A Fifty-Year Outlook," presentation at World Future Society 9th General Assembly, Washington, DC, 1999.

19. Strauss, William, and Howe, Neil, *The Fourth Turning...An American Prophesy*, Broadway Books, a division of Bantam Doubleday Dell Publishing Group, Inc., NY, 1998, p. 3.

20. "GI Generation (Born 1912-1927)," The Intergenerational Center, Temple University. http://cil.templecil.org/node/35.

21. Brokaw, Tom, *The Greatest Generation*, Random House, NY, 1998, p. xx.

22. *Rolling Stone,* "Frank Sinatra." www.rollingstone.com.

23. Halberstam, David, *The Fifties*, The Amateurs Limited, Fawcett Columbine book, Ballantine Books, 1993, pp. 456-457.

24. Barry, Dan, "Boomers Hit New Self-Absorption Milestone: Age 65," *New York Times*, Dec. 31, 2010. http://www.nytimes.com/2011/01/01/us/01boomers.html?_r=0&pagewanted=print.

25. Pew Research Center, "Baby Boomers Retire," Dec. 29, 2010. http://www.pewresearch.org/daily-number/baby-boomers-retire/.

26. Park, Alice, "Baby Boomers: Not the Healthiest Generation," *Time, Health & Family*, Feb. 5, 2013. http://healthland.time.com/2013/02/05/baby-boomers-not-the-healthiest-generation/.

27. "Boomers: Well-Known People, Famous Celebrities," *Boomers Life*. http://www.boomerslife.org/boomers_social_leaders_actors_musicians_list.htm.

28. "Generation Jones." http://www.generationjones.com.

29. Williams, Alex, "For Gen Xers, A Wake-Up Call," *The New York Times*, Aug. 24, 2012. http://www.nytimes.com/2012/08/26/fashion/paul-ryan-a-wake-up-call-for-gen-xers-cultural-studies.html?pagewanted=all&_r=0.

30. Thielfoldt, Diane, and Scheef, Devon, "Generation X and the Millennials: What You Need to Know About Mentoring the New Generations," *Law Practice Today*, Nov. 2005, American Bar Association. http://apps.americanbar.org/lpm/lpt/articles/mgt08044.html.

31. "Bridging the Generation Gap," unattributed to individual but consistent with other research and observations.

http://www3.ag.purdue.edu/counties/newton/Extension%20Homemakers/Bridging%20the%20Generational%20Gap.pdf.

32. Greenberg, Eric, with Weber, Karl, *Generation We*, Pachatusan, Emeryville, CA. 2008, p. 6.

33. Graves, Tanisha, "Managing Unique Segments of the Workforce," Ithaca College, *Operations Management*, Spring 2009. http://staff.ithaca.edu/ngraves/docs/publications/generations.pdf.

34. Strauss, William and Howe, Neil, *Millennials Rising*, Vintage Books, NY, 2000, pp. 88-95.

35. Novak, Jill, "The Six Living Generations in America," University of Phoenix, Texas A&M University, *Marketing Teacher.com*. http://www.marketingteacher.com/lesson-store/lesson-six-living-generations.html.

36. *The Future According to Kids*, Nickelodeon Television Network, Viacom International, Inc., Roundtable Press, 1999, front dust cover leaf.

37. Pew Research Center for the People & the Press, "A Portrait of Generation Next, How Young People View Their Lives, Futures, and Politics," Summary of Findings, Released Jan. 9, 2007. http://www.people-press.org/2007/01/09/a-portrait-of-generation-next/.

38. Pew Research Center, "The Millennials: Confident, Connected, Open to Change," *Millennials, A Portrait of Generation Next*, Feb. 24, 2010. http://www.pewresearch.org/millennials/.

39. Pew Research Center, "Confident, Connected, Open to Change," *Millennials: A Portrait of a Generation*, Feb. 2010. http://www.pewsocialtrends.org/files/2010/10/millennials-confident-connected-open-to-change.pdf.

40. Honisch, Marty; Leaf, Margaret; and Ryan, Rebecca, "Where are the Next Cities?" *Area Development Online*, (drawing from observations of Richard Florida). http://www.areadevelopment.com/siteSelection/august09/next-generation-cities-knowledge-workers.shtml.

41. Fischler, Marcelle Sussman, "8 Amazingly-Tiny Micro-Apartments," *Forbes*, Places & Spaces, March 17, 2013. http://www.forbes.com/sites/marcellefischler/2013/03/17/8-amazingly-tiny-micro-apartments/.

42. Zollo, Peter, *Getting Wiser to Teens*, New Strategist Publications, Inc., NY, 20004, pp. 146, 205.

43. Weinberger, Hannah, "Where Are All the Millennial Feminists?" *CNN Living*, Nov. 10, 2012. http://www.cnn.com/2012/11/09/living/millennials-feminism.

44. "The Millennial Generation," (slides). http://www.indiana.edu/~oem/Retreat%20Handouts/Millennial%20Generation/Millennial%20OEM%20Retreat.pdf.

45. Strauss, William, and Howe, Neil, *Millennials Rising...The Next Great Generation*, Vintage Books, a division of Random House, NY, 2000, p. 324.

46. Friese, Lauren, and Jowett, Cassandra, of TalentEgg, "The Six Ways Generation Y Will Transform the Workplace,"*Globe and Mail*, Canada, March 12, 2013. http://www.theglobeandmail.com/report-on-business/careers/the-future-of-work/the-six-ways-generation-y-will-transform-the-workplace/article9615027/. Also see http://testegg.ca/.

47. "Restless Generation is a Challenge...and a Huge Opportunity for Employers," *deloitte.com*. http://www.deloitte.com/view/en_US/us/Services/consulting/human-capital/organization-and-talent/a90f49642dff0210VgnVCM100000ba42f00aRCRD.htm.

48. Matchar, Emily, "How Those Spoiled Millennials Will Make the Workplace Better for Everyone," *The Washington Post*, Aug. 16, 2012. http://articles.washingtonpost.com/2012-08-16/opinions/35490487_1_boomerang-kids-modern-workplace-privileged-kids,.

49. Clark, Dorie, "3 Indispensable Tips for Leaders Under 30," *Forbes*, Leadership, Jan. 24, 2013, article based on suggestions from Annmarie Neil. http://www.forbes.com/sites/dorieclark/2013/01/24/3-indispensable-tips-for-leaders-under-30/.

50. Strauss, William, and Howe, Neil, *Millennials Rising...The Next Great Generation*, Vintage Books, a division of Random House, NY, 2000, p. 320.

51. "How to Manage Different Generations," *Wall Street Journal*, April 7, 2009. http://guides.wsj.com/management/managing-your-people/how-to-manage-different-generations/tab/print/.

52. Gandhi, Mahatma, quote, Brainy Quote. http://www.brainyquote.com/quotes/quotes/m/mahatmagan160841.html.

53. Mead, Margaret quote, *Search Quote*. http://www.searchquotes.com/quotation/We_are_now_at_a_point_where_we_must_educate_our_children_in_what_no_one_knew_yesterday,_and_prepare_/15891/.

54. The Intergenerational Center, Temple University,*"Capturing Experience,"* a series of information and study modules covering generations. http://cil.templecil.org/node/5.

55. Burnstein, David, "Fast Future: The Rise of the Millennial Generation," *TED Talk* by the founder of Generation 18, a youth political engagement group, posted Nov. 30, 2012. http://macromon.wordpress.com/2012/11/30/fast-future-the-rise-of-the-millennial-generation-ted-talks/.

56. Yang Lan, "The Generation That's Remaking China," featuring, China's equivalent of Oprah, *TED Talk*, posted Oct. 2011, (17:46). http://www.ted.com/talks/yang_lan.html.

57. Strauss, William, and Howe, Neil, *The fourth turning . . . An American prophesy*. New York: Bantam Doubleday, 1998.

58. Greenberg, Eric, with Weber, Karl, *Generation We*, (2008), Pachatusan, Emeryville, CA.

59. Friese, Lauren, and Jowett, Cassandra, of TalentEgg, "The Six Ways Generation Y Will Transform the Workplace,"*Globe and Mail*, Canada, March 12, 2013. http://www.theglobeandmail.com/

report-on-business/careers/the-future-of-work/the-six-ways-generation-y-will-transform-the-workplace/article9615027/. Also see http://testegg.ca/.

60. *Wall Street Journal* Interview with Facebook founder and Millennial Mark Zuckerberg, posted June 10, 2010. http://live.wsj.com/video/d8-facebook-ceo-mark-zuckerberg-full-length-video/29CC1557-56A9-4484-90B4-539E282F6F9A.html#!29CC1557-56A9-4484-90B4-539E282F6F9A.

61. Yen, Hope, "White Population to Lose Majority Status in 2043," Associated Press, The Boston Globe, Dec. 13, 2012. http://www.bostonglobe.com/news/nation/2012/12/13/census-whites-longer-majority/cldoCAQfjIWT34hnhmXFAM/story.html.

62. Frey, William H., "Census Projects New "Majority Minority" Tipping Points, Brookings Institution, *Opinion, State of Metropolitan America*, No. 58 of 61, data used in two illustrations, "Year When Non-Hispanic White Become a Minority, By Age Group" and "Projected U.S. General Population by Race/Ethnicity, 2012 and 2060: Total Population, Youth Under 18, Seniors 65+, Percentages." http://www.brookings.edu/research/opinions/2012/12/13-census-race-projections-frey.

63. "U.S. Whites Now Losing Majority in Under-5 Age Group as Deaths Outnumber Births for First Time," *Washington Post* Business, *Associated Press*, June 13, 2013. http://www.washingtonpost.com/local/white-deaths-outnumber-births-for-first-time/2013/06/13/3bb1017c-d388-11e2-a73e-826d299ff459_story.html.

64. Roberts, Sam, "Census Benchmark for White Americans: More Deaths Than Births," *New York Times*, June 13, 2013. http://www.nytimes.com/2013/06/13/us/census-benchmark-for-white-americans-more-deaths-than-births.html?_r=0.

65. U.S. Census Bureau, "U.S. Census Bureau Projections Show a Slower Growing, Older, More Diverse Nation a Half Century from Now," News Release, Dec. 12, 2012. https://www.census.gov/newsroom/releases/archives/population/cb12-243.html.

66. Yen, Hope, "White Population to Lose Majority Status in 2043," Associated Press, *The Boston Globe*, Dec. 13, 2012. http://www.bostonglobe.com/news/nation/2012/12/13/census-whites-longer-majority/cldoCAQfjIWT34hnhmXFAM/story.html.

67. Bernstein, Robert, "Most Children Younger than Age 1 are Minorities," Census Bureau Reports, U.S. Census Bureau News Release, May 17 2012. http://www.census.gov/newsroom/releases/archives/population/cb12-90.html.

68. Siek, Stephanie, and Sterling, Joe, "Most U.S. Children Under 1 Are Minorities, Census Says," *CNN,* May 17, 2012. http://www.cnn.com/2012/05/17/us/census-population-diversity.

69. "Definitions of Diversity," New Mexico State University, Aug. 1, 2010, includes NMSU definition and definitions developed by others. Included are items developed by NMSU, Iowa State University (ISU) and Kansas State University (KSU). http://www.nmsu.edu/diversity/diversity-defined.html.

70. "Cambridge Becomes Latest District to Integrate by Income," *Education Week*, January 9, 2002, p. 11.

71. Armstrong, Thomas, *Neurodiversity in the Classroom*, ASCD, Alexandria, VA., 2012.

72. U.S. Census Bureau: *World Almanac* Books, 2004. Foreign-born population: Top countries of origin, 1920, 1960, 2000. In *World Almanac* (p. 378), NY, p. 378.

73. Camarota, Steven A., "Immigrants in the U.S., 2010: A Profile of America's Foreign-Born Population," Center for Immigration Studies, Aug. 2012, Table 5." Top 20 Immigrant-Sending Countries, 1990, 2000, 2010, reflecting the 2010 American Community Survey. http://www.cis.org/2012-profile-of-americas-foreign-born-population.

74. Camarota, Steven A., "Immigrants in the U.S., 2010: A Profile of America's Foreign-Born Population," Center for Immigration Studies, Aug. 2012, Table 1." State Immigrant Population in 2010 by Year of Arrival," drawn from 2010 American Community Survey by U.S. Census Bureau. http://www.cis.

org/2012-profile-of-americas-foreign-born-population.

75. U.S. Census Bureau, "The Foreign-Born Population in the United States: 2010," *American Community Survey Reports*, May 2012. http://www.census.gov/prod/2012pubs/acs-19.pdf.

76. Passel, Jeffrey S., and Cohn, D'Vera, "U.S. Foreign-Born Population: How much Change from 2009 to 2010?" Pew Hispanic Center, Jan 9, 2012. http://www.pewhispanic.org/files/2012/01/Foreign-Born-Population.pdf.

77. U.S. Census Bureau, "Sources of Population Growth," March 2004, Online: www.npg.org/popfacts.htm.

78. Bernstein, Robert, "Most Children Younger than Age 1 are Minorities," Census Bureau Reports, U.S. Census Bureau News Release, May 17 2012. http://www.census.gov/newsroom/releases/archives/population/cb12-90.html.

79. Cohn, D'Vera, "Census Bureau Lowers U.S. Growth Forecast, Mainly Due to Reduced Immigration and Births," *Pew Research Social & Demographic* Trends, Dec. 14, 2012. http://www.pewsocialtrends.org/2012/12/14/census-bureau-lowers-u-s-growth-forecast-mainly-due-to-reduced-immigration-and-births/, March 11, 2013.

80. U.S. Census Bureau, International Data Base, Comparing 2000 and 2050. http://www.census.gov/population/international/data/idb/informationGateway.php.

81. U.S. Census Bureau, "Population Projections, 2000 National Population Projections: Summary Tables, 2050-2070 and 2075-2100." http://www.census.gov/population/projections/data/national/natsum.html.

82. Passel, Jeffrey; Livingston, Gretchen; and Cohn, D'Vera, "Explaining Why Minority Births Now Outnumber White Births," *Pew Research Social & Demographic Trends*, May 17, 2012. http://www.pewsocialtrends.org/2012/05/17/explaining-why-minority-births-now-outnumber-white-births/.

83. "What is the Relationship Between Race and Achievement in Our Schools? Minority Student Achievement Network Statement of Purpose," Minority Student Achievement Network, Adopted June 2003. http://msan.wceruw.org/about/statement.aspx.

84. "Achievement Gap," *Education Week*, updated July 7, 2011. http://www.edweek.org/ew/issues/achievement-gap/.

85. "Child Poverty," National Center for Children in Poverty (NCCP). http://www.nccp.org/topics/childpoverty.html.

86. "Frequently Asked Questions for Test-Takers," GED Testing Service. http://www.gedtestingservice.com/testers/faqs-test-taker#number_people.

87. Layton, Lyndsey, "National Public High School Graduation Rate at a Four-Decade High, *The Washington Post*, Jan. 22, 2013. http://articles.washingtonpost.com/2013-01-22/local/36472838_1_graduation-rate-dropout-rate-asian-students.

88. Stillwell, Robert, and Sable, Jennifer, U.S. Department of Education, "Public School Graduates and Dropouts from the Common Core of Data: School Year 2009-10, First Look (Provisional Data), National Center for Education Statistics (NCES), issued Jan. 2013. http://nces.ed.gov/pubs2013/2013309/index.asp and "Graduation Rate Hits Record High for High School Students: Government Report," *Huffington Post*, updated Feb. 1, 2013. http://www.huffingtonpost.com/2013/01/22/graduation-rate-record-high-school-students_n_2522128.html.

89. Camarota, Steven A., "Immigrants in the U.S., 2010: A Profile of America's Foreign-Born Population," Center for Immigration Studies, Aug. 2012, Executive Summary, Overall Numbers. http://www.cis.org/2012-profile-of-americas-foreign-born-population.

90. Lewis, John, "The Civil Rights Movement Past and Present," Buskin Lecture, Education Writers

Association Conference, 2000, p. 16.

91. Ryan, Kevin, Remarks at conference devoted to "Solving Ethno-national Conflicts in Europe," Tutzing, Germany, April 2013.

92. Gould, J.J., "Slavery's Global Comeback." *The Atlantic*, Dec. 2012. http://www.theatlantic.com/international/archive/2012/12/slaverys-global-comeback/266354/.

93. Booth, William, "The Myth of the Melting Pot, One Nation, Indivisible: Is It History?" *Washington Post*, February 22, 1998, p. A1. http://www.washingtonpost.com/wp-srv/national/longterm/melting-pot/melt0222.htm.

94. "Bending Toward Justice: Unfinished Legacy of Brown v. Board," *PEN Weekly NewsBlast*, summarized in *American School Board Journal* article, April, 2004. http://www.asbj.com/MainMenuCategory/Archive/2004/April/Bending-toward-Justice.html?DID=275568.

95. Booth, William, "The Myth of the Melting Pot, One Nation, Indivisible: Is It History?" *Washington Post*, February 22, 1998, p. 2. http://www.washingtonpost.com/wp-srv/national/longterm/meltingpot/melt0222.htm.

96. Marx, Gary, *Ten Trends…Educating Children for a Profoundly Different Future*, Educational Research Service, 2000, p. 16.

97. Bloch, Matthew; Carter, Shan; and McLean, Alan, "Mapping the 2010 U.S. Census," *The New York Times*, Census Bureau; socialexplorer.com, by enables access to certain demographic information for each county in the U.S. http://projects.nytimes.com/census/2010/map.

98. Frey, William H., "Census Projects New "Majority Minority" Tipping Points., Brookings Institution, *Opinion, State of Metropolitan America*, No. 58 of 61. http://www.brookings.edu/research/opinions/2012/12/13-census-race-projections-frey.

99. Kotkin, Joel, *Tribes…How Race, Religion, and Identity Determine Success in the New Global Economy*, Random House, Inc., NY,, 1992, pp. 3-4, 255.

100. Yen, Hope, "White Population to Lose Majority Status by 2043,"*Associated Press, Boston Globe*, Dec. 13, 2012. http://www.bostonglobe.com/news/nation/2012/12/13/census-whites-longer-majority/cldoCAQfjIWT34hnhmXFAM/story.html.

101. Bernstein, Robert, "Most Children Younger than Age 1 are Minorities, Census Bureau Reports," U.S. Census Bureau News Release, May 17 2012. http://www.census.gov/newsroom/releases/archives/population/cb12-90.html.

102. Maxwell, Lesli A., "Push is On To Identify English Language Learners," *Education Week*, Feb. 26, 2013. http://www.edweek.org/ew/articles/2013/02/27/22ell_ep.h32.html?tkn=XTMFx2T7Flxt-K47NSVxMBmYBCf3%2BcrSNQq7y&print=1

103. "Whites are a Minority in 1 in 10 Counties in the U.S., according to Census Bureau," *Associated Press, Philadelphia Inquirer*, Aug. 9, 2007. (hard copy)

104. Kimmel, Michael, "Solving the 'Boy Crisis' in Schools," *Huffington Post*, April 30, 2013. http://www.huffingtonpost.com/michael-kimmel/solving-the-boy-crisis-in_b_3126379.html.

105. Deruy, Emily, "Arizona Ruling Blocks Ethnic Studies," *ABC News*, March 13, 2013. http://abcnews.go.com/ABC_Univision/News/ethnic-studies-programs-minority-kids/story?id=18720280.

106. Parker, Ryan, "Denver Forum on Immigration Reform Draws Packed House," *The Denver Post*, Feb. 2, 2013. http://www.denverpost.com/breakingnews/ci_22507295/denver-forum-immigration-reform-draws-packed-house.

107. Hayes, Dianne, "Report: Despite Outpacing Men in Educational Attainment, Women's Pay Still Lagging," *Diverse Issues in Higher Education*. http://diverseeducation.com/article/48716/.

108. Southern Poverty Law Center, "The Top 5 Diversities Stories of 2010," Teaching Tolerance, Jan. 6, 2011. http://www.tolerance.org/blog/top-5-diversity-stories-2010.

109. Bloch, Matthew; Carter, Shan; and McLean, Alan, "Mapping the U.S. Census," interactive U.S. map. Source U.S. Census Bureau, socialexplorer.com. http://projects.nytimes.com/census/2010/map.

110. Abraham Lincoln's Emancipation Proclamation presented January 1, 1863, Smithsonian National Museum of American History. http://americanhistory.si.edu/documentsgallery/exhibitions/document_transcripts/transcript_emancipation_proclamation.pdf.

111. PBS site devoted to *Human Diversity*. http://www.pbs.org/race/004_HumanDiversity/004_00-home.htm.

112. *Precious Children, Diversity in the Classroom*, Teaching Activities and Related Readings, PBS, includes teaching suggestions, materials, and videos. http://www.pbs.org/kcts/preciouschildren/diversity/index.html.

113. *Precious Children, Diversity in the Classroom*, Teaching Activities and Related Readings, PBS, includes teaching suggestions, materials, and videos. http://www.pbs.org/kcts/preciouschildren/diversity/index.html.

114. Kotkin, Joel, *The Next Hundred Million: America in 2050*. NY, Penguin Books. 2010; and Kotkin, Joel, *Tribes . . . How Race, Religion, and Identity Determine Success in the New Global Economy*. Random House, Inc., NY, 1992.

115. U.S. Census Bureau, "The Foreign-Born Population in the United States: 2010," *American Community Survey Reports*, issued May 2012. http://www.census.gov/prod/2012pubs/acs-19.pdf.

116. Chevalier, Maurice quotes, *Brainy Quotes*. http://www.brainyquote.com/quotes/authors/m/maurice_chevalier.html.

117. This item is based on multiple references included in Tables T2.1, T3.1, T3.2, and T3.3.

118. "Hail to the Old Man," Presidents Day 2011. http://www.bfdg.com/other-thoughts/life-span-usa-presidents.html. Life expectancy for men in 1789, less than 35, considering infant mortality rate. Life expectancy for women would have been somewhat higher.

119. U.S. Census Data, Info Please, Pearson Education, Inc., "Life Expectancy at Birth by Race and Sex, 1930-2010," Both Sexes, and "Life Expectancy by Age, 1850-2004," sexes listed separately, 2011. http://www.infoplease.com/ipa/A0005148.html and Sullivan, Patricia, *Washington Post*, "U.S. Life Expectancy Rises, But Bad Habits Persist," July 11, 2013, p. A-6

120. U.S. Census Data, Info Please, Pearson Education, Inc., "Life Expectancy at Birth by Race and Sex, 1930-2010," Both Sexes, and "Life Expectancy by Age, 1850-2004," sexes listed separately, 2011. http://www.infoplease.com/ipa/A0005148.html. Compared with "Life Expectancy in the USA, 1900-98, Men and Women." http://demog.berkeley.edu/~andrew/1918/figure2.html and Sullivan, Patricia, *Washington Post*, "U.S. Life Expectancy Rises, But Bad Habits Persist," July 11, 2013, p. A-6.

121. Gayle, Damien, "How Long Will You Live?" *Mail Online*, Nov. 30, 2012, worldwide life expectancies based on report in *CIA World Factbook*. http://www.dailymail.co.uk/news/article-2240855/How-does-nation-rank-world-map-life-expectancy.html. U.S. ranking drawn from this report and most recent U.S. life expectancy drawn from 2010 U.S. Centers for Disease Control data.

122. U.S. Census Data, Info Please, Pearson Education, Inc., "Life Expectancy at Birth by Race and Sex, 1930-2010," Both Sexes, and "Life Expectancy by Age, 1850-2004," sexes listed separately, 2011. http://www.infoplease.com/ipa/A0005148.html. Compared with "Life Expectancy in the USA, 1900-98, Men and Women." http://demog.berkeley.edu/~andrew/1918/figure2.html.

123. Wallace, Paul, *Agequake: Riding the Demographic Rollercoaster Shaking Business, Finance, and Our World*, Nicholas Brealey Publishing, Ltd., London, 1999, p. 5-6.

124. U.S. Census Bureau, "Population Projections, 2000 National Population Projections, Summary Tables." http://www.census.gov/population/projections/data/national/natsum.html, click on appropriate individual tables.

125. United Nations, "Table 1.4. Median Age in Years for the World and Major Areas," *World Population Prospects, The 2010 Revision, Vol. 1: Comprehensive Tables*, NY, 2011. http://esa.un.org/unpd/wpp/ Documentation/pdf/WPP2010_Volume-I_Comprehensive-Tables.pdf.

126. *The New York Times Almanac,* "Births and Deaths in the U.S., 1910-2008." Birth Rate, per thousand reduced to per hundred by author, 2011, New York Times Company, NY, p.294.

127. U.S. Census Bureau, "Live Births, Deaths, Marriages, and Divorces: 1960 to 2007, Table 78,", *Statistical Abstract of the U.S., 2011.* http://www.census.gov/compendia/statab/2011/tables/11s0078.pdf.

128. "Countries with the Highest Birth Rates." http://www.aneki.com/birth.html. "Countries with the Lowest Birth Rates." http://www.aneki.com/lowest_birth.html. Birth rates also drawn from *Index Mundi.* http://www.indexmundi.com/g/r.aspx?v=25. Information attributed to *CIA Factbook,* as of Jan. 2012.

129. Haub, Carl, "World Population Data Sheet 2012," Population Reference Bureau. http://www.prb. org/Publications/Datasheets/2012/world-population-data-sheet/fact-sheet-world-population.aspx.

130. "Fertility Rates, Michigan and U.S. Residents, Selected Years, 1900-2011. http://www.mdch.state. mi.us/pha/osr/natality/tab1.3.asp; and National Vital Statistics Reports, Vol. 61, No. 1, Aug. 28, 2012. http://www.cdc.gov/nchs/data/nvsr/nvsr61/nvsr61_01.pdf.

131. Central Intelligence Agency (CIA), "Country Comparison: Total Fertility Rate," *The CIA World Factbook.* https://www.cia.gov/library/publications/the-world-factbook/rankorder/2127rank.html.

132. Centers for Diseases Control and Prevention (CDC), "Deaths and Mortality." http://www.cdc. gov/nchs/fastats/deaths.htm.

133. Table 1, *National Center for Health Statistics 2000.* In Marx, Gary, *Ten Trends,* Educational Research Service, Arlington, VA, 2000. p. 6.

134. U.S. Census Bureau, International Data Base, Demographic Overview, Crude Death Rate per 1,000 Population. http://www.census.gov/population/international/data/idb/informationGateway.php, Feb. 28, 2012.

135. National Vital Statistics Reports, Vol. 61, No. 1, and Vol. 60, No. 4, "Births: Final Data for 2010." http://www.census.gov/compendia/statab/2012/tables/12s0078.pdf , and "Deaths: Preliminary Data for 2010." http://www.cdc.gov/nchs/data/nvsr/nvsr60/nvsr60_04.pdf.

136. U.S. Census Bureau, International Data Base, "World Births, Deaths, and Population Growth, 2010." *2011 New York Times Almanac,* edited by John W. Wright and editors of the *New York Times,* 2010, New York Times Company, p 484.

137. U.S. Census Bureau , U.S. PopClock Projection. http://www.census.gov/population/www/pop-clockus.html, March 1, 2013.

138. National Center for Health Statistics, Life Expectancy (Data for the U.S. in 2001), "More Than Just Numbers," http://www.cdc.gov/nchs/data/nvsr/nvsr60/nvsr60_09.pdf, Centers for Disease Control and "Automakers Use Design to Adapt to AARP Generation," Guyette, James E., SearchAutoParts.com, http://www.searchautoparts.com/aftermarket-business/operations-distribution/automakers-use-de-sign-adapt-aarp-generation.

139. "Hail to the Old Man," Presidents Day 2011. http://www.bfdg.com/other-thoughts/life-span-usa-presidents.html. Life expectancy for men in 1789, less than 35, considering infant mortality rate. Life expectancy for women would have been somewhat higher.

140. U.S. Census Data, Info Please, Pearson Education, Inc., "Life Expectancy at Birth by Race and Sex,

1930-2010," Both Sexes, and "Life Expectancy by Age, 1850-2004," sexes listed separately, 2011. http://www.infoplease.com/ipa/A0005148.html and Sullivan, Patricia, *Washington Post*, "U.S. Life Expectancy Rises, But Bad Habits Persist," July 11, 2013.

141. Centers for Disease Control and Prevention (CDC), "Life Expectancy," U.S. Final 2010 Data. http://www.cdc.gov/nchs/fastats/lifexpec.htm and Sullivan, Patricia, *Washington Post*, "U.S. Life Expectancy Rises, But Bad Habits Persist," July 11, 2013, p. A-6.

142. U.S. Census Bureau, "Expectations of Life at Birth, 1970 to 2008, and Projections, 2010 to 2020," *Statistical Abstract of the U.S., 2012*. http://www.census.gov/compendia/statab/2012/tables/12s0105.pdf.

143. Gayle, Damien, "How Long Will You Live?" *Mail Online*, Nov. 30, 2012, Worldwide life expectancies based on report in *CIA World Factbook*. http://www.dailymail.co.uk/news/article-2240855/How-does-nation-rank-world-map-life-expectancy.html. U.S. ranking drawn from this report and most recent U.S. life expectancy drawn from 2010 U.S. Centers for Disease Control data, reported earlier.

144. Wallace, Paul, *Agequake: Riding the Demographic Rollercoaster Shaking Business, Finance, and Our World*, Nicholas Brealey Publishing Limited, London, 1999, p. 6. U.S. Census Bureau, "National Population Projections, Summary Profiles, Total Population by Age, Sex, Race, and Hispanic Origin, Middle Series, 1999-2100," Washington, DC, http://www.census.gov/population/projections/data/national/natsum.html, Feb. 28, 2013.

145. Haines, Michael R., "Median Age of the Population by Sex for the U.S.: 1800 to 2010," Figure 5.1, (2006c), Historical Statistics of the United States: Millennial Edition, Vol. 1, pp. 1040 to 1-101, especially pp. 1-71; and the U.S. Census Bureau, decennial census publications, Sept. 22, 2012. http://www.demographicchartbook.com/Chartbook/images/chapters/gibson05.pdf.

146. "Population Projections, 2000 National Population Projections: Summary Tables," 2050-2070 and 2075-2100. http://www.census.gov/population/projections/data/national/natsum.html.

147. United Nations, "Table 1.4. Median Age in Years for the World and Major Areas," *World Population Prospects, The 2010 Revision, Vol. 1: Comprehensive Tables*, NY, 2011. http://esa.un.org/unpd/wpp/Documentation/pdf/WPP2010_Volume-I_Comprehensive-Tables.pdf.

148. "Births and Deaths in the U.S., 1910-2008," Birth Rate, per thousand reduced to per hundred by author, *The NewYork Times Almanac*, 2011, New York Times Company, NY, p. 294.

149. National Center for Health Statistics, U.S. Dept. of Health and Human Services, "Birth Rates and Fertility Rates by Age of Mother, 1950 to 2010, *World Almanac 2013*, World Almanac Books, NY, pp. 201-202.

150. U.S. Census Bureau, *Statistical Abstract of the U.S., 2012*, "Live Births, Deaths, Marriages, and Divorces: 1960 to 2008, Table 78." http://www.census.gov/compendia/statab/2011/tables/11s0078.pdf.

151. U.S. Census Bureau, *Statistical Abstract of the U.S., 2011*, "Live Births, Deaths, Marriages, and Divorces: 1960 to 2007, Table 78." http://www.census.gov/compendia/statab/2011/tables/11s0078.pdf.

152. "Countries with the Highest Birth Rates." http://www.aneki.com/birth.html. "Countries with the Lowest Birth Rates." http://www.aneki.com/lowest_birth.html. Information attributed to *CIA Factbook*, as of Jan. 2012.

153. National Center for Health Statistics, "Fertility Rates, Michigan and U.S. Residents, Selected Years, 1900-2011." U.S. numbers converted from per thousand to per hundred by author. http://www.mdch.state.mi.us/pha/osr/natality/tab1.3.asp.

154. *World Almanac, Book of Facts, 2008*, "Birth Rates; Fertility Rates, by Age of Mother, 1950-2005," World Almanac Education Group, NY, 2008, National Center for Health Statistics, U.S. Department of Health and Human Services, p. 159.

155. Centers for Disease Control (CDC), *National Vital Statistics Report*, Vol. 61, No. 1, Aug. 28, 2012,

"General Fertility Rate for 2010 in U.S.," converted from thousands to hundreds by author, p. 2. http://www.cdc.gov/nchs/data/nvsr/nvsr61/nvsr61_01.pdf.

156. U.S. Census Bureau, Statistical Abstract of the U.S., 1999, "No. 1420, Live Births, Deaths, Infant Deaths, and Maternal Deaths, 1900 to 1997," 20th Century Statistics, Section 31, p 874. http://www.census.gov/prod/99pubs/99statab/sec31.pdf.

157. Centers for Disease Control (CDC), "Deaths and Mortality," Final Data for 2010. http://www.cdc.gov/nchs/fastats/deaths.htm.

158. The World Bank "Death Rate, Crude (per 1,000 people), 2010 figures, listing two relatively high countries and two relatively low. http://data.worldbank.org/indicator/SP.DYN.CDRT.IN/countries.

159. National Vital Statistics Reports, Vol. 61, No. 1, and Vol. 60, No. 4, "Births: Final Data for 2010." http://www.census.gov/compendia/statab/2012/tables/12s0078.pdf , and "Deaths: Preliminary Data for 2010." http://www.cdc.gov/nchs/data/nvsr/nvsr60/nvsr60_04.pdf.

160. U.S. Census Bureau, International Data Base, "World Births, Deaths, and Population Growth, 2010." *2011 New York Times Almanac*, edited by John W. Wright and editors of the *New York Times*, New York Times Company, p. 484.

161. "Table 1, Population by Hispanic or Latino Origin and by Race for the United States: 2000 and 2010," Overview of Race and Hispanic Origin, 2010 Census Briefs, issued March 2011, p 4. http://www.census.gov/prod/cen2010/briefs/c2010br-02.pdf.

162. U.S. Census Bureau, "Table 2. Projections of the Population by Selected Age Groups and Sex for the U.S.: 2015-2060." http://www.census.gov/population/projections/files/summary/NP2012-T2.xls.

163. U.S. Census Bureau, "NP-T5-H, Projections of the Resident Population by Race, Hispanic Origin, and Nativity: Middle Series, 2075-2100," 2100 Data, released Jan. 13, 2000. http://www.census.gov/population/projections/files/natproj/summary/np-t5-h.pdf.

164. U.S. Census Bureau, "Population Projections, 2000 National Population Projections, Summary Tables." http://www.census.gov/population/projections/data/national/natsum.html, click on appropriate individual tables.

165. U.S. Census Bureau, International Data Base, "Total Mid-Year Population of the World, 1950-2050," http://www.census.gov/population/international/data/idb/worldpoptotal.php.,U.S. Census Bureau, "Total Midyear Population for the World: 1950-2050," U.S. Pop Clock, Worldometer. http://www.census.gov/population/www/popclockus.html.

166. Cohn, D'Vera, "Census Bureau Lowers U.S. Growth Forecast, Mainly Due to Reduced Immigration and Births," Pew Research Social & Demographic Trends, Dec. 14, 2012. http://www.pewsocial-trends.org/2012/12/14/census-bureau-lowers-u-s-growth-forecast-mainly-due-to-reduced-immigration-and-births/.

167. Camarota, Steven A., "Immigrants in the U.S., 2010: A Profile of America's Foreign-Born Population," Aug. 2012, Center for Immigration Studies. http://www.cis.org/2012-profile-of-americas-foreign-born-population#execsum.

168. Passel, Jeffrey and Cohn, D'Vera, "U.S. Population Projections: 2005-2050," *Pew Research Social & Demographic Trends*, Feb. 11, 2008. http://www.pewsocialtrends.org/2008/02/11/us-population-projections-2005-2050/.

169. Shrestha, Laura B., and Heisler, Elayne J., "The Changing Demographic Profile of the U.S.," Table 1: Demographic Features of the More Developed Regions and the Less Developed Regions, 1950-2050, according to the United Nations' Medium Projection 1998 (United Nations 1999), Congressional Research Service, March 31, 2011. http://www.fas.org/sgp/crs/misc/RL32701.pdf.

170. "2011 Technical Panel on Assumptions and Methods, Report to the Social Security Advisory Board," Sept. 2011. http://www.ssab.gov/Reports/2011_TPAM_Final_Report.pdf; "Ratio of Covered

Workers to Beneficiaries." http://www.ssa.gov/history/ratios.html; "Life Expectancy at Birth by Race and Sex, 1930 to 2010." http://www.infoplease.com/ipa/A0005148.html; and "Life Expectancy for Social Security, Social Security Administration. http://www.ssa.gov/history/li.eexpect.html.

171. "2011 Technical Panel on Assumptions and Methods, Report to the Social Security Advisory Board," Sept. 2011. http://www.ssab.gov/Reports/2011_TPAM_Final_Report.pdf. "Ratio of Covered Workers to Beneficiaries." http://www.ssa.gov/history/ratios.html. "Life Expectancy at Birth by Race and Sex, 1930 to 2010." http://www.infoplease.com/ipa/A0005148.html. and "Life Expectancy for Social Security," Social Security Administration. http://www.ssa.gov/history/lifeexpect.html.

172. National Vital Statistics Reports, Vol. 61, No. 1, and Vol. 60, No. 4, "Births: Final Data for 2010." http://www.census.gov/compendia/statab/2012/tables/12s0078.pdf , and "Deaths: Preliminary Data for 2010.." http://www.cdc.gov/nchs/data/nvsr/nvsr60/nvsr60_04.pdf..

173. 2010 Census Briefs "Table 1, Population by Hispanic or Latino Origin and by Race for the United States: 2000 and 2010," Overview of Race and Hispanic Origin, issued March 2011, p 4. http://www.census.gov/prod/cen2010/briefs/c2010br-02.pdf.

174. U.S. Census Bureau, Data for 2015 through 2060 drawn from "Table 2. Projections of the Population by Selected Age Groups and Sex for the U.S.: 2015-2060," released Dec. 2012, Population Division. http://www.census.gov/population/projections/files/summary/NP2012-T2.xls. Some groups determined by calculating with help of "Single Year Age Groups-Custom Region-United States" for years 2000, 2010, 2015, 2020, 2030, 2040, and 2050." http://www.census.gov/population/international/data/idb/informationGateway.php, select and click for data; and "U.S. Census Bureau Projections of the Total Resident Population by 5-Year Age Groups, 2100." http://www.census.gov/population/projections/files/natproj/summary/np-t3-h.pdf, scroll down to 2100.

175. U.S. Census Bureau, "NP-T5-H, Projections of the Resident Population by Race, Hispanic Origin, and Nativity: Middle Series, 2075-2100," 2100 Data, released Jan. 13, 2000. http://www.census.gov/population/projections/files/natproj/summary/np-t5-h.pdf.

176. U.S. Census Bureau, International Data Base, "Population Pyramids," U.S. Census Bureau, International Data Base, Mid-year Population by Single Year Age Groups-Custom Region-United States for years 2000, 2010, 2015, 2020, 2030, 2040, and 2050. http://www.census.gov/population/international/data/idb/informationGateway.php, select Population Pyramids, select year or years, select U.S.

177. U.S. Census Bureau, International Data Base, "Total Mid-Year Population of the World, 1950-2050." http://www.census.gov/population/international/data/idb/worldpoptotal.php., U.S. Census Bureau, "Total Midyear Population for the World: 1950-2050," U.S. Pop Clock, Worldometer. http://www.census.gov/population/www/popclockus.html.

178. U.S. Census Bureau, International Data Base, Comparing 2000 and 2050. http://www.census.gov/population/international/data/idb/informationGateway.php, click on selection of years and name of country, as directed, also on Overview, Mid-Year Population of Youth Age Groups, Mid-Year Population of Older Age Groups, and Mid-Year Population by Five-Year Age Groups,". Note that U.S. Life Expectancy for 2000, at 76.8, is from U.S. Census Bureau, *Statistical Abstract of the U.S.: 2012*, Table 104. http://www.census.gov/compendia/statab/2012/tables/12s0104.pdf.

179. U.S. Census Bureau, Population Division, U.S. Information revised with data for total population and 2050, drawn from "Table 2. Projections of the Population by Selected Age Groups and Sex for the U.S.: 2015-2060," released Dec. 2012. http://www.census.gov/population/projections/files/summary/NP2012-T2.xls. Birth to 19 for U.S. refers to birth to 18 in this case, since specific information for the 19-year-old group for a comparable year was not found.

180. U.S. Department of Education, National Center for Education Statistics (2012). *Digest of Education Statistics, 2011* (NCES 2012-001), Chapter 1, "Fast Facts, Enrollment Trends." http://nces.ed.gov/fastfacts/display.asp?id=65.

181. U.S. Department of Education, "Projections of Education Statistics to 2020," National Cen-

ter for Education Statistics, "Actual and Projected Numbers for Enrollment in All Postsecondary Degree-Granting Institutions: Fall 1995 through fall 2020." http://nces.ed.gov/programs/projections/projections2020/tables/table_22.asp?referrer=list.

182. U.S. Bureau of Labor Statistics (BLS), "Kindergarten and Elementary School Teachers," Summary. http://www.bls.gov/ooh/Education-Training-and-Library/Kindergarten-and-elementary-school-teachers.htm.

183. Bureau of Labor Statistics, "Employment Outlook 2010-2020, Occupational Employment Projections to 2020," *Monthly Labor Review*, Jan. 2012, Education, Training, and Library occupations, p. 94, and "Employment and Wages of Occupations with the Largest Numeric Grow in Jobs, 2010-2010," p. 101. http://www.bls.gov/opub/mlr/2012/01/art5full.pdf.

184. Khallash, Sally. "Staff Retention Key to M&A Success Finds Mercer Study," Global Talent Strategy, Feb. 25, 2013. http://globaltalentstrategy.com/en/article/staff-retention-key-to-ma-success-finds-mercer-study-348.

185. Hamilton, Craig, "Chasing Immortality, An Interview with Ray Kurzweil," *Enlightenment Magazine*. http://www.singularity.com/WIEnlightenment_KurzweilAritcle.pdf

186. Barrow, Becky, "One in Six Will Live to be 100, Official Figures Predict 10 m. Alive Today Will get Royal Telegram,", *Mail Online*, Dec. 31, 2010. http://www.dailymail.co.uk/health/article-1342589/1-6-live-100-Official-figures-predict-10m-alive-today-royal-telegram.html.

187. Brownstein, Joseph, "Most Babies Born Today May Live Past 100," *ABC News* Medical Unit, *ABC News*, Oct. 1, 2009. http://abcnews.go.com/Health/WellnessNews/half-todays-babies-expected-live-past-100/story?id=8724273.

188. U.S. Census Bureau, "Older Americans Month: May 2012," Profile America Facts for Features, March 1, 2012. http://www.census.gov/newsroom/releases/archives/facts_for_features_special_editions/cb12-ff07.html.

189. Marx, Gary, *Ten Trends…Educating Children for a Profoundly Different Future*, Educational Research Service, Alexandria, Virginia, 2000, pp. 10-11.

190. Marx, Gary, *Sixteen Trends…Their Profound Impact on Our Future*, 2006, quotes from Arnold Fege and Keith Marty, Educational Research Service and *Education Week* Press, p. 31.

191. Moyers, Bill, "On Not Growing Old," In the Know Opinion, *aarp.org/bulletin*, April 2012. http://pubs.aarp.org/aarpbulletin/201204_DC?pg=34#pg34.

192. Rosling, Hans, "Let My Dataset Change Your Mindset," *TED Talk* video, (19:40), addressing audience at U.S. State Department, June 2009. http://www.ted.com/talks/hans_rosling_at_state.html.

193. Worldometer. http://www.worldometers.info/world-population/.

194. Christensen, Clayton M., Allworth, James, and Dillon, Karen, *"How Will You Measure Your Life?"* Harper-Collins, NY, 2012.

195. Hamilton, Craig, Suggested Readings: "Chasing Immortality, An Interview with Ray Kurzweil," *Enlightenment Magazine*. http://www.singularity.com/WIEnlightenment_KurzweilAritcle.pdf and Kurzweil, Ray, and Grossman Terry, *Transcend: Nine Steps to Living Well Forever* (Paperback), Rodale, Inc., NY, 2010.

196. Markoff, John, "Computer Wins on 'Jeopardy!': Trivial, It's Not," *New York Times*, Feb. 16, 2011. http://www.nytimes.com/2011/02/17/science/17jeopardy-watson.html?pagewanted=all&_moc.semityn.www.

197. Merron, Jeff, "The List…Biggest Falls from Grace," *ESPN Page 2*, story about Joe Jackson, accused of throwing the World Series in the 1919 Black Sox scandal. http://espn.go.com/page2/s/list/fallfromgrace.html.

198. Markoff, John, "Computer Wins on 'Jeopardy!': Trivial, It's Not," *New York Times*, Feb. 16, 2011. http://www.nytimes.com/2011/02/17/science/17jeopardy-watson.html?pagewanted=all&_moc. semityn.www.

199. "Seven Themes for the Coming Decade," *The Futurist*, World Future Society, James Lee's Blog, Aug. 28, 2012. http://www.wfs.org/blogs/james-lee/seven-themes-for-coming-decade.

200. "Zettabyte," International Data Corporation. http://en.wikipedia.org/wiki/Zettabyte.

201. Richey, Kevin W., Virginia Tech, "The ENIAC." http://ei.cs.vt.edu/~history/ENIAC.Richey. HTML.

202. "IBM System/360. On April 7, 1964, the entire concept of computers changed," IBM 1964 promotional announcement/brochure. http://www-03.ibm.com/ibm/history/ibm100/us/en/icons/system360/words/.

203. "Apollo 11: The Computers That Put a Man on the Moon," *Computer Weekly.com*. http://www.computerweekly.com/feature/Apollo-11-The-computers-that-put-man-on-the-moon.

204. Long, Tony, "Jan. 24, 1984: Birth of the Cool (Computer, That Is)," Wired online, Jan. 24, 2008. http://www.wired.com/science/discoveries/news/2008/01/dayintech_0124.

205. Lohr, Steve, "The Age of Big Data," *The New York Times,* Feb. 11, 2012. http://www.nytimes.com/2012/02/12/sunday-review/big-datas-impact-in-the-world.html?pagewanted=all

206. "Cukier, Kenneth, and Mayer-Schoenberger, Viktor, "The Rise of Big Data," *Foreign Affairs*, May/June 2013, p. 28.

207. Strickland, Jonathan, "How Cloud Computing Works," How Stuff Works, A Discovery Company. http://computer.howstuffworks.com/cloud-computing/cloud-computing.htm.

208. "It's a Small World After All, Nemo Disney Reopens Beloved Ride," Associated Press, Feb. 5, 2009. www.msnbc.msn.com/id/29019618/ns/travel-family/t-its-small-world-after-all-nemo.

209. National Science and Technology Council, Committee on Technology, Interagency Working Group on Nanoscience, Engineering, and Technology, *Nanotechnology…Shaping the World Atom by Atom*, September 1999. http://www.wtec.org/loyola/nano/IWGN.Public.Brochure/IWGN.Nanotechnology.Brochure.pdf.

210. University of Oxford, Oxford Martin School, News, "The Next Technological Revolutions," report on presentation by Eric Drexler of the Foresight Institute. http://www.oxfordmartin.ox.ac.uk/news/201111-news-FutureTechLaunch.

211. Drexler, Eric, "Why Care about Nanotechnology?" *Foresight Institute Nanotechnology Newsletter*, March 31, 2004. http://www.foresight.org/nano/EricDrexler.html.

212. Deardorff, Julie, "Scientists: Nanotech-Based Products Offer Great Potential but Unknown Risks," *Chicago Tribune*, July 10, 2012. http://articles.chicagotribune.com/2012-07-10/health/ct-met-nanotechnology-20120710_1_nanoparticles-sunscreens-chad-mirkin.

213. Wilkins, Alasdair, "What Today's Higgs Boson Discovery Really Means," *We Come from the Future, Physics*, July 4, 2012. http://io9.com/5923494/what-todays-higgs-boson-discovery-really-means.

214. Cain, Fraser, "How Many Galaxies in the Universe," *Universe Today*, May 4, 2009. http://www.universetoday.com/30305/how-many-galaxies-in-the-universe/.

215. Jha, Alok, "Two New Planets are Most Earth-like Ever Seen—But Hot as Hell," *The Guardian*, Dec. 20, 2011. http://www.guardian.co.uk/science/2011/dec/20/planets-earth-like-exoplanet-solar-system.

216. Jet Propulsion Laboratory, "Voyager, The Interstellar Mission." http://voyager.jpl.nasa.gov/mission/fastfacts.html.

217. "How Space Tourism Works," How Stuff Works.com. http://science.howstuffworks.com/space-tourism.htm.

218. Achenbach, Joel, "A Distant Object Sheds Light on Our Universe," *Washington Post*, Style Section, March 17, 2004, p. C-1. http://www.washingtonpost.com/wp-dyn/articles/A64910-2004Mar16.html.

219. "Pando Networks Releases Global Internet Speed Study. http://www.pandonetworks.com/Pando-Networks-Releases-Global-Internet-Speed-Study.

220. ibid

221. Ookla.com, "Household Download Index, All Countries," graph period: May 9, 2011-Nov. 7, 2013, updated Nov. 7, 2013 http://www.netindex.com/download/allcountries.
Federal Communications Commission (FCC), "Universal Service for Schools and Libraries." http://transition.fcc.gov/wcb/tapd/universal_service/schoolsandlibs.html.

222. Eder, Steve, "Teachers Fight Online Slams," *The Wall Street Journal*, Sept. 17, 2012. http://online.wsj.com/article/SB10000872396390443779404577644032386310506.html.

223. Fraser-Thill, Rebecca, "Cyberbullying Facts—How Common is Cyberbullying," *Parenting Tweens, About.com Tweens*. http://tweenparenting.about.com/od/behaviordiscipline/a/Cyberbullying-Facts.htm.

224. "How Hackers Steal Your Personal Data," AOL online promotion, Aug. 10, 2012.

225. Guterl, Fred, *The Fate of the Species*, Bloomsbury USA, NY, 2012, brief comments based on Guterl's observations, pp. 132-143.

226. Muller, Joann, "With Driverless Cars, Once Again It Is California Leading the Way," *Forbes*, Sept. 26, 2012. http://www.forbes.com/sites/joannmuller/2012/09/26/with-driverless-cars-once-again-it-is-california-leading-the-way/print/.

227. Waldron, Ben "Hypersonic Plane Could Revolutionize Commercial Flight," *ABC News*, Oct. 25, 2012. http://abcnews.go.com/blogs/technology/2012/10/hypersonic-plane-could-revolutionize-commercial-flight/.

228. Enns, Don, "Opinion: B.C. Can Be Leader in Ensuring Biotech Advances...Help Improve Health Care Delivery," *Vancouver Sun*, Canada, Sept. 17, 2012. http://www.vancouversun.com/Opinion/Op-Ed/Opinion+leader+ensuring+biotech+advances+help+improve/7256240/story.html.

229. "Gene Therapy," *The Free Dictionary* online. http://medical-dictionary.thefreedictionary.com/DNA+surgery.

230. Burton, Jeff, Kinter, Marcia, and Marx, Dan, *2005 Guide to Digital Imaging*, 2004, Specialty Graphic Imaging Association/Digital Printing and Imaging Association, Fairfax, VA, 2004, pp. 1, 7-9.

231. "3D Printing, Wikipedia. http://en.wikipedia.org/wiki/3D_printing.

232. "Could 3D Printing Change the World?" Strategic Foresight Report, Atlantic Council, Oct. 2011. http://www.acus.org/files/publication_pdfs/403/101711_ACUS_3DPrinting.PDF.

233. Quillen, Ian, "Can Technology Replace Teachers?" *Education Week*, August 8, 2012, online August 7, 2012. http://www.edweek.org/ew/articles/2012/08/08/37replace_ep.h31.html?print=1.

234. "Using 21st Century Tools for College Success, This Week's Top Educator Resource," promotion announcement, eSchool *News Online*.

235. Dede, Chris, Presentation and Comments during Harvard Future of Learning Institute, Cambridge, Mass., Aug. 2, 2012.

236. William Ferriter, "Are Kids Really Motivated by Technology," *Smart Blog*, Aug. 17, 2012. http://smartblogs.com/education/2012/08/17/are-kids-really-motivated-technology/.

237. FitzGerald, Eileen, "The Digital Divide in Greater Danbury's Schools," Aug. 14, 2012. http://

www.newstimes.com/local/article/The-digital-divide-in-Greater-Danbury-s-schools-3781704.php.

238. Plummer, Sarah, "Technology Will Generate Benefits and Challenges for Schools," Beckley, WV, *Register-Herald*, Aug. 9, 2012. http://www.register-herald.com/todaysfrontpage/x1402350278/Technology-will-generate-benefits-and-challenges-for-schools.

239. Bonvillian, Crystal, "Chapman Students Show Off Wi-Fi on Huntsville School Buses," *The Huntsville Times*, Aug. 22, 2012. http://blog.al.com/breaking/2012/08/chapman_students_show_off_wi-f.html.

240. Andersen, Erin, "Fear Giving Way to App-Friendly Classrooms," *Lincoln Journal & Star*, Nebr., Aug. 9, 2012. http://journalstar.com/lifestyles/family/fear-giving-way-to-app-friendly-classrooms/article_c3da06af-9ac4-5dfa-ab57-541203dae2f8.html.

241. FitzGerald, Eileen, "The Digital Divide in Greater Danbury's Schools, Aug. 14, 2012. http://www.newstimes.com/local/article/The-digital-divide-in-Greater-Danbury-s-schools-3781704.php.

242. *Fairfax County Public Schools Handbook,* 2013-14, "A Guide for FCPS Parents, Students, and Staff Members, http://www.fcps.edu/cco/pubs/handbook.pdf, p.2.

243. Millstone, Jessica, "Teacher Attitudes about Digital Games in the Classroom," The Joan Ganz Cooney Center at Sesame Workshop in collaboration with BrainPOP, May 2, 2012, slide show. http://www.joanganzcooneycenter.org/images/presentation/jgcc_teacher_survey.pdf.

244. Devaney, Laura, "Sesame Street Boosts Early Childhood Learning,," *eSchool News*, May 6, 2013. http://www.eschoolnews.com/2013/05/06/study-sesame-street-boosts-early-childhood-learning/.

245. Brenner, Joanna, "Pew Internet: Teens," Aug. 27, 2012. http://pewinternet.org/Commentary/2012/April/Pew-Internet-Teens.aspx#.

246. World Bank, *Information and Communication Technologies*, "Maximizing Mobile Report Highlights Development Potential of Mobile Communications," July 27, 2012. http://web.worldbank.org/WBSITE/EXTERNAL/TOPICS/EXTINFORMATIONANDCOMMUNICATIONANDTECHNOLOGIES/0,,contentMDK:23242711~pagePK:210058~piPK:210062~theSitePK:282823,00.html,,

247. Peckham, Matt, "The Collapse of Moore's Law: Physicist Says It's Already Happening," *Time Techland*, Innovation, May 1, 2012. http://techland.time.com/2012/05/01/the-collapse-of-moores-law-physicist-says-its-already-happening/.

248. Ford, Dale, "Intel's Semiconductor Market Share Surges to More Than 10-Year High in 2011," HIS Supply Market Research, March 26, 2012.

249. Johnson, Brian D., Intel Corp., Presentation to World Future Society Annual Conference, Toronto, Ontario, Canada, July 24, 2012.

250. "Petaflop Definition," *Petaflop Web Site.* http://www.petaflop.info/.

251. Waugh, Rob, "No Speed Limit: IBM Scientists on Verge of Creating 'Quantum Computers' Faster Than Any Supercomputer on Earth," *Daily Mail*, U.K., March 1, 2012. http://www.dailymail.co.uk/sciencetech/article-2108160/Quantum-computers-IBM-verge-creating-machine-faster-supercomputer.html.

252. Peckham, Matt, "The Collapse of Moore's Law: Physicist Says It's Already Happening," *Time Techland*, Innovation, May 1, 2012. http://techland.time.com/2012/05/01/the-collapse-of-moores-law-physicist-says-its-already-happening/.

253. Peckham, Matt, "The Collapse of Moore's Law: Physicist Says It's Already Happening," *Time Techland*, Innovation, May 1, 2012. http://techland.time.com/2012/05/01/the-collapse-of-moores-law-physicist-says-its-already-happening/.

254. "Quantum Leap: The World's Biggest R&D Labs Are Racing to Build a Quantum Computer. Geordi Rose Thinks He Can Beat Them. Is He Just Blowing Smoke?" *CNN Money*, Aug. 1, 2004.

http://money.cnn.com/magazines/business2/business2_archive/2004/08/01/377387/index.htm.

255. "USC, Lockheed Martin, D-Wave to Launch Quantum Computing Center," *HTC Wire*, Oct. 28, 2011. http://www.hpcwire.com/hpcwire/2011-10-28/usc,_lockheed_martin,_d-wave_to_launch_quantum_computing_center.html.

256. Marklein, Mary Beth, "College May Never Be the Same," *USA Today*, Sept. 12, 2012. http://www.usatoday.com/news/nation/story/2012/09/12/college-may-never-be-the-same/57752972/1.

257. Rominiecki, Joe, "Desperately Seeking Social Media ROI," *Associations Now*, April 5, 2013. http://associationsnow.com/2013/04/desperately-seeking-social-media-roi/.

258. Lazarro, Helen, Flipped Learning Network, Presentation during Education Summit, World Future Society Annual Conference, Chicago, July 19, 2013.

259. November, Alan, "What's Your Plan for Making Every Home a Center of Learning?" *eSchool News*, March 30, 2012. http://www.eschoolnews.com/2012/03/30/alan-november-whats-your-plan-for-making-every-home-a-center-of-learning/.

260. *New Media Literacies*, materials and video, University of Southern California, http://www.newmedialiteracies.org/. Also see white paper, *Confronting the Challenges of Participatory Culture: Media Education for the 21st Century* (Jenkins et al., 2006, produced with support from the MacArthur Foundation. http://www.newmedialiteracies.org/wp-content/uploads/pdfs/NMLWhitePaper.pdf.

261. Bloxham, Andy, "Internet Addition is a 'Clinical Disorder,'" *The Telegraph*, June 18, 2008. http://www.telegraph.co.uk/news/uknews/2152972/Internet-addiction-is-a-clinical-disorder.html.

262. Dede, Chris, "MindShift on Cyberlearning and Games," *YouTube* video. http://www.youtube.com/watch?v=nNAZXB0DnT4&feature=related.

263. "Steve Jobs: How to live before you die," *TED Talk*, Commencement Address at Stanford University, June 2005. http://www.ted.com/talks/steve_jobs_how_to_live_before_you_die.html.

264. U.S. Department of Education, Office of Technology, *Transforming American Education, Learning Powered by Technology, National Education Technology Plan 2010*. http://www.ed.gov/sites/default/files/NETP-2010-final-report.pdf.

265. *New Media Literacies*, Materials and Video, online, University of Southern California. http://www.newmedialiteracies.org/.

266. *Confronting the Challenges of Participatory Culture: Media Education for the 21st Century* (Jenkins et al., 2006, produced with support from the MacArthur Foundation. http://www.newmedialiteracies.org/wp-content/uploads/pdfs/NMLWhitePaper.pdf.

267. Drexler, Eric,Blog, "Metamodern," Drexler is CEO of The Foresight Insititute. Particular emphasis on nanotechnology, http://metamodern.com/.

268. Video of a 30-story Chinese hotel being built in 15 days, from *BSB, Broad Sustainable Building*. https://www.youtube.com/embed/GVUsIlwWWM8?rel=0.

269. Whitfield, Norman, and Strong, Barrett, "I Heard It Through the Grapevine," chorus from song written by for Motown Records in 1966. Recorded by, among others, Marvin Gaye, Smokey Robinson & the Miracles, and Gladys Knight and the Pips, *ST Lyrics*. http://www.stlyrics.com/lyrics/thebigchill/ihearditthroughthegrapevine.htm, and Wikipedia, http://en.wikipedia.org/wiki/I_Heard_It_Through_the_Grapevine.

270. Heller, Joseph quotes. http://www.goodreads.com/author/quotes/3167.Joseph_Heller.

271. Turkle, Sherry, "Connected, but alone?" *TED Talk*, Sherry Turkle, posted April 2012. http://www.ted.com/talks/sherry_turkle_alone_together.html.

272. Richtel, Matt, "Silicon Valley Says Step Away From the Device," *New York Times*, July 23, 2012.

http://www.nytimes.com/2012/07/24/technology/silicon-valley-worries-about-addiction-to-devices.html.

273. Vance, Ashlee, "Facebook: The Making of 1 Billion Users," *Bloomberg Businessweek*, Oct. 4, 2012. http://www.businessweek.com/printer/articles/74602-facebook-the-making-of-1-billion-users.

274. Timberg, Craig, "Our Cars Say So Much about Us," *Washington Post*, March 6, 2013. http://www.pressdisplay.com/pressdisplay/viewer.aspx, hard copy day of publication.

275. Beckford, Martin, "More Than 200 Schools Have CCTV in Toilets and Changing Rooms," *The Telegraph*, Sept. 12, 2012. http://www.telegraph.co.uk/education/9535931/More-than-200-schools-have-CCTV-in-toilets-and-changing-rooms.html, access Sept. 13, 2012.

276. *Shaping Tomorrow*, "DIY Internet, What's Changing?" Oct. 2, 2013 online issue, www.shapingtomorrow.com.

277. Fisher, Marc, and Timberg, Craig, "Surveillance: A Fear, A Comfort," *Washington Post*, Dec. 22, 2013, pp.1,8, http://www.washington post.com/world/national-security/Americans-uneasy-about-surveillance-but-often-use-snooping-tools-post-poll-finds/2013/12/21/cal5e990-67f9-11e3-ae56-22de072140a2_story.html.

278. "Cloud Security Alliance Sets Sights on Privacy, Big Data," *SimplySecurity.com*. http://www.simplysecurity.com/2012/09/07/cloud-security-alliance-sets-sights-on-privacy-big-data/.

279. "Empowering Consumers, Protecting Privacy," Privacy Rights Clearinghouse. https://www.privacyrights.org/fs/fs29-education.htm.

280. U.S. Department of Education, "Family Educational Rights and Privacy Act (FERPA)." http://www2.ed.gov/policy/gen/guid/fpco/ferpa/index.html.

281. U.S. Department of Education, "FERPA Facts." http://faculty.irsc.edu/DEPT/AdultEducation/adult_ed/E-Learning/employees/teacher%20links/ferpa_fast_facts.pdf.

282. Privacy Rights Clearinghouse, "Empowering Consumers, Protecting Privacy." https://www.privacyrights.org/fs/fs29-education.htm.

283. "What is FOIA?" FOIA.gov. http://www.foia.gov.

284. Gormley, Ken, "One Hundred Years of Privacy," *Wisconsin Law Review*, University of Wisconsin, 1992. http://cyber.law.harvard.edu/privacy/Gormley--100%20Years%20of%20Privacy.htm.

285. U.S. Department of Health and Human Services (HHS), "Health Information Privacy." http://www.hhs.gov/ocr/privacy/index.html.

286. "Spotlight on Social Media in the Classroom," *Education Week*. http://www.edweek.org/ew/marketplace/products/spotlight-social-media-in-the-classroom.html.

287. Barseghian, Tina, "Online Privacy: Kids Know More Than You Think," *Media Shift*, April 2, 2012. http://www.pbs.org/mediashift/2012/04/online-privacy-kids-know-more-than-you-think-093.html.

288. "Kids Rules for Online Safety (for pre-teens)," *SafeKids.com*, "Online Safety & Civility." http://www.safekids.com/kids-rules-for-online-safety/.

289. Davis, Michelle, "'Safe' Social Networking Tailored for K-12 Schools," *Education Week*, June 15, 2011, (print), and June 13, 2011 (online). http://www.edweek.org/ew/articles/2011/06/15/35mm-social.h30.html.

290. "The Antisocial Effects of Social Media," *New York Times*, Opinion Pages, To the Editor, Housman, Donna, Weston, MA, April 25, 2012. http://www.nytimes.com/2012/04/26/opinion/the-antisocial-effects-of-social-media.html.

291. Vance, Ashlee, "The Very Rich Perils of Rich Kids on Social Networks," *Business Week,* August 10,

2012. http://www.businessweek.com/articles/2012-08-10/the-very-real-perils-of-rich-kids-on-social-networks.

292. Federal Trade Commission, "About Identity Theft: Deter, Detect, Defend, and Avoid ID Theft." http://www.ftc.gov/bcp/edu/microsites/idtheft/consumers/about-identity-theft.html.

293. "Consumer Privacy 'Bill of Rights' Seeks to Give Web Users More Control Over Their Data," *Huffington Post*, Tech, posted Feb. 23, 2012. http://www.huffingtonpost.com/2012/02/23/web-privacy-bill-of-rights_n_1294230

294. "White House Pushing Online Privacy Bill," *ASAE Inroads*, Vol. 27, No. 8, Feb. 23, 2012.

295. "How Companies Are 'Defining Your Worth' Online," Story on an NPR interview with Joseph Turow, Feb. 22, 2012. http://www.npr.org/2012/02/22/147189154/how-companies-are-defining-your-worth-online.

296. Turkle, Sherry, "Connected, but alone?" *TED Talk*, posted April 2012. http://www.ted.com/talks/sherry_turkle_alone_together.html.

297. *Consumer Data Privacy in a Networked World*, The White House, Washington, DC, Feb. 2012. http://www.whitehouse.gov/sites/default/files/privacy-final.pdf.

298. "Silicon Valley Says, 'Step Away From the Device,'" *New York Times*, July 23, 2012. http://www.nytimes.com/2012/07/24/technology/silicon-valley-worries-about-addiction-to- devices.html

299. "Computer Game Addiction," Australian Broadcasting Company (ABC), April 26, 2012. http://www.abc.net.au/catalyst/stories/3488130.htm.

300. "Our Space: Being a Responsible Citizen of the Digital World," a collaboration of The GoodPlay Project at Harvard and Project New Media Literacies at USC, 2011. http://www.goodworkproject.org/ourspacefiles/0__Introductory_Materials_Our_Space.pdf.

301. Gunther, Marc, "Immelt: We Are in an Emotional, Social, Economic Reset," *GreenBiz.com*, Nov. 8, 2008. http://www.greenbiz.com/blog/2008/11/08/immelt-we-are-emotional-social-economic-reset.

302. American Society of Civil Engineers (ASCE), "Report Card for American Infrastructure." To view and order: http://www.infrastructurereportcard.org/report-cards and http://www.asce.org/reportcard/.

303. "Economic Report of the President," Transmitted to the Congress, Feb. 2012, Together with the "Annual Report of the Council of Economic Advisers," U.S. Government Printing Office, Washington, DC, 2012, p 4. http://www.nber.org/erp/ERP_2012_Complete.pdf.

304. Zakaria, Fareed, "Can America Be Fixed?" *Foreign Affairs*, Jan.-Feb 2013, p 26.

305. Florida, Richard L. *The Great Reset*, Harper-Collins, NY, 2010/2011. pp. viii, ix.

306. Rifkin, Jeremy, *The Third Industrial Revolution,* Palgrave Macmillan, NY, 2011, pp. 234-235, pp. 36-37. Also included is background from "Jeremy Rifkin." Bio. http://en.wikipedia.org/wiki/Jeremy_Rifkin.

307. Harvard Business School, "The Forgotten Real Estate Boom of the 1920s," *Bubbles, Panics, & Crashes, Historical Collections*, Presidents and Fellows of Harvard College, 2012. http://www.library.hbs.edu/hc/crises/forgotten.html.

308. "A New Economic Analysis of Infrastructure Investment," A Report Prepared by the Department of the Treasury with the Council of Economic Advisers, March 23, 2012. http://www.treasury.gov/resource-center/economic-policy/Documents/20120323InfrastructureReport.pdf.

309. Population Reference Bureau, "The U.S. Census Tradition." http://www.prb.org/Articles/2009/censustradition.aspx?p=1.

310. "Stages of Urbanization." http://www.answers.com/topic/urbanization.

311. *The New York Times Almanac 2011*, Edited by John W. Wright, New York Times Company, Penguin Books, NY, 2010.

312. Florida, Richard L., *The Great Reset*, Harper-Collins, NY, 2010/2011, p. 21.

313. "The First Measured Society," hosted by Ben Wattenberg, Population: Urban, Rural, Suburban, from PBS program, "Think Tank with Ben Wattenberg." http://www.pbs.org/fmc/book/1population6. htm. Wattenberg wrote a book by the same name with fellow authors Theodore Caplow and Louis Hicks. The AEI Press, Washington, DC

314. Florida, Richard L., *The Great Reset*, Harper-Collins, NY, 2010/2011, pp. 160-161.

315. Geron, Tomio, "How People Make Cash in the Share Economy," *Forbes*, Jan. 23, 2013. http://www.forbes.com/sites/tomiogeron/2013/01/23/how-people-make-cash-in-the-share-economy/.

316. Fleming, Thomas, "Fleming: What Life Was Like in 1776," Wall Street Journal Opinion, July 2, 2012. http://online.wsj.com/article/SB10001424052702303561504577496620544901322.html#printMode.

317. Levine, Linda, "An Analysis of the Distribution of Wealth Across Households, 1989-2010," Table 2, Congressional Research Service (CRS), July 17, 2012. http://www.fas.org/sgp/crs/misc/RL33433.pdf

318. Stiglitz, Joseph E., "Inequality is Holding Back the Recovery," *New York Times Opinionator*, Jan. 19, 2013. http://opinionator.blogs.nytimes.com/2013/01/19/inequality-is-holding-back-the-recovery/

319. Krugman, Paul, "White Collars Turn Blue," *New York Times*, Sept. 29, 1996. http://mit.edu/krugman/www/BACKWRD2.htmlaccessed, Dec. 21, 2012.

320. "Leading Lights, An Interview with Tom Stewart," *Knowledge Inc.*, 1997. http://www.brint.com/members/online/120205/stewart.html.

321. "Bring Me Sunshine, The Sudden Popularity of a Controversial American Economist," *The Economist*, Nov. 11, 2010. http://www.economist.com/node/17468554/ or http://www.amazon.com/The-Rise-Creative-Class-Transforming/dp/0465024769.

322. Gibson, Rowan, *Rethinking the Future*, "Becoming a Leader of Leaders," chapter written by Warren Bennis, Nicholas Brealey Publishing, London, 2002, p. 149.

323. Putnam, Robert, address to a White House Conference on the New Economy, April 5, 2000. Article, "Leadership Skills: Networking vs. Building Social Capital," Cook, Marian, quoting Putnam, WITI, Women in Technology International, WITI Leadership. http://www.witi.com/wire/articles/58/Leadership-Skills:-Networking-vs.-Building-Social-Capital-(Part-1-of-3)/print/.

324. Putnam, Robert, "Social Capital: Measurement and Consequences," Organization for Economic Cooperation and Development (OECD), 1995. http://www.oecd.org/edu/country-studies/1825848.pdf. Putnam's book, *Bowling Alone*, was published in 2000.

325. Fukuyama, Francis, *Trust...The Social Virtues and The Creation of Prosperity*, Francis Fukuyama, Free Press Paperbacks, NY, p. 10.

326. Thurow, Lester C., *Building Wealth...The New Rules for Individuals, Companies, and Nations in a Knowledge-Based Economy*, Lester C. Thurow, HarperCollins Publishers, Inc., NY, 1999, p. 92.

327. Marx, Gary, *Ten Trends...Educating Children for a Profoundly Different Future*, Educational Research Service, Arlington, Virginia, 2000, p. 23.

328. Clinton, Bill, former U.S. President, addressing White House Conference on the New Economy, April 5, 2000. http://www.gpo.gov/fdsys/pkg/PPP-2000-book1/pdf/PPP-2000-book1-doc-pg625.pdf.

329. Siegel Bernard, Tara, "Working Financial Literacy in With the Three R's," *New York Times*, April 9, 2010. http://www.nytimes.com/2010/04/10/your-money/10money.html.

330. "Advocacy Tool Kit, Council on Economic Education, see section for Oklahoma. http://www.

councilforeconed.org/about/policy-and-advocacy/toolkit/.

331. Marx, Gary, *Ten Trends…Educating Children for a Profoundly Different Future*, Educational Research Service, Arlington, Virginia, 2000, p. 25.

332. "Congressional Research Service Disproves Trickle-Down Economics," *Neon Tommy*, Annenberg Digital News, Nov. 6, 2012. http://www.neontommy.com/news/2012/11/congressional-research-service-disproves-trickle-down-economics.

333. Florida, Richard, presentation "The Rise of the Creative Class," based on his book of the same title, UCTV, program (59:23) presented in San Diego. http://www.youtube.com/watch?v=iLstkIZ5t8g.

334. Florida, Richard, presentation "The Rise of the Creative Class," based on his book of the same title, UCTV, program (59:23) presented in San Diego. http://www.youtube.com/watch?v=iLstkIZ5t8g.

335. "Economic Report of the President," Transmitted to Congress, Feb. 2012, plus Annual Report of the Council of Economic Advisers. http://www.nber.org/erp/ERP_2012_Complete.pdf.

336. Rifkin, Jeremy, presentation *TEDx Talk*, video, (11:06), "Leading the Way to *The Third Industrial Resolution*, June 8, 2012. http://www.youtube.com/watch?v=snsb3Pc_C4M.

337. "The Federal Reserve and You," Video, (23.34), Produced by the Federal Reserve. http://2www.phil.frb.org/education/federal-reserve-and-you/.

338. "A New Economic Analysis of Infrastructure Investment," A report prepared by the Department of the Treasury with the Council of Economic Advisers, March 23, 2012. http://www.treasury.gov/resource-center/economic-policy/Documents/20120323InfrastructureReport.pdf.

339. American Society of Civil Engineers (ASCE), "Report Card for American Infrastructure." To view and order: http://www.infrastructurereportcard.org/report-cards and http://www.asce.org/reportcard/.

340. Frey, Thomas, "2011 in Review, 55 Jobs of the Future. http://www.futuristspeaker.com/2011/12/year-in-review-top-10-articles-on-futuristspeaker-com/.

341. Popcorn, Faith, and Hanft, Adam, *Dictionary of the Future*, Hyperion, NY, 2001, p. 299.

342. U.S. Bureau of Labor Statistics (BLS), "National Longitudinal Surveys, Frequently Asked Questions." http://www.bls.gov/nls/nlsfaqs.htm#anch41, modified March 8, 2012.

343. U.S. Bureau of Labor Statistics (BLS), "Usual Weekly Earnings of Wage and Salary Workers, Third Quarter, 2012," Table 5. http://www.bls.gov/news.release/pdf/wkyeng.pdf.

344. Krugman, Paul, "White Collars Turn Blue," *New York Times*, Sept. 29, 1996. http://mit.edu/krugman/www/BACKWRD2.html.

345. Florida, Richard, "America's Leading Creative Class Metros," *The Atlantic*, July 9, 2012. http://www.theatlanticcities.com/jobs-and-economy/2012/07/americas-leading-creative-class-metros/2233/#slide1.

346. Florida, Richard, "Creatives and the Crisis," *The Atlantic*, Oct. 22, 2012. http://www.theatlanticcities.com/jobs-and-economy/2012/10/creatives-and-crisis/1727/.

347. Harden, Blaine, "Brain-Gain Cities Attract Educated Young," Urban Warfare/American Cities Fight for Talent Series, *The Washington Post*, November 9, 2003, pp. 1, 14-15. http://lists.ibiblio.org/pipermail/internetworkers/2003-November/012572.html.

348. Lee, James H., "Hard at Work in the Jobless Future," World Future Society. http://www.wfs.org/book/export/html/2477.

349. Gallman, Robert E., "Trends in the Location of Population, Industry, and Employment, Ohio State University, Table 1. http://ageconsearch.umn.edu/bitstream/17629/1/ar610017.pdf, based on information from Historical Statistics of the United States, Bureau of the Census, 1960, p 74.

350. U.S. Bureau of Labor Statistics, "Employment Projections. Employment by Major Industry Sector," Table 2.1. http://data.bls.gov/cgi-bin/print.pl/emp/ep_table_201.htm.

351. *The World of Agriculture*, Unit 2, p 15. http://www.okcareertech.org/educators/cimc/new-products/clusters/ag-cluster/pdf-files/8gradeagteacher.pdf. Information corroborated by *Today's Farmer*.

352. U.S. Environmental Protection Agency, "Demographics," Ag 101, based on U.S. Census, updated June 27, 2012. http://www.epa.gov/oecaagct/ag101/demographics.html.

353. Gallman, Robert E., "Trends in the Location of Population, Industry, and Employment," Ohio State University, Table 1. http://ageconsearch.umn.edu/bitstream/17629/ar610017.pdf, based on information from Historical Statistics of the United States, Bureau of the Census, 1960, p 74.

354. U.S. Bureau of Labor Statistics, "Employment Projections. Employment by Major Industry Sector," Table 2.1. http://data.bls.gov/cgi-bin/print.pl/emp/ep_table_201.htm.

355. Beard, Charles A. Mary R., and William, *The Beards' New Basic History of the United States*, Doubleday, NY, 1960, pp 189-190.

356. "Manufacturing Surprise: The U.S. Still Leads in Making Things," *CBS News*, June 21, 2010, based on information from the Bureau of Economic Analysis. http://www.cbsnews.com/8301-505123_162-36742134/manufacturing-surprise58-the-us-still-leads-in-making-things/.

357. Royte, Elizabeth, "The Printed World,", *Smithsonian Magazine*, May 2013, p 52.

358. Florida, Richard L., *The Great Reset*, Harper-Collins, NY, 2010/2011, pp 116-128.

359. Fishman, Charles "The Insourcing Boom," *The Atlantic*. http://www.theatlantic.com/magazine/print/2012/12/the-insourcing-boom/309166/.

360. "U.S. Companies Begin Insourcing Jobs from China, Japan, & Mexico; India Could Be Next," The *Economic Times*, Jan. 12, 2012. http://articles.economictimes.indiatimes.com/2012-01-12/news/30616823_1_assembly-plant-galaxe-solutions-new-jobs.

361. "11 Key Trends in Global Mobility (Study)," Global Talent Strategy, Dec. 31, 2012. http://globaltalentstrategy.com/en/article/11-key-trends-in-global-mobility-study-317.

362. U.S. Department of Labor, Bureau of Labor Statistics, "Fastest Growing Occupations, 2002-2012" and "Occupations with the Largest Job Declines, 2002-2012," *Monthly Labor Review*, February 2004. http://www.ait.org.tw/infousa/enus/economy/workforce/docs/art5full.pdf.

363. U.S Bureau of Labor Statistics "Employment Outlook: 2010-2020, Industry Employment and Output Projections to 2020," *Monthly Labor Review*, Jan. 2012. http://www.bls.gov/opub/mlr/2012/01/art4full.pdf.

364. Briode, Blaire, "Five Reasons Boomers Will Go Bust," based on "Money in Your 50s and 60s," *The Fiscal Times*, Aug. 23, 2012. http://money.msn.com/baby-boomers/5-reasons-boomers-will-go-bust-fiscaltimes.aspx.

365. Fisher, Anne, "10 Hot Careers for 2012-and Beyond," *Fortune*, Reported by *CNN Money*, Dec. 27, 2011. http://management.fortune.cnn.com/2011/12/27/10-hot-careers-for-2012-and-beyond/.

366. U.S Bureau of Labor Statistics, "Projections Overview," Table 1, Occupations with the Fastest Growth, Projected 2010-2020,"and "Table 3, Occupations with the Fastest Decline, Projected 2010-2020." http://www.bls.gov/ooh/about/print/projections-overview.htm.

367. U.S. Bureau of Labor Statistics, "Projections Overview," Table 7, Percent Change in Employment, by Education Category, 2010-2020," and "Table 8, New Jobs by Education Category, 2010-2020." http://www.bls.gov/ooh/about/print/projections-overview.htm.

368. Gordon, Ed, presentation, World Future Society Annual Conference, Chicago, Ill., July 20, 2013. Gordon is author of *Future Jobs: Solving the Employment and Skills Crisis*, Praeger, Chicago, Sept. 2013.

369. U.S Bureau of Labor Statistics, "Projections Overview, Table 9, Number of Jobs Due to Growth and Replacement Needs, by Occupational Group, 2010-2020. http://www.bls.gov/ooh/about/print/projections-overview.htm.

370. Bascuas, Katie, "So You've Decided to Implement Teleworking. Now What?" *Associations Now,* May 2, 2013. http://associationsnow.com/2013/05/so-youve-decided-to-implement-teleworking-now-what/.

371. Athitakis, Mark, "Warning: Engaged Employees," *Associations Now,* April 8, 2013. http://associationsnow.com/2013/04/warning-engaged-employees/.

372. Wagner, Cynthia G., "Emerging Careers and How to Create Them," *The Futurist,* Jan.-Feb. 2011, p 30.

373. "About the Society for Neuroscience." http://www.sfn.org/index.aspx?pagename=about_sfn.

374. "Nanotechnology: Big Opportunities for Those Who Think Small," IEEE News Source. http://theinstitute.ieee.org/career-and-education/career-guidance/big-opportunities-for-those-who-think-small.

375. University of Houston, College of Technology, "Career Opportunities: Possibilities for Careers with a Degree in Biotechnology." http://tech.uh.edu/programs/undergraduate/biotechnology/career-opportunities/.

376. American Academy of Forensic Science "Forensic Science Careers." http://www.all-about-forensic-science.com/forensic-science-careers.html.

377. Houghton, Stuart, "Google Glass: What You Need to Know," *techradar.AV.* http://www.techradar.com/us/news/video/google-glass-what-you-need-to-know-1078114.

378. Frey, Thomas "55 Jobs of the Future." http://www.futuristspeaker.com/2011/11/55-jobs-of-the-future/.

379. Hamlisch, Marvin, "A Chorus Line," Broadway musical, music by Marvin Hamlisch, lyrics by Edward Kleban, book by James Kirkwood and Nicholas Dante, conception of Michael Bennett. http://www.imagi-nation.com/moonstruck/albm56.html, and http://www.careerhubblog.com/main/2006/11/who_am_i_anyway.html.

380. "Employability Skills: SCANS Profile." http://www.learningconnections.org/ss/pdfs/biotechnologybiomedical/bio4.pdf.

381. Caprino, Kathy, "What You Don't Know Will Hurt You: The Top 8 Skills Professionals Need to Master," *Forbes,* April 27, 2012. http://www.forbes.com/sites/kathycaprino/2012/04/27/what-you-dont-know-will-hurt-you-the-top-8-skills-professionals-need-to-master/.

382. The Conference Board of Canada, "Employability Skills 2000+." http://www.conferenceboard.ca/topics/education/learning-tools/employability-skills.aspx?pf=true, Jan. 12, 2013.

383. The White House, "Education: Knowledge and Skills for the Jobs of the Future, Reform for the Future." http://www.whitehouse.gov/issues/education/reform.

384. Pennsylvania Department of Education, "Career Education & Work (CEW) Standards Toolkit." http://www.pacareerstandards.com/.

385. National Association of State Directors of Career Technical Education Consortium, "The 16 Career Clusters." http://www.careertech.org/career-clusters/glance/careerclusters.html.

386. Fishman, Charles, "The Insourcing Boom," *The Atlantic.* http://www.theatlantic.com/magazine/print/2012/12/the-insourcing-boom/309166/.

387. Robinson, Kenneth, "How Do We Prepare Students for Jobs That Don't Exist Yet?" *Edudemic,* video, Oct. 25, 2011. http://edudemic.com/2011/10/students-of-the-future/.

388. Lee, James H., "Hard at Work in the Jobless Future," World Future Society. *The Futurist*, March-April 2012. http://www.wfs.org/book/export/html/2477.

389. "Nanotechnology-related careers." http://trynano.org/.

390. Goldman, Jordan, "Hot Jobs of the Future," Good to Know, *ABC News* Video, (13.41), featuring and interview with Jordan Goldman, founder and CEO of *Unigo.com*, Dec. 21, 2009. http://abcnews.go.com/Business/video/hot-jobs-future-9396730.

391. Travis, Merle, "16 Tons," song written by Merle Travis, made popular by Tennessee Ernie Ford, 1955, CowboyLyrics.com, AnthemCollege.edu. http://www.cowboylyrics.com/lyrics/classic-country/sixteen-tons---tennessee-ernie-ford-14930.html.

392. *Worldchanging: A User's Guide for the 21st Century*, edited by Steffen, Alex; introduction by McKibben, Bill, Abrams, NY, 2011.

393. U.S. Energy Information Administration, "What is Energy? Explained, Basics, Washington, DC http://www.eia.gov/energyexplained/print.cfm?page=about_home,.

394. "Energy," *About.com. Physics*. http://physics.about.com/od/glossary/g/energy.htm.

395. U.S. Energy Information Administration, "What is Energy? Explained, Basics, Washington, DC http://www.eia.gov/energyexplained/print.cfm?page=about_home.

396. "2012, The Outlook for Energy: A View to 2040, Taking on the World's Toughest Energy Challenges," Exxon Mobil. http://www.exxonmobil.com/corporate/files/news_pub_eo2012.pdf.

397. U.S. Department of Energy, "Understanding Earth's Energy Sources," Energy Education and Workforce Development, PowerPoint slide presentation for Grades 9-12 energy education, Karen Harrell and Dora Moore, Denver Public Schools, 2006, "Oil." http://www1.eere.energy.gov/education/pdfs/acts_harrell_understandingfossilfuels_307.pdf.

398. "Oil Sands." http://en.wikipedia.org/wiki/Oil_sands.

399. Institute for Energy Research, "Oil Shale." http://www.instituteforenergyresearch.org/energy-overview/oil-shale/.

400. State of Connecticut Department of Energy & Environmental Protection, Public Utilities Regulatory Authority, "How was natural gas formed? http://www.naturalgas.org/overview/background.asp.

401. "Typical Composition of Natural Gas," NaturalGas.org, a web site identified as reflecting information from the natural gas industry. http://www.naturalgas.org/overview/background.asp.

402. "Fracking: Definition Varies," *About.com. Energy*. http://energy.about.com/b/2012/02/19/fracking-definition-varies.htm, posted Feb. 19, 2012.

403. World Nuclear Organization, "The Cosmic Origins of Uranium." http://www.world-nuclear.org/info/inf78.html.

404. U.S. Department of Energy, "Renewable Energy." http://www.eere.energy.gov/topics/renewable_energy.html.

405. Union of Concerned Scientists, "How Geothermal Energy Works." http://www.ucsusa.org/clean_energy/our-energy-choices/renewable-energy/how-geothermal-energy-works.html.

406. U.S. Department of Energy, "Energy Basics, Biomass Technologies." http://www.eere.energy.gov/topics/biomass.html, updated Aug. 12, 2011.

407. U.S. Department of Energy, "Energy Basics, Biofuels." http://www.eere.energy.gov/basics/renewable_energy/biofuels.html, updated Aug. 12, 2011.

408. American Coal Council, "Biomass Co-Firing with Coal as an Emissions Reduction Strategy." http://www.americancoalcouncil.org/displaycommon.cfm?an=1&subarticlenbr=162.

409. U.S. Department of Energy, "Energy Basics, Biopower." http://www.eere.energy.gov/basics/renewable_energy/biopower.html, updated Aug. 12, 2011.

410. U.S. Department of Energy, "Energy Basics, Bio-Based Products." http://www.eere.energy.gov/basics/renewable_energy/biobased_products.html, updated Aug. 12, 2011.

411. Wocken, Chad, "The Power of Algae, *Biomass Magazine*. http://biomassmagazine.com/articles/3096/the-power-of-algae.

412. U.S. Energy Information Administration, "U.S. Energy Facts, Explained, Consumption & Production, Americans Use Many Type of Energy." http://www.eia.gov/energyexplained/print.cfm?page=us_energy_home, updated Oct. 15, 2012.

413. Advanced Power & Energy, University of California, Irvine, 2007, "Building Templates: Commercial Building: Education Buildings." http://www.apep.uci.edu/der/buildingintegration/2/BuildingTemplates/School.aspx

414. U.S. Energy Information Administration, "U.S. Energy Facts, Explained, Consumption & Production, Americans Use Many Type of Energy." http://www.eia.gov/energyexplained/print.cfm?page=us_energy_home, updated Oct. 15, 2012.

415. Goodstein, David, *Out of Gas, The End of the Age of Oil*, W.W. Norton & Companies, Inc., NY, 2004, pp. 123-126.

416. Smil, Vaclav, "Power Density Primer," *Market Resource* energy blog, May 13, 2010. http://www.masterresource.org/2010/05/smil-density-new-renewables-iv/.

417. Gates, Sara, "Two-Thirds of Solar Panels Installed Worldwide in 2011 Were in Europe, Report Says," *Huffington Post*, Sept. 25, 2012. http://www.huffingtonpost.com/2012/09/25/solar-panels-2011-europe_n_1912999.html.

418. University of California at San Diego, "Space-Based Solar Power, Do the Math." http://physics.ucsd.edu/do-the-math/2012/03/space-based-solar-power/, posted March 20, 2012.

419. Whittington, Mark, "David Criswell's Big Idea: Lunar-Based Solar Power," *Examiner.com*. http://www.examiner.com/article/david-criswell-s-big-idea-lunar-based-solar-power, posted June 23, 2010.

420. Wesoff, Eric, "Sources: Bloom Box Costs $12.50 Per Watt," *greentechmedia*, posted Jan. 4, 2011. http://www.wired.com/business/2011/01/bloom-box-cost/.

421. U.S. Department of Energy, "An Update on Advanced Battery Manufacturing," Oct. 16, 2012. http://energy.gov/articles/update-advanced-battery-manufacturing.

422. DaVinci Institute, "2050 and the Future of Transportation." http://www.davinciinstitute.com/papers/2050-and-the-future-of-transportation/.

423. U.S. Energy Information Administration, "Frequently Asked Questions, How much electricity is lost in transmission and distribution in the U.S?"http://www.eia.gov/tools/faqs/faq.cfm?id=105&t=3.

424. NOVA Science NOW, "Smart Grid," video, hosted by Neil deGrasse Tyson, *PBS*. http://video.pbs.org/video/1801235533/.

425. Ford, Matt, "Can a Smart Grid Turn Us On To Energy Efficiency?" CNN Tech, March 2, 2009. http://edition.cnn.com/2009/TECH/03/01/eco.smartgrid/index.html.

426. "Wichita Lineman," *CowboyLyrics.com*. http://www.cowboylyrics.com/lyrics/campbell-glen/wichita-lineman-622.html.

427. Green, Miranda, "How to Stop Blackouts: Smart-Grid Technology Could Ease Storms' Aftermath," *The Daily Beast*. http://www.thedailybeast.com/articles/2012/11/04/how-to-stop-blackouts-smart-grid-technology-could-ease-storms-aftermath.print.html, posted Nov. 4, 2012.

428. "USA's 20th Century Power Grid Fails Its 21st Century Economy," *USA Today*, Opinion, Our View,

Nov. 13, 2012, p. 8a.

429. "The Smart Grid." *Smartgrid.com*. http://www.smartgrid.gov/the_smart_grid.

430. Cobb, Kurt, "Our Current Infrastructure Was Build for a Different World," *OilPrice.com*, July 30, 2012. http://oilprice.com/Energy/Energy-General/Our-Current-Infrastructure-was-Built-for-a-Different-World.html.

431. "Advanced Batteries for Energy Storage will Represent a Market of Nearly $30 Billion by 2022, Forecasts Pike Research," *Business Wire*, Aug. 12, 2012. http://www.marketwatch.com/story/advanced-batteries-for-energy-storage-will-represent-a-market-of-nearly-30-billion-by-2022-forecasts-pike-research-2012-08-07.

432. Electrochemical Society, "Large Scale Energy Storage for Smart Grid Applications," *ecstransactions*, Vol. 41, No. 23. http://ecst.ecsdl.org/content/41/23/local/front-matter.pdf.

433. The World Bank, "Energy and the World Bank." http://web.worldbank.org/WBSITE/EXTERNAL/.OPICS/EXTENERGY2/0,,print:Y-isCURL:Y-contentMDK:22858145-pagePK:210058-piPK:210062-theSitePK:4114200,00.html.

434. "Global Energy Statistical Yearbook 2012, Enerdata." http://yearbook.enerdata.net/.

435. U.S. Department of Energy, "Energy Literacy…Essential Principles and Fundamental Concepts for Energy Education, A framework for Energy Education for Learners of All Ages." http://www1.eere.energy.gov/education/pdfs/energy_literacy_1_0_low_res.pdf.

436. U.S. Department of Energy, "Lesson Plans, Teach and Learn, Energy Education & Workforce Development." http://www1.eere.energy.gov/education/lessonplans/default.aspx.

437. U.S. Department of Energy, "Energy Literacy…Essential Principles and Fundamental Concepts for Energy Education…A Framework for Energy Education for Learners of All Ages." http://www1.eere.energy.gov/education/pdfs/energy_literacy_1_0_low_res.pdf.

438. U.S. Department of Energy, "Lesson Plans, Teach and Learn, Energy Education & Workforce Development." http://www1.eere.energy.gov/education/lessonplans/default.aspx.

439. "Energy Independence and Security Act of 2007: Summary of Provisions."

http://www.eia.gov/oiaf/aeo/otheranalysis/aeo_2008analysispapers/eisa.html.

440. "Smart Grid," *NOVA Science Now*, *PBS* Video. http://video.pbs.org/video/1801235533/.

441. EIA Technical Review Guidelines: Energy Generation and Transmission, Volume II-Appendices, U.S. Energy Information Agency in collaboration with USAID, EPA, and CCAD. http://www.epa.gov/international/regions/lac/eia-guidelines/energyvol2.pdf.

442. National Geographic Society, "Solar Power," video. http://video.nationalgeographic.com/video/environment/energy-environment/solar-power/.

443. National Geographic Society, "Great Energy Challenge." http://video.nationalgeographic.com/video/environment/energy-environment/great-energy-challenge/.

444. U.S. Energy Information Agency, "U.S. Energy Facts Explained, Consumption and Production." http://www.eia.gov/energyexplained/print.cfm?page=us_energy_home.

445. NIST (National Institute for Standards and Technology), "Smart Grid: A Beginner's Guide." http://www.nist.gov/smartgrid/beginnersguide.cfm post created July 17, 2012.

446. Goodall, Jane, interviewed on *Bill Moyers Journal, PBS*, Nov. 27, 2009. http://www.pbs.org/moyers/journal/11272009/watch.html.

447. Gillis, Justin, "Heat-Trapping Gas Passes Milestone, Raises Fears," *New York Times*, May 10, 2013. http://www.nytimes.com/2013/05/11/science/earth/carbon-dioxide-level-passes-long-feared-milestone.

html?pagewanted=all&_r=0.

448. Walsh, Bryan, and *Time* Correspondents, *Planet in Peril*, Time Books, NY, 2012.

449. Based on Huitt, W., Maslow's Hierarchy of Needs. *Educational Psychology Interactive*, Valdosta State University, Valdosta, GA, 2007. http://www.edpsycinteractive.org/topics/regsys/maslow.html and Parkinson-Hardman, Linda, "Maslow's Hierarchy of Needs Theory," Mowgli Foundation. http://mowgli.org.uk/3792/maslows-hierarchy-of-needs-theory.html.

450. "Global Warming & Climate Change (Doha Talks, 2012)," *New York Times*, Science, posted/updated Dec. 2, 2012. http://topics.nytimes.com/top/news/science/topics/globalwarming/index.html#.

451. "Background on the UNFCCC: The International Response to Climate Change," United National Framework Convention on Climate Change. http://unfccc.int/essential_background/items/6031.php.

452. "About Six Degrees Could Change the World," National Geographic television program. http://natgeotv.com/ca/six_degrees/about.

453. McKibben, Bill, "Global Warming's Terrifying New Math," *Rolling Stone*, July 19, 2012. http://www.rollingstone.com/politics/news/global-warmings-terrifying-new-math-20120719?print=true.

454. National Center for Atmospheric Research, "Future Warming Likely to Be on High Side of Climate Projections, Analysis Finds," *Atmos News*, Nov. 8, 2012. https://www2.ucar.edu/atmosnews/news/8264/future-warming-likely-be-high-side-climate-projections-analysis-finds.

455. National Center for Atmospheric Research, "Who Needs Glaciers?" *Atmos News*, Aug. 13, 2012. https://www2.ucar.edu/atmosnews/water-futures/who-needs-glaciers.

456. Vedantam, Shankar, "Report on Global Ecosystems Calls for Radical Changes," *Washington Post*, March 30, 2005, p. A2. http://www.washingtonpost.com/ac2/wp-dyn/A10966-2005Mar29?language=printer.

457. IPCC Report, "G8+5 Academies Joint Statement: Climate Change and the Transformation of Energy Technologies for a Low Carbon Future." http://www.nationalacademies.org/includes/G8+5energy-climate09.pdf.

458. Broder, John M., "Climate Change Doubt is Tea Party Article of Faith," *New York Times*, Oct. 20, 2010. http://www.nytimes.com/2010/10/21/us/politics/21climate.html.

459. "Global Warming & Climate Change Myths." http://www.skepticalscience.com/argument.php.

460. Environmental Defense Fund, "Scientific Consensus on Global Warming, Peer Review Ensures Sound Science." http://www.edf.org/climate/scientific-consensus.

461. Leiserowitz, Anthony; Maibach, Edward; Roser-Renouf, Connie; & Hmielowski, Jay, (2012) *Global Warming's Six America's in March 2012 and November 2011*, Yale University and George Mason University. New Haven, CT: Yale Project on Climate Change Communication. http://environment.yale.edu/climate/files/Six-Americas-March-2012.pdf.

462. Walsh, Bryan, "The Kyoto Accords-and Hope-Are Expiring," *Time Magazine*, Nov. 8, 2011. http://www.time.com/time/printout/0,8816,2098887,00.html#.

463. Guterl, Fred, "Searching for Clues to Calamity," *New York Times*, July 20, 2012, http://www.nytimes.com/2012/07/21/opinion/the-climate-change-tipping-point.html.

464. Guterl, Fred, *The Fate of the Species…Why The Human Race May Cause Its Own Extinction and How We Can Stop It*, Bloomsbury USA, NY, 2012, p. 58. Subquote from Hansen, James, *Storms of My Grandchildren*, Bloomsbury USA, NY, 2009, ch. 7.

465. "Venus: Fast Facts and Images." http://nai.arc.nasa.gov/astrotech/candidates/venus.htm.

466. Naam, Ramez, "Can Innovation Save the Planet?" Presentation, World Future Society Annual Conference, Toronto, Canada, July 28, 2012.

467. Jin, Zhouying, "Systemic Solution to Achieving Green Economic Growth and Sustainable Development," focused on China, July 28, 2012, World Future Society Conference, Toronto, Canada. Presentation delivered with Kenneth Hunter, senior fellow, Maryland China Initiative, University of Maryland, and board chair, World Future Society,

468. "The White House, "Energy, Climate Change, and Our Environment." http://www.whitehouse. gov/energy.

469. National Highway Traffic Safety Administration (NHTSA), "CAFÉ—Fuel Economy." http:// www.nhtsa.gov/fuel-economy.

470. "The Blueprint for A Secure Energy Future: Progress Report," President Barack Obama, March 2012. http://www.whitehouse.gov/sites/default/files/email-files/the_blueprint_for_a_secure_energy_future_oneyear_progress_report.pdf.

471. "About EPA, Our Mission and What We Do." http://www.epa.gov/aboutepa/whatwedo.html.

472. Pontin, Jason, "Why We Can't Solve Big Problems," *MIT Technology Review*, November-December 2012. http://www.technologyreview.com/featuredstory/429690/why-we-cant-solve-big-problems/.

473. Rifkin, Jeremy, *The Third Industrial Revolution*, Palgrave Macmillan, NY, 2012, p 13.

474. Accuweather, "Top 5 Most Expensive Natural Disasters in History," March 30, 2011. http://www.accuweather.com/en/weather-news/top-5-most-expensive-natural-d/47459#.

475. Schwab, Martin, "Future of Planetary Defense," Homeland Defense Institute, World Future Society presentation at Capital Science 2004, Arlington, Virginia, March 20, 2004, at National Science Foundation.

476. Pearlman, Robert, "NASA Simulates Asteroid Mission for Potential 2025 Flight," *SPACE.com*, Aug. 31, 2012. http://www.space.com/17412-nasa-mock-asteroid-mission-2025-tests.html.

477. "The World's Richest and Poorest Countries," *Global Finance*. http://www.gfmag.com/tools/global-database/economic-data/11934-richest-poorest-countries.html#axzz2Eg6KpUts.

478. Renner, Michael, and Sheehan, Molly O., "Poverty and Inequality Block Progress," Sarin, Radhika, "Rich-Poor Divide Growing," and Halweil, Brian, "Harvesting of Illegal Drugs Remains High," *Vital Signs 2003*, The Worldwatch Institute, Washington, DC, 2003, pp.18, 88, 98, and 138, and *World Development Report 2000/2001*, World Bank, Oxford University Press, NY, 2001, p. 51.

479. Joanna Zelman, "50 Million Environmental Refugees by 2020, Experts Predict," *Huffington Post*, May 25, 2011. http://www.huffingtonpost.com/2011/02/22/environmental-refugees-50_n_826488.html?view=print&comm_ref=false.

480. Russell, Jack, "Carbon Emissions on the Rise But Policies Growing Too," *Vital Signs 2009*, The Worldwatch Institute, Washington, DC, pp. 59-61.

481. "World Military Spending Levels Out After 13 Years of Increases, says SIPRI, Stockholm International Peace Research Institute, April 17, 2012. http://www.sipri.org/media/pressreleases/17-april-2012-world-military-spending-levels-out-after-13-years-of-increases-says-sipri.

482. Singer, Peter W., "Comparing Defense Budgets, apples to Apples," *Time*. http://nation.time.com/2012/09/25/comparing-defense-budgets-apples-to-apples/.

483. Halweil, Brian, "Harvesting of Illegal Drugs Remains High," *Vital Signs 2003*, The Worldwatch Institute, Washington, DC, 2003, pp. 98-99.

484. World Health Organization, "Almost a Quarter of All Disease Caused by Environmental Exposure," June 16, 2006, Geneva. http://www.who.int/mediacentre/news/releases/2006/pr32/en/index.html.

485. Environmental Protection Agency, "Asthma Triggers: Gain Control." http://www.epa.gov/asthma/

triggers.html.

486. "Environmental Factors in Autism Initiative," *Autism Speaks*. http://www.autismspeaks.org/science/initiatives/environmental-factors-autism-initiative and "Autism Rates: New High in U.S. Inspires Renewed Debate," Reuters, May 29, 2012, http://www.huffingtonpost.com/2012/03/29/autism-rates-high_n_1388342.html.

487. Centers for Disease Control, "Data & Statistics Prevalence of Autism." http://www.cdc.gov/ncbddd/autism/data.html.

488. Lavelle, Marianne, and Kurlantzick, Joshua, "Water, Tapped Out, The Coming Water Crisis," *U.S. News and World Report*, Special Edition, The Future of Earth, Morton B. Zuckerman, Executive Committee and Editor-in -Chief, Summer 2004, p. 58.

489. Markels, Alex, "The War Over Water," *U.S. News and World Report*, Special Edition, The Future of Earth, Morton B. Zuckerman, Executive Committee and Editor-in -Chief, Summer 2004, p. 64.

490. Fuller, R. Buckminster, *Operating Manual for Spaceship Earth*, Southern Illinois University Press, Touchstone Book published by Simon and Schuster, NY, 1969, p. 2 of ch. 5.

491. Marx, Gary, *Sixteen Trends…Their Profound Impact on our Future*, Educational Research Service, 2006-2008, Editorial Projects in Education, 2012, p 230.

492. Environmental Protection Agency (EPA), "Environmental Education (EE)." http://www.epa.gov/envrroed.

493. U.S. Environmental Protection Agency (EPA), "Definition of Green Building." http://www.epa.gov/greenbuilding/pubs/about.htm.

494. U.S. Environmental Protection Agency, "Teacher Resources and Lesson Plans" for teaching about the environment. http://www.epa.gov/students/teachers.html.

495. National Geographic Society, "This Bulb." http://video.nationalgeographic.com/video/kids/green-kids/this-bulb-kids/.

496. National Center for Atmospheric Research, "Learning Center: Climate Change Courses & Consequences," information plus teaching resources. http://spark.ucar.edu/climate-change-causes-and-consequences.

497. "Alex Steffen: The Route to a Sustainable Future." *TED Talk*, 2005, Oxford, U.K. http://climatechange.worldbank.org/sites/default/files/Turn_Down_the_heat_Why_a_4_degree_centrigrade_warmer_world_must_be_avoided.pdf.

498. World Bank *Turn Down the Heat…Why a 4 degree C. Warmer World Must Be Avoided*, A World Bank Report by the Potsdam Institute for Climate Impact Research and Climate Analytics, Washington, DC, Nov. 2012. http://climatechange.worldbank.org/sites/default/files/Turn_Down_the_heat_Why_a_4_degree_centrigrade_warmer_world_must_be_avoided.pdf.

499. Walsh, Bryan, *Global Warming: The Causes, The Perils, The Solutions*, Time Home Entertainment, Inc., *TIME Books*, NY 2012.

500. Goodall, Jane, interviewed on *Bill Moyers Journal*, PBS, Nov. 27, 2009. http://www.pbs.org/moyers/journal/11272009/watch.html.

501. National Geographic Society, "Global Warming, A Way Forward: Facing Climate Change," video, Washington, DC, with support from the United Nations Foundation. http://video.nationalgeographic.com/video/environment/global-warming-environment/way-forward-climate/.

502. MIT Video, Environment. http://video.mit.edu/channel/environment/.

503. Fuller, R. Buckminster, *Operating Manual for Spaceship Earth*, Southern Illinois University Press, Touchstone Book published by Simon and Schuster, NY, 1969.

504. Guterl, Fred, *The Fate of the Species*, Bloomsbury USA, NY, 2012.

505. Wilber, Ken, *A Brief History of Everything*, Shambhala Publications, Boston, MA, 2000.

506. Thoreau, Henry, quote, *Great Quotes.com*. http://www.great-quotes.com/quotes/category/Sustainability.

507. Fiksel, Joseph; Eason, Tasha; and Frederickson, Herbert, "A Framework for Sustainability, Indicators at EPA," National Risk Management Research Laboratory, Office of Research and Development, U.S. Environmental Protection Agency (EPA), Oct. 2012. http://www.epa.gov/sustainability/docs/framework-for-sustainability-indicators-at-epa.pdf, pp 9-10.

508. Rifkin, Jeremy, *The Third Industrial Revolution, How Lateral Power is Transforming Energy, The Economy, and The World*, Palgrave Macmillan, NY, 2011, pp. 236-239.

509. "Sustainable," definition, synonyms and antonyms, *Merriam-Webster Online Dictionary*. http://www.merriam-webster.com/dictionary/sustainable.

510. Mr. Y (pseudonym for Captain Wayne Porter, U.S. Navy, and Colonel Mark "Puck" Mykleby, U.S. Marine Corps "A National Strategic Narrative," published by the Woodrow Wilson Center, Washington, DC, with a preface by Anne-Marie Slaughter, professor of politics and international affairs, Princeton University. http://www.wilsoncenter.org/sites/default/files/A%20National%20Strategic%20Narrative.pdf.

511. Sustainable Cities Institute, "Land Use" and "Economic Development: Environmental Challenges." http://www.sustainablecitiesinstitute.org/view/page.basic/class/tag.topic/land_use.

512. "Chicago Launches Green Building Retrofit Program," *Sustainable Business.com*, Aug. 18, 2011. http://www.sustainablebusiness.com/index.cfm/go/news.printerfriendly/id/22682.

513. Kaye, Leon, "Chicago, My Kind of Sustainable Business Town," *Triple Pundit.com*, March 5, 2013. http://www.triplepundit.com/2013/03/chicago-sustainable-business/.

514. City of Toronto, "What the City Doing to Shrink Its Footprint?" http://www.toronto.ca/environment/initiatives/energy_retrofits.htm.

515. Burr, Andrew C., "Sears Tower to Undergo Historic Green Retrofit," *The CoStar Group*, June 28, 2009. http://www.costar.com/News/Article/Sears-Tower-to-Undergo-Historic-Green-Retrofit/113203.

516. "Empire State Building Efficiency Retrofit Model Rolls Out Across U.S.," *The Energy Collective*, July 3, 2013. http://theenergycollective.com/silviomarcacci/244691/empire-state-building-efficiency-retrofit-model-rolls-out-across-us.

517. Ford, Daniel, "U.K. Restaurant Association Launches Sustainability Rating System,*"Associations Now*, May 6, 2013. http://associationsnow.com/2013/05/u-k-restaurant-association-launches-sustainability-rating-system/.

518. Porter, Michael E., and Kramer, Mark R., "Creating Shared Value," *Harvard Business Review*, Jan. 2011. http://hbr.org/2011/01/the-big-idea-creating-shared-value/.

519. "Business Sustainability," *Financial Times, Lexicon*. http://lexicon.ft.com/Term?term=business-sustainability.

520. Rouse, Margaret, "LEED (Leadership in Energy and Environmental Design), Definition," April 2010. http://searchdatacenter.techtarget.com/definition/LEED-Leadership-in-Energy-and-Environmental-Design.

521. Howard, Brian Clark, "10 of the Greenest Colleges in America," *The Daily Green*. http://www.thedailygreen.com/environmental-news/latest/10-greenest-colleges-460708#slide-1.

522. Howard, Brian Clark, "12 of the Most Healthy and Sustainable College Cafeterias," *The Daily Green*. http://www.thedailygreen.com/environmental-news/latest/greenest-college-cafete-

rias-4608093#slide-1.

523. University of South Dakota, "Proposal to the Board of Regents for Establishment of Sustainability Degree Program at the University of South Dakota," Nov. 16, 2011, p. 3. See news release at http://www.usd.edu/press/news/news.cfm?nid=2326, Dec. 21, 2011.

524. Dallas County Community College District, "Sustainable Schools and College Track." http://www.dcccd.edu/AU/Chancellor/OrgLearn/Councils/WG/DCCCDGreen/2012SS/Pages/Schools2012.aspx.

525. Ian Symmonds & Associates, "Seven Essential Characteristics of Sustainable Schools and Colleges". http://static.squarespace.com/static/500ee7a3c4aaf86e468510a3/t/5242c0a0e4b0c4a-5c2a3128e/1380106400878/Seven%20Essential%20Characteristics%20of%20Sustainable%20Schools%20and%20Colleges.pdf.

526. Gutman, Robert, quote, *Good Reads.com.* http://www.goodreads.com/quotes/tag/sustainability.

527. Samuel, Henry, "Hypersonic Jet to Travel Paris to Tokyo in Two-and-One-Half Hours Travel," June 19, 2011. http://www.telegraph.co.uk/travel/travelnews/8585139/Hypersonic-jet-to-travel-from-Paris-to-Tokyo-in-two-and-a-half-hours.html, (includes video).

528. Wilson, Edward O., *Letters to a Young Scientist*, Liveright, W.W. Norton & Company, NY, 2013. Ordering information at http://www.amazon.com/Letters-Young-Scientist-Edward-Wilson/dp/0871403773 and Smithsonian Associates program announcements, April 13, 2013, p 16.

529. Fiksel, Joseph; Eason, Tasha; and Frederickson, Herbert, "A Framework for Sustainability, Indicators at EPA," National Risk Management Research Laboratory, Office of Research and Development, U.S. Environmental Protection Agency (EPA), Oct. 2012. http://www.epa.gov/sustainability/docs/framework-for-sustainability-indicators-at-epa.pdf.

530. Portland State University Graduate School of Education, Review "Sustainability Education Resource," Review numerous resources that include teaching materials and videos. http://www.pdx.edu/elp/sustainability-education-resources.

531. Mr. Y is pseudonym for Captain Wayne Porter, U.S. Navy, and Colonel Mark "Puck" Mykleby, U.S. Marine Corps, "A National Strategic Narrative," *Mr. Y*, published by the Woodrow Wilson Center, Washington, DC, 2011, (with a preface by Anne-Marie Slaughter, professor of politics and international affairs, Princeton University. http://www.wilsoncenter.org/sites/default/files/A%20National%20Strategic%20Narrative.pdf.

532. *Science Daily's* web site, including its informational and teaching materials, including videos. http://www.sciencedaily.com/videos/earth_climate/.

533. View "Ecosystem Markets Task Force Video: A New Lens for Business; A New Economy," March 2013, University of Cambridge, U.K., business executives discuss a sustainable economy, (5:00). http://www.cpsl.cam.ac.uk/Business-Platforms/Natural-Capital-Leaders-Platform.aspx?#fragment-2.

534. Wilson, Edward O., (2013), *Letters to a Young Scientist*, Liveright, W.W. Norton & Company, NY. .

535. Robert Engelman, *Vital Signs, 2011*, The Worldwatch Institute, Washington, DC In section devoted to "World Population Growth Slows Modestly, Still on Track for 7 Billion in Late 2011, Table 1, Regional Population Basics for 2010, Share of Absolute Global Population. Data from U.N. Population Division, http://esa.un.org/unup/.

536. Matt Rosenberg, "Basic Earth Facts," *About.com*, Geography, About.com Guide. http://geography.about.com/od/learnabouttheearth/a/earthfacts.htm.

537. Washington, Jesse, "Rodney King Death: 'Can We All Get Along?' Plea Measures His Lasting Moments," *Huffington Post*, June 17, 2012. http://www.huffingtonpost.com/2012/06/17/rodney-king-death-can-we-all-get-along_n_1604450.html?view=print&comm_ref=false.

538. Luers, William H., "A Message from UNA-USA's President, Looking Forward," email letter, November 10, 2004.

539. Friedman, Thomas L., *The World is Flat...A Brief History of the 21ˢᵗ Century*, 2005-2006-2007, Straus & Giroux, NY.

540. Harvard Graduate School of Education, "Future of Learning: Strand 3: Globalization" introductory materials, Future of Learning Institute, Aug. 3, 2012.

541. "Measuring Globalization," *Foreign Policy Magazine*, Jan.-Feb., 2001, Issue 122, pp. 56-65, from online abstract. http://www.columbia.edu/itc/sipa/U6347/client_edit/Measuring%20globalization. html.

542. Pink, Daniel, "The Changing Workplace, An Interview with Daniel Pink," Global Issues, The Challenges of Globalization," *eJournal USA*, Feb. 2006. p. 20. http://iipdigital.usembassy.gov/st/english/publication/2012/07/20120724145054su0.4112316.html#axzz2a1moYsXT.

543. Caselli, Marco, "On the Nature of Globalization and Its Measurement," UNU-CRIS Occasional Papers, United Nations University, Comparative Regional Integration Studies, O-2006/3. http://www.cris.unu.edu/fileadmin/workingpapers/20060220113557.O-2006-3.pdf.

544. Gibson, Rowan, *Rethinking the Future*, "Rethinking Business," Rowan Gibson, Nicholas Brealey Publishing, 2002, pp. 4-7, 10-11.

545. Winchester, Simon, *Krakatoa*, HarperCollins Publishers, Inc., NY, 2004, p. 193.

546. Suarez-Orozco, Marcello M., Comments by Marcello Suarez Orozco, Harvard University Future of Learning Institute, Session devoted to "Globalization and the Future of Learning," Aug. 3, 2012.

547. U.S. Census Bureau, International Data Base, "World Population: 1950-2050." http://www.census. gov/population/international/data/idb/worldpopgraph.php.

548. U.S. Census Bureau, International Data Base, "World Population Growth Rates, 1950-2050." http://www.census.gov/population/international/data/worldpop/graph_growthrate.php and "Negative Population Growth Facts & Figures, Total Midyear World Population, 1950-2050." http://www.npg. org/facts/world_pop_year.htm.

549. U.S. Census Bureau, "World Population by Region and Development Category: 1950-2050, Table A-1," *Global Population Profile*, first released on Internet March 22, 2004. http://www.7bn.net/ipc/prod/wp02/tabA-01.pdf.

550. "Measuring Globalization, Economic Reversals, Forward Momentum," study conducted and copyrighted by A.T. Kearney, Inc., and the Carnegie Endowment for International Peace, *Foreign Policy Magazine*, March-April 2004, based on article on pp. 54-60, including Global Index Chart on p. 57.

551. "The Globalization Index," *Foreign Policy Magazine*, Nov.-Dec 2007, A.T. Kearney, Inc., and *Foreign Policy*, owned by the Carnegie Endowment for International Peace, pp. 70-71.

552. "Measuring Globalization, The Global Top 20," *Foreign Policy* Magazine, May-June 2005, A.T. Kearney, Inc., and *Foreign Policy*, owned by the Carnegie Endowment for International Peace, p. 55.

553. ibid

554. Appiah, Anthony, Comments by Anthony Appiah, Harvard University Future of Learning Institute, Session devoted to "Globalization and the Future of Learning," Aug. 3, 2012.

555. "Economics Focus, Drain or Gain? Poor Countries Can End Up Benefiting When Their Brightest Citizens Emigrate," *The Economist*, May 26, 2011. http://www.economist.com/node/18741763/print.

556. Boeri, Tito: Brucker, Herbert: et al. (eds) "Book Review: Brain Drain and Brain Gain: The Global Competition to Attract High-Skilled Migrants," Oxford University Press. July 2012. http://blogs.lse. ac.uk/lsereviewofbooks/2012/10/31/book-review-brain-drain-and-brain-gain-the-global-competition-

to-attract-high-skilled-migrants/.

557. "European Education and Training Systems in the Second Decennium of the Lisbon Strategy," European Commission, Network of Experts in the Social Sciences of Education & Training (NESSE), and European Export Network on Economics in Education, June 2008. http://www.nesse.fr/nesse/activities/reports/challenges-for-european-education-pdf.

558. Khallash, Sally, "11 Key Trends in Global Mobility (Study)," *Global Talent Strategy*, Dec. 31, 2012. http://globaltalentstrategy.com/en/article/11-key-trends-in-global-mobility-study-317 and "Towers Watson's Global Survey Report 2012," *Global Talent Strategy*, Khallash, Sally, Jan. 7, 2013. http://globaltalentstrategy.com/en/article/towers-watsons-global-pay-survey-report-2012-321.

559. Khallash, Sally, "Only 8% of Scientists Would Relocate in China," Global Talent Strategy, Jan. 7, 2013. http://globaltalentstrategy.com/en/article/only-8-of-scientists-would-relocate-to-china-320.

560. Batalova, Jeanne, and Lee, Alicia, "US in Focus," Migration Information Source, Migration Policy Institute, March 2012. http://www.migrationinformation.org/usfocus/display.cfm?ID=886#7.

561. Camarota, Steven A. and Zeigler, Karen, "Jobs Americans Won't Do? A Detailed Look at Immigrant Employment by Occupation," Center for Immigration Studies (CIS), August 2009. http://www.cis.org/illegalImmigration-employment.

562. "European Education and Training Systems in the Second Decennium of the Lisbon Strategy," European Commission, Network of Experts in the Social Sciences of Education & Training (NESSE), and European Export Network on Economics in Education, June 2008. http://www.nesse.fr/nesse/activities/reports/challenges-for-european-education-pdf.

563. Tavares, Laura, Notes from Presentation by Laura Tavares, "What Do We Do with Difference?" Harvard Future of Learning Institute, Aug. 2, 2012, Cambridge, MA.

564. United Nations, "Universal Declaration of Human Rights," Adopted by the General Assembly as Resolutions 217 A(III) of December 10, 1948. http://www.un.org/en/documents/udhr/.

565. Marmolejo, Francisco, "Deficiency in Foreign Language Competency: What is Wrong with the U.S. Educational System," *The Chronicle of Higher Education*, Nov. 9, 2012. http://chronicle.com/blogs/worldwise/deficiency-in-foreign-language-competency-what-is-wrong-with-the-u-s-educational-system/27558.

566. "Internet World Users by Language, Top Ten Languages, Internet," *World Stats*. http://www.internetworldstats.com/stats7.htm, updated May 31, 2011.

567. "International Education Exchanges are at All-Time High, Strengthening Economies and Societies around the World," Press Release, "Open Doors 2012: "International Student Enrollment Increased by 6 Percent," Nov. 12, 2012. http://www.iie.org/Who-We-Are/News-and-Events/Press-Center/Press-Releases/2012/2012-11-13-Open-Doors-International-Students#.UtV7D_1Lc8M, and http://www.iie.org/en/Research-and-Publications/Open-Doors/Data/Fact-Sheets-by-Country.

568. European Commission, "From 6 to 28 Members." http://ec.europa.eu/enlargement/policy/from-6-to-28-members/index_en.htm.

569. "North American Free Trade Agreement." http://en.wikipedia.org/wiki/North_American_Free_Trade_Agreement.

570. "Overview, Association of Southeast Asian Nations." http://www.asean.org/asean/about-asean/history.

571. "The Twelfth Meeting of ASEAN Plus Three." http://www.asean.org/.

572. "Mercosur Trade Center. http://www.mercosurtc.com/.

573. African Union, "AU in a Nutshell." http://www.au.int/en/about/nutshell.

574. Gardner, Lauren, "Worldwide Urbanization," *World Politics Review*, June 28, 2007. http://www. worldpoliticsreview.com/trend-lines/895/worldwide-urbanization.

575. United Nations Department of Economic and Social Affairs, Population Division, "World Urbanization Prospects, The 2011 Revision," Table 3. Population of Urban Agglomerations with 10 Million Inhabitants or More, 1950, 1975, 2011, and 2025 (Millions), New York, 2012, pp. 6-7. http://esa. un.org/unup/pdf/FINAL-FINAL_REPORT%20WUP2011_Annextables_01Aug2012_Final.pdf.

576. U.S. Census Bureau, International Data Base, "Total Midyear Population for the World: 1950-2050." http://www.census.gov/population/international/data/idb/worldpoptotal.php.

577. DeCapua, Joe, "Urbanization Grows Worldwide," Voice of America (VOA), Aug. 31, 2012. http://www.voanews.com/content/world-urban-forum-31aug12/1499111.html.

578. Legarde, Christine, "A New Global Economy for a New Generation," speech by IMF Managing Director in Davos, Switzerland, Jan. 23, 2013. An "Introduction: Priorities for 2013." http://www.imf.org/external/np/speeches/2013/012313.htm.

579. "About TIMSS and PIRLS," International Study Center, one web site provides an overview of TIMSS and PIRLS. http://timssandpirls.bc.edu/home/pdf/TP_About.pdf . "TIMSS International Results in Science" describes the TIMSS approach to science, http://timss.bc.edu/timss2011/international-results-science.html. "TIMSS International Results in Mathematics" describes the TIMMS approach to mathematics. http://timss.bc.edu/timss2011/international-results-mathematics.html.

580. "OECD Program for International Student Assessment (PISA), Organization for Economic Cooperation and Development. http://www.oecd.org/pisa/.

581. "OECD Program for International Student Assessment (PISA), Organization for Economic Cooperation and Development. http://www.oecd.org/pisa/aboutpisa/.

582. Reimers, Fernando M., "Leading for Global Competency," *Educational Leadership*, Sept. 2009, Volume 67, Number 1. http://www.ascd.org/publications/educational-leadership/sept09/vol67/num01/Leading-for-Global-Competency.aspx.

583. Smith, Andrew F., and Czarra, Frederick R., "Teaching in a Global Context," *ASCD INFObrief.* http://www.ascd.org/publications/newsletters/policy-priorities/jan03/num32/toc.aspx.

584. "International Economic Development Policy, Georgetown University (Syllabus), Center for Global Development, Washington, DC. http://www.cgdev.org/content/publications/detail/1424345/.

585. "Putting the World into World-Class Education, State Innovations and Opportunities," CCSSO and the Asia Society. http://asiasociety.org/files/stateinnovations.pdf

586. Perrier, Craig, Blog, "The Global, History Educator." http://cperrier.edublogs.org/global-competencies/. Update on global competencies, Feb. 9, 2013.

587. "Methodology-Teaching Diplomacy," Diplo Foundation, Malta and Geneva, 1999-2005. Diplomacy Foundation web site is at http://www.diplomacy.edu/. Original information appeared in the *Sixteen Trends* book by Gary Marx, pp. 251-252, but no longer online, from http://www.diplomacy.edu/Edu/Methodology/teaching.asp.

588. "Putting the World into World-Class Education, State Innovations and Opportunities," CCSSO and the Asia Society. http://asiasociety.org/files/stateinnovations.pdf.

589. Mead, Walter Russell, "America's Sticky Power," *Foreign Policy* Magazine, March-April 2004, pp. 46, 51.

590. Marx, Gary, *Sixteen Trends…Their Profound Impact on our Future*, Educational Research Service, 2006-2008, Editorial Projects in Education, 2012, p. 264.

591. Institute for International Education, "Open Doors 2012" Background Information, Nov. 13, 2012. http://www.iie.org/en/Research-and-Publications/Open-Doors.

592. Ashdown, Paddy, "The Global Power Shift," (video) a controversial *TED Talk* by Paddy Ashdown, a former member of the British Parliament and a diplomat with a lifelong commitment to cooperation, recorded Dec. 2011. http://www.ted.com/talks/lang/en/paddy_ashdown_the_global_power_shift.html.

593. Friedman, Thomas L., *The World is Flat*, Picador, Trade Paperback Edition, 2007. Available for Purchase from: http://www.amazon.com/World-Flat-3-0-History-Twenty-first/dp/0312425074/.

594. Fuller, R. Buckminster, *Operating Manual for Spaceship Earth*, Simon and Schuster, NY, 1969.

595. United Nations, "Universal Declaration of Human Rights," adopted by the General Assembly as Resolution 217 (AIII) of December 10, 1948. Available from http://www.un.org/en/documents/udhr/.

596. "The Globalization Index," *Foreign Policy* Magazine, Nov.-Dec 2007, A.T. Kearney, Inc., and *Foreign Policy*, owned by the Carnegie Endowment for International Peace, pp. 68-71.

597. Brown, Gordon, "Wiring the Web for Global Good," *TED Talk*, featuring Gordon Brown, U.N. Special Envoy for Global Education and former U.K. Prime Minister, taped July 2009, (16:46). http://www.ted.com/talks/lang/en/gordon_brown.html.

598. Reimers, Fernando M., "Leading for Global Competency," *Educational Leadership*, Sept. 2009, Volume 67, Number 1. http://www.ascd.org/publications/educational-leadership/sept09/vol67/num01/Leading-for-Global-Competency.aspx.

599. Suarez-Orozco, Marcello M., *Learning in the Global Era: International Perspectives on Globalization and Education*, University of California Press, 2007. http://www.amazon.com/Learning-Global-Era-International-Globalization/dp/0520254368, ordering information.600. Friedman, George, *The Next 100 Years, A Forecast for the 21ˢᵗ Century*, Anchor Books, a division of Random House, NY, 2009.

601. *Edudemic*, "Dr. Seuss Quotes That Can Change Your Life." http://www.edudemic.com/2012/12/30-dr-seuss-quotes-you-should-never-forget/.

602. Perkins, David, Presentation during Harvard University Future of Learning Institute, Cambridge, MA, July 31, 2012.

603. Personalized Learning. http://en.wikipedia.org/wiki/Personalized_learning.

604. U.S Department of Education, "Individualized, Personalized, and Differentiated Instruction." http://www.ed.gov/technology/draft-netp-2010/individualized-personalized-differentiated-instruction,

605. ibid

606. "Learning as Transformation," Transformative Learning Theory web site. http://transformative-learningtheory.com.

607. Casey, Katherine, and Grossman, Francesca, "Teaching Values: Let's End Our Misguided Approach," Kaplan, *Education Week*, Feb. 27, 2013. http://www.edweek.org/tm/articles/2013/02/25/fp_casey_grossman.html.

608. BusinessDictionary.com, "Market of One,".

609. Strauss, Valerie, "A School Brings Brain Research to the Center of Its Curriculum," *Washington Post*, March 5, 2013. http://articles.washingtonpost.com/2013-03-05/local/37459205_1_brain-research-teachers-education.

610. Marberry, Wendy, in Events and Conferences, Professional Development, "You don't need a computer," Pearson OLE (Online Learning Exchange), Community Blog, Quote from Steve Jobs. http://www.olecommunity.com/steve-jobs-said-you-dont-need-a-computer/. November 8, 2011.

611. Common Core Standards, "Frequently Asked Questions." http://www.corestandards.org/resources/frequently-asked-questions.

612. Perkins, David, Presentation during Harvard University Future of Learning Institute, Cambridge, MA, July 31, 2012.

613. Hymes, Donald, with Chafin, Ann, and Gonder, Peggy, *The Changing Face of Testing and Assessment*, American Association of School Administrators, 1991, p. 4.

614. Kohn, Alfie, "Confusing Harder with Better," *Education Week*, September 15, 1999. http://www.edweek.org/ew/articles/1999/09/15/02kohn.h19.html.

615. Marx, Gary, "Key Questions about Standards," *Sixteen Trends...Their Profound Impact on our Future*, Educational Research Service, 2006, 2008, Editorial Projects in Education, 2011, p. 146.

616. Megan, Kathleen, "State to Replace Mastery Test with Computerized, Personalized Test in Two Years," *Hartford Courant, Courant.com*, July 15, 2012. http://articles.courant.com/2012-07-15/news/hc-mastery-testing-ending-0713-20120713_1_new-tests-questions-student.

617. Sandler, Susan, "People vs. Personalization," *Education Week*, Feb. 29, 2012, pp. 20, 22.

618. Immordino-Yang, M.H. and Fischer, K.W., "Neuroscience Bases of Learning," Elsevier, Ltd., 2012, page 310, reading for 2012 Harvard University Future of Learning Institute. Immordino-Yang,, M.H., & Fischer, K.W. (2009, in press). Neuroscience bases of learning. In V.G. Aukrust (Ed.), *International Encyclopedia of Education, 3rd Edition*, Section on Learning and Cognition, Oxford, England: Elsevier. http://www-bcf.usc.edu/~immordin/papers/Immordino-Yang+Fischer_2009_NeuroscienceBasesof-Learning.pdf

619. Szalavitz, Maia, "Performance-Enhancing Drugs O.K. in School, but Not in Sports, Students Say," *Time Healthland*. http://healthland.time.com/2012/05/09/performance-enhancing-drugs-o-k-in-school-but-not-in-sports-students-say/. May 9, 2012.

620. Sparks, Sarah, "Smart Pills' Promising, Problematic," *Education Week*, Oct. 24, 2012, p 1, 16.

621. Kurzweil, Ray, "The Accelerating Power of Technology," *TED Talk*, Posted Nov. 2006, TED 2005. http://www.ted.com/talks/ray_kurzweil_on_how_technology_will_transform_us.html.

622. CAST (2011). Universal Design for Learning Guidelines version 2.0. Wakefield, MA: Author. *Universal Design for Learning web site*, "About UDL." http://www.cast.org/udl/ and http://www.udlcenter.org/aboutudl/udlguidelines.

623. Gardner, Howard, *Multiple Intelligences, The Theory in Practice*, BasicBooks, a division of Harper-Collins Publishers, Inc., 1993, drawn from pp. 15-27.

624. Item based on stories/news releases from SIIA and ASCD, including: "District Race to the Top Appropriately Prioritizes Personalized Learning." http://www.siia.net/blog/index.php/2012/05/district-race-to-the-top-appropriately-prioritizes-personalized-learning/ and "Education Leaders Identify Top 10 Components of Personalized Learning." http://www.ascd.org/news-media/Press-Room/News-Releases/Education-Leaders-Identify-Top-10-Components-of-Personalized-Learning.aspx.

625. Education Writers Association, "High School Reform," *Education Reform*, Backgrounder 15, January 2001, p. 1.

626. "Join the Coalition of Essential Schools," website promotion. http://www.essentialschools.org/join.

627. Stansbury, Meris, "Mass Customized Learning: The Key to Education Reform," *eSchool News*, Feb. 21, 2012. http://www.eschoolnews.com/2012/02/21/mass-customized-learning-the-key-to-education-reform/.

628. Denvir, Daniel, "School: It's Way More Boring Than When You Were There," *Salon.com*, Sept. 14, 2011. http://www.salon.com/2011/09/14/denvir_school/.

629. Guterl, F., *The Fate of the Species*, Bloomsbury USA, NY, 2012, p. 122.

630. Perkins, David, Presentation during Harvard University Future of Learning Institute, Cambridge, MA, July 31, 2012.

631. Marx, Gary, *Sixteen Trends...Their Profound Impact on our Future*, Educational Research Service,

2006-2008, Editorial Projects in Education, 2012, pp. 160-161.

632. Immordino-Yang, "Embodied Brains, Social Minds," *TEDx* Manhattan Beach, Manhattan Beach Education Foundation, 2009, *YouTube*, http://www.youtube.com/watch?v=RViuTHBIOq8,

633. CAST (2011). Universal Design for Learning Guidelines version 2.0. Wakefield, MA: Author. http://www.udlcenter.org.

634. Gardner, Howard, *Multiple Intelligences: The Theory in Practice*, Basic Books, Harper Collins, NY, 1993.

635. Immordino-Yang, M.H., & Fischer, K.W. (2009, in press). Neuroscience bases of learning. In V.G. Aukrust (Ed.), *International Encyclopedia of Education, 3rd Edition*, Section on Learning and Cognition, Oxford, England: Elsevier. http://www-bcf.usc.edu/~immordin/papers/Immordino-Yang+Fischer_2009_NeuroscienceBasesofLearning.pdf

636. Council of Chief State School Officers, *Common Core State Standards: Implementation Tools and Resources*, August 2012. http://www.ccsso.org/documents/2012/common_core_resources.pdf

637. Ravitch, Diane, *The Death and Life of the Great American School System...How testing and Choice are Undermining Education*, Basic Books, NY, 2010.

638. Einstein, Albert, quotes, Quote from interview with G.S. Viereck, Oct. 26, 1929. Reprinted in "Glimpses of the Great (1930). http://Einstein.biz/quotes.

639. Definition of ingenuity, *Merriam-Webster's Collegiate Dictionary*, Eleventh Edition, Merriam-Webster, Inc., 2003, p. 642.

640. Definition of creativity, *Dictionary.com*.

641. Definition of imagination, *Merriam-Webster Dictionary online*.

642. ibid

643. Marx, Gary, *Ten Trends...Educating Children for a Profoundly Different Future*, Educational Research Service, 2000, p. 62.

644. Florida, Richard, *The Rise of the Creative Class*, Basic Books, a member of the Perseus Books Group, 2002, p. 33.

645. Wilson, Edward O., *Consilience...The Unity of Knowledge*, Borzoi Book, published by Alfred A. Knopf, Inc., NY, 1998, pp. 8-9.

646. Wilber, Ken, *A Brief History of Everything*, Shambhala Publications, Inc., Boston, 1996 and 2000, pp. 15-17.

647. Robinson, Kenneth, *TED Talk*, http://www.ted.com/speakers/sir_ken_robinson.html.

648. Robinson, Kenneth, "Creativity in the Classroom, Innovation in the Workplace," *Principal Voices, online*. http://www.principalvoices.com/voices/ken-robinson-white-paper.html.

649. Olson, Max, "Fumbling the Future at Xerox PARC," *Future Blind*, May 16, 2011.http://www.futureblind.com/2011/05/fumbling-the-future-at-xerox-parc/.

650. Gladwell, Malcolm, "Creation Myth," *The New Yorker*, May 18, 2011. http://www.newyorker.com/reporting/2011/05/16/110516fa_fact_gladwell.

651. Thomas, Douglas, and Brown, John Seely, "Cultivating the Imagination: Building Learning Environments for Innovation," *Teachers College Record*, Date Published: Feb. 17, 2011. http://www.tcrecord. orgID Number 16341. http://www.newcultureoflearning.com/TCR.pdf.

652. Sparks, Sarah, "Studies Explore How to Nurture Students' Creativity," *Education Week*, Dec. 13, 2011. http://www.edweek.org/ew/articles/2011/12/14/14creative.h31.html.

653. "Howard Gardner: Principal Publications." http://www.old-pz.gse.harvard.edu/PIs/HGpubs.htm.

654. Gardner, Howard, Presentation during Harvard University Future of Learning Institute, Cambridge, MA, July 31, 2012.

655. Sparks, Sarah D., "Studies Explore How to Nurture Students' Creativity," *Education Week*, Dec. 13, 2011. http://www.edweek.org/ew/articles/2011/12/14/14creative.h31.html.

656. Johnson, Brian, "Waking Up the Algorhythm," presentation at World Future Society Annual Conference, July 27, 2012, Toronto, Ontario, Canada.

657. Herlin, Niko. "Crowdsourcing Weak Signals Collection in Corporate Foresight," presentation at World Future Society Annual Conference, July 29, 2012, Toronto, Ontario, Canada.

658. Amabile, T.M.; Hadley, C.N.; and Kramer, S.J., "Creativity Under the Gun," *Harvard Business Review* on The Innovative Enterprise, Harvard Business School Press, Boston, 2003, pp. 1-25.

659. Reference: "Elvis Presley," http://www.history-of-rock.com/elvis_presley.htm.

660. Vanderbilt, Tom, "Better Living Through Imitation," *Smithsonian Magazine*, September 2012, p. 51.

661. da Vinci, Leonardo, *The Quotation Page*, Leonardo da Vinci quote. http://www.quotationspage.com/quote/24512.html.

662. Presley, Elvis, http://www.history-of-rock.com/elvis_presley.htm. "Elvis", "Elvis Presley" and "Graceland" are Registered Trademarks of Elvis Presley Enterprises, Inc. © 1996 E.P.E., Inc.

663. "National Endowment for the Arts (NEA) Announces Upcoming Education Leaders Institute," *Art Works*, NEA web site, April 9, 2010. http://www.nea.gov/news/news10/ELI.html.

664. Amundson, Kristen, *Performing Together…The Arts and Education*, John F. Kennedy Center for the Performing Arts and American Association of School Administrators, 1985.

665. Eisner, Elliot, "Ten Lessons the Arts Teach." http://www.arteducators.org/advocacy/10-lessons-the-arts-teach.

666. Winner, Ellen, "Distinguishing Between Research and Advocacy: Examining Claims about the Effects of Brain Stimulation and the Arts," Presentation at the Future of Learning Institute, Harvard University, Aug. 1, 2012.

667. Morris, Holly J.; Kulman, Linda; Satchell, Michael; Lord, Lewis; Curry, Andrew; Tolson, Jay; Sklaroff, Sara,; Gilgoff, Dan, *U.S. News and World Report, Special Collector's Edition, American Ingenuity…The Culture of Creativity That Made a Nation Great*, "Making Music," Spring 2003, pp. 32, 39, 40, 47, 48.

668. Amundson, Kristen, *Performing Together…The Arts and Education*, John F. Kennedy Center for the Performing Arts and American Association of School Administrators, 1985.

669. Perkins, David, Presentation during Harvard University Future of Learning Institute, Cambridge, MA, July 31, 2012.

670. Perkins, David, and Tishman, Shari, "Patterns of Thinking," Project Zero, Harvard University. http://www.pz.gse.harvard.edu/patterns_of_thinking.php.

671. Costa, Arthur L., "The Thought-Filled Curriculum," *Educational Leadership*, ASCD, Feb.2008, pp. 20-24. http://www.ascd.org/publications/educational_leadership/feb08/vol65/num05/The_Thought-Filled_Curriculum.aspx.

672. Gladwell, Malcolm, *Blink…The Power of Thinking Without Thinking*, Little, Brown, and Company, NY, 2005, p. 69.

673. MacArthur Foundation Fellows Program, FAQ. http://www.macfound.org/media/files/FELLOWSQA.PDF.

674. "Raising Classroom Standards Means Ramping Up Non-Fiction," WNYC News, Oct. 11, 2010. http://www.wnyc.org/articles/wnyc-news/2010/oct/11/raising-classroom-standards-means-ramping-non-fiction/.

675. Qi, Lin Wang, "Activating for 21ˢᵗ Century Learning," Blog, Nov. 23, 2011. http://www.qi-global.com/blog/tag/patrick-newell.

676. Florida, Richard, *The Great Reset*, Harper Paperback, NY, 2011, p.18.

677. Immordino-Yang, M.H. and Fischer, K.W., "Neuroscience Bases of Learning," Elsevier, Ltd., 2012, page 310, reading for 2012 Harvard University Future of Learning Institute. Immordino-Yang,, M.H., & Fischer, K.W. (2009, in press). Neuroscience bases of learning. In V.G. Aukrust (Ed.), *International Encyclopedia of Education, 3ʳᵈ Edition*, Section on Learning and Cognition, Oxford, England: Elsevier. http://www-bcf.usc.edu/~immordin/papers/Immordino-Yang+Fischer_2009_NeuroscienceBasesof-Learning.pdf

678. Illes, Judy (Editor), *Neuroethics: Defining the Issues in Theory, Practice, and Policy*, Oxford Scholarship Online, Sept. 2009, Chapter 18, "Neuroethics in Education," Sheridan, Kimberley; Zinchenko, Elena; and Gardner, Howard, pp. 265-275.

679. Gardner, Howard, *Five Minds for the Future.*(paperback).Harvard Business School Press, Cambridge, MA, 2009.

680. Perkins, D. (2010). *Making Learning Whole…How Seven Principles of Teaching Can Transform Education*. San Francisco, CA, Jossey-Bass.

681. Wilson, Edward O., *Consilience. .. The Unity of Knowledge*. Borzoi Books, Alfred A. Knopf, NY, 1998.

682. Robinson, Kenneth, TED presentation, creativity. http://www.ted.com/speakers/sir_ken_robinson.html.

683. Costa, Arthur, *Habits of Mind Across the Curriculum: Practical and Creative Strategies for Teachers* (paperback).ASCD, Alexandria, VA, 2009.

684. Newell, Patrick, *21:21 The Movie, Aligning 21ˢᵗ Century Learning with 21ˢᵗ Century Learners,"* (21:21), 21 Foundation, Tokyo International School. http://www.21foundation.com/2121-the-movie/.

685. ASCD's *Whole Child Initiative web page*. http://www.wholechildeducation.org/.

686. Tomlinson, Stephen, "Edward Lee Thorndike and John Dewey on the Science of Education,", *Oxford Review of Education*, Vol. 23, No. 3 (Sept. 1997), pp. 365-383. http://www.wou.edu/~girodm/611/Thorndike_vs_Dewey.pdf.

687. Gibboney, Richard A., "Centennial Reflections, Intelligence by Design: Thorndike versus Dewey," *Phi Delta Kappan*, Oct. 2006. http://www.pdkmembers.org/members_online/publications/Archive/pdf/k0610cen.pdf.

688. Robelen, Erik, "Most Teachers See the Curriculum Narrowing, Survey Finds," *Education Week*, Sec. 8, 2011. http://blogs.edweek.org/edweek/curriculum/2011/12/most_teachers_see_the_curricul.html.

689. "Reinforcement Theory of Motivation," *Management Study Guide*. http://www.managementstudyguide.com/reinforcement-theory-motivation.htm.

690. Rumberger, Russell, "Solving the Nation's Dropout Crisis," *Education Week* Commentary, Oct. 26. 2011.

691. Boyd Pitts, Annette, "Conversations with Annette Boyd Pitts," *Civitas in Romania…External On-Site Evaluation*, Marx, Gary, Center for Civic Education, May-June 2011.

692. Gelb, Michael J., *How to Think Like Leonardo da Vinci*, Dell Publishing, a division of Random

House, NY, 1998, p. 17.

693. Hutchins, Alex, "What is the Half Life of Education?" Examiner.com, Knoxville, June 1, 2011. http://www.examiner.com/article/what-is-the-half-life-of-education.

694. "Kurzweil is Right. People Don't Understand Exponential Growth," Kurzweil Accelerated Intelligence Forums. http://www.kurzweilai.net/forums/topic/kurzweil-is-right-people-dont-understand-exponential-growth.

695. The European Graduate School, "Plato-Biography." http://www.egs.edu/library/plato/biography/.

696. Moen, Matthew, *Liberal Arts & Sciences FAQs*, Council of Colleges of Arts & Sciences (brochure), 2011.

697. Badolato, Robert, "The Educational Theory of Horace Mann," New Foundations, 2002. http://www.newfoundations.com/GALLERY/Mann.html. See also the Horace Mann League web site, http://www.hmleague.org.

698. Gibboney, Richard A., "Centennial Reflections, Intelligence by Design: Thorndike versus Dewey," *Phi Delta Kappan*, Oct. 2006. http://www.pdkmembers.org/members_online/publications/Archive/pdf/k0610cen.pdf.

699. Introduction, WEBDuBois.org.

700. Du Bois, W.E.B., "The Talented Tenth," from *The Negro Problem: A Series of Articles by Representative Negroes of Today*, New York, 1903.

701. *The Story of American Public Education*, Master Timeline, *PBS*, 2001. http://www.pbs.org/kcet/publicschool/roots_in_history/choice_master3.html

702. Adler, Mortimer, *The Paideia Proposal...An Educational Manifesto*. Institute for Philosophical Research, Chicago, 1982. New York, 1998, first Touchstone edition, p. 43. Also http://www.economyprofessor.com/theorists/mortimeradler.php and http://en.wikipedia.org/wiki/Paideia_Proposal.

703. U.S. Department of Education, *A Nation at Risk: The Imperative for Educational Reform, National Commission on Excellence in Education*, Washington, DC, pp. 5-6, 24-27.

704. Hoffman, David, and Broder, David, "Summit Sets 7 Main Goals for Education," *The Washington Post*, September 29, 1989, p. A3.

705. "History, 1989 to Present," National Education Goals Panel, Item-Feb. 1990. http://govinfo.library.unt.edu/negp/page1-7.htm.

706. Uchida, Donna, with Cetron, Marvin, and McKenzie, Floretta, *Preparing Students for the 21st Century*, American Association of School Administrators, Arlington, VA, 1996, drawn from items, pp. 1-75.

707. Gardner, Howard, *Multiple Intelligences: New Horizons*, Basic Books, NY, 2006. Earlier version, *Multiple Intelligences: The Theory in Practice, A Reader*, NY, 1993, also Basic Books.

708. Gardner, Howard, *Five Minds for the Future*. Harvard Business School Press, Cambridge, MA, 2007.

709. U.S. Department of Education, "Stronger Accountability, Testing for Results Helping Families, Schools, and Communities Understand and Improve Student Achievement," Sept. 16, 2004. http://www2.ed.gov/nclb/accountability/ayp/testingforresults.html.

710. U.S. Department of Education, "Key Policy Letters Signed by the Education Secretary or Deputy Secretary," July 24, 2001. http://www2.ed.gov/print/policy/elsec/guid/secletter/020724.html.

711. "The Whole Child Initiative, ASCD, Background and Tenets, http://www.ascd.org/whole-child.aspx.

712. Partnership for 21st Century Skills, web site, http://www.p21.org/.

713. The White House, Office of the Press Secretary, "President Obama Launches 'Educate to Innovate' Campaign for Excellence in Science, Technology, Engineering, & Math (STEM) Education," Nov. 23, 2009. http://www.whitehouse.gov/the-press-office/president-obama-launches-educate-innovate-campaign-excellence-science-technology-en, accesses, Sept. 4, 2012.

714. "Introduction to the Common Core State Standards." June 10, 2010. http://www.corestandards.org/assets/ccssi-introduction.pdf.

715. "How to Teach: Toward a Philosophy of Meaning," Coverage vs. Postholing, *studentsfriend.com, a guide to teaching world history and geography*. http://www.studentsfriend.com/onhist/how.html.

716. "Scaffolding," Glossary of Education, http://www.education.com/definition/scaffolding.

717. Brownsburg High School Senior Academy. http://www.brownsburg.k12.in.us/senioracademy/sa/.

718. Moen, Matthew, Liberal *Arts & Sciences FAQs*, Council of Colleges of Arts & Sciences (brochure), 2011.

719. Gentzel, Thomas J., "The Question," *PSBA Bulletin*, Closing Thoughts, April 2012. p. 48.

720. Perkins, David, Presentation during Harvard University Future of Learning Institute, Cambridge, MA, July 31, 2012.

721. Gove, Michael, "GCSEs Have Too Much Breadth, Not Enough Depth," BBC interview with U.K. Education Secretary Michael Gove. http://www.bbc.co.uk/news/education-18530504.

722. Robelen, Erik, "Most Teachers See the Curriculum Narrowing, Survey Finds," *Education Week*, Sec. 8, 2011. http://blogs.edweek.org/edweek/curriculum/2011/12/most_teachers_see_the_curricul.html.

723. Gibboney, Richard A., "Centennial Reflections, Intelligence by Design: Thorndike versus Dewey," *Phi Delta Kappan*, Oct. 2006. http://www.pdkmembers.org/members_online/publications/Archive/pdf/k0610cen.pdf.

724. Mathews, Jay, "Will Depth Replace Breadth in Schools?" Class Struggle, *The Washington Post*, Feb. 27, 2009. http://voices.washingtonpost.com/class-struggle/2009/02/will_depth_replace_breadth_in.html.

725. Moen, Matthew, Liberal *Arts & Sciences FAQs*, Council of Colleges of Arts & Sciences (brochure), 2011.

726. Hirsch, Jr., E.D.; Kett, Joseph F.; and Trefil, James, *The Dictionary of Cultural Literacy…What Every American Needs to Know*, Houghton Mifflin Company, Boston, 1988, p. 300.

727. Page, Susan, "Public Service Valued; Politics Not So Much," *USA Today*, July 22, 2013, pp. 1,7a.

728. Cleveland, Harlan, *Nobody in Charge…Essays on the Future of Leadership*, Jossey-Bass, A Wiley Company, San Francisco, 2002, p. 161.

729. Bartholomew, Dave and King, Pearl, "I Hear You Knockin' but You Can't Come In," song written by Dave Bartholomew and Pearl King, published in 1955, EMI Music Publishing. Hit song recorded by many artists. http://en.wikipedia.org/wiki/I_Hear_You_Knocking.

730. "How Political Polarization Hurts The Economy," *USA Today*, Sept. 11, 2012, p 10 A.

731. Timpany, Katherine, pastor of the First Congregational Church UCC in Sioux Falls, South Dakota, *Rev It Up*, Reflections on Faith and Life (newsletter), Nov. 4, 2009.

732. Edwards, Mickey "How to Turn Republicans and Democrats into Americans," *The Atlantic*, July-Aug. 2011. http://www.theatlantic.com/magazine/archive/2011/07/how-to-turn-republicans-and-democrats-into-americans/308521/.

733. Hamilton, Lee H., "How Should the Winners Govern?" *Comments on Congress*, The Center on

Congress at Indiana University," June 13, 2012. http://congress.indiana.edu/how-should-the-winners-govern.

734. Hamilton, Lee H., "Why We're So Polarized.," *Comments on Congress*, Oct. 29, 2010. http://congress.indiana.edu/why-were-so-polarized.

735. Friedman, Thomas L., "Advice from Grandma," *The New York Times*, Nov. 22, 2009. http://www.nytimes.com/2009/11/22/opinion/22friedman.htm.

736. Riley, Sean, "Charley Reese's 'Final" Column on How Washington, DC, Operates," *Examiner.com*, Feb. 15, 2012, based on re-emergence of Charley Reese's last column for the *Orlando Sentinel* on Feb. 3, 1984. http://www.examiner.com/article/charley-reese-s-final-column-on-how-washington-d-c-operates.

737. Ward, Barbara, *Spaceship Earth*, Columbia University Press, NY, Hamish Hamilton, Great Britain, 1966, p. 15.

738. Hirsch, Jr., E.D.; Kett, Joseph F.; and Trefil, James, *The Dictionary of Cultural Literacy…What Every American Needs to Know*, Houghton Mifflin Company, 1988, p. 460.

739. Madison, James, Hamilton, Alexander, and Jay, John, *The Federalist…A Commentary on The Constitution of the United States*, Bicentennial Edition, Robert B. Luce, Inc., Washington, DC, and NY, 1976, pp. 54-55.

740. National Institute for Civil Discourse, Purpose, "The Breakdown of Civil Discourse is Putting America at Risk." http://nicd.arizona.edu/purpose.

741. Pecquet, Julian, "Monday's Global Agenda: Washington Hosts AIDS Summit," *The Hill,* July 23, 2012. http://thehill.com/blogs/global-affairs/foreign-aid/239401-mondays-global-agenda-washington-hosts-aids-summit.

742. "Impacts World 2013-International Conference on Climate Change Effects, Potsdam, May 27-30." http://thehill.com/blogs/global-affairs/foreign-aid/239401-mondays-global-agenda-washington-hosts-aids-summit.

743. Zakaria, Fareed, *CNN*, "Why Political Polarization has Gone Wild in American (and What to Do About It)," July 24, 2011. http://globalpublicsquare.blogs.cnn.com/2011/07/24/why-political-polarization-has-gone-wild/.

744. Huntington, Samuel P., *The Clash of Civilizations, Remaking of World Order*, Samuel P. Huntington, Touchstone, a registered trademark of Simon and Schuster, Inc., NY, 1997, pp. 13-14.

745. Marx, Gary, *Sixteen Trends…Their Profound Impact on Our Future*, Educational Research Service and Education Week Press, 2006, p. 245.

746. Viadero, Debra, "Researchers Try to Promote Students' Ability to Argue," *Education Week* Eye on Research, Sept. 16, 2009. http://www.edweek.org/ew/articles/2009/09/16/03argue.h29.html.

747. Friedman, Thomas L., "Advice from Grandma," *New York Times,* Nov. 22, 2009. http://www.nytimes.com/2009/11/22/opinion/22friedman.html?_r=0.

748. Haidt, Jonathan, "How Common Threats Can Make Common (Political) Ground," *TED Talk*. http://www.ted.com/talks/jonathan_haidt_how_common_threats_can_make_common_political_ground.html, (20:02) posted Jan. 2013.

749. Edwards, Mickey, "How to Turn Republicans and Democrats into Americans," *The Atlantic*, July-Aug. 2011. http://www.theatlantic.com/magazine/archive/2011/07/how-to-turn-republicans-and-democrats-into-americans/308521/.

750. Cleveland, Harlan, *Nobody in Charge . . . Essays on the Future of Leadership.* Jossey-Bass, A Wiley Co., San Francisco, 2002.

751. Reich, John, ASU Emeritus Psychology Professor, Arizona State University YouTube video, (9:53).

http://www.youtube.com/watch?v=sBSJ4F-p1YE.

752. Stengel, Richard, "Mandela: His 8 Lessons of Leadership," *Time*, July 9, 2008. http://www.time.com/time/magazine/article/0,9171,1821659,00.html.

753. "Definitions of Authority,, Authoritarian," *Merriam-Webster Dictionary Online*. http://www.merriam-webster.com/dictionary/authority, http://www.merriam-webster.com/dictionary/authoritative, http://www.merriam-webster.com/dictionary/authoritarian.

754. "Pericles." http://www.fordham.edu/halsall/ancient/pericles-funeralspeech.asp.

755. Wheeler, L. Kip, "Niccoló Machiavelli and 'The Prince,'" updated March 6, 2013. http://web.cn.edu/kwheeler/machiavelli.html.

756. "Hobbes, Locke, Montesquieu, and Rousseau on Government," Constitutional Rights Foundation, Bill of Rights in Action, Spring 2004. http://www.crf-usa.org/bill-of-rights-in-action/bria-20-2-c-hobbes-locke-montesquieu-and-rousseau-on-government.html.

757. Center for Civic Education, "*Foundations of Democracy*, High School, How Should We Choose People for Positions of Authority?" excerpt from the Center's Foundations of Democracy curriculum. http://new.civiced.org/fod-high-lesson-4.

758. Center for Civic Education, "We the People," Constitution Day, Introduction to concept of Authority for lesson plans. http://new.civiced.org/resources/curriculum/constitution-day-and-citizenship-day. Further information from www.civiced.org.

759. Myatt, Mike "10 Reasons Your Top Talent Will Leave You," *Forbes*, Dec. 13, 2012. http://www.forbes.com/sites/mikemyatt/2012/12/13/10-reasons-your-top-talent-will-leave-you/print/.

760. Turley, Cari, "The 8 Elements of Employee Engagement," *Achievers.com*. http://blog.achievers.com/2013/04/8-elements-employee-engagement.

761. Center for Civic Education, Suggested "Center for Civic Education Lesson Plans." http://new.civiced.org/resources/curriculum/lesson-plans, and Lesson Plans devoted to *What is Authority?* http://new.civiced.org/resources/curriculum/constitution-day-and-citizenship-day.

762. Center for Civic Education, Ordering information for *We the People: The Citizen and the Constitution*; *We the People: Project Citizen*; and *Foundations of Democracy*. http://new.civiced.org/resources/publications/ebooks/new-enhanced-ebook.

763. Constitutional Rights Foundation, Ordering information and learning resources. http://www.crf-usa.org/publications/.

764. Rifkin, Jeremy, *The Third Industrial Revolution*, Palgrave Macmillan, NY, 2012, Part 3, Chapters 7, 8, and 9, pp. 193-270.

765. "Global Power Shifts," Historian, Diplomat, and former head of Harvard's Kennedy School of Government Joseph Nye, *TEDGlobal*, Oxford, England, U.K., posted Oct. 2010, video (18:16), http://www.ted.com/talks/joseph_nye_on_global_power_shifts.html.

766. Sipe, James W., and Frick, Don M., "Leads with Moral Authority." http://www.leaderswhoserve.com/index_files/Pillar7.htm, excerpt from *Seven Pillars of Servant Leadership*, 2009, Paulist Press, N.J. available from http://www.amazon.com/Seven-Pillars-Servant-Leadership-Practicing/dp/080914560X.

767. Samuelson, Robert J., "Rethinking the Great Recession," *The Wilson Quarterly*, Winter 2011. http://www.wilsonquarterly.com/article.cfm?AID=1768.

768. *Webster's New Collegiate Dictionary*, Merriam-Webster, Inc., NY, 1983, p. 426.

769. "The Whistleblowers," *Time*, "Persons of the Year" cover story, December 30, 2002.

770. Kidder, Rushworth, quote from Rushworth Kidder, founder of the Institute for Global Ethics, from his book, *Moral Courage*, William Morrow Paperbacks, NY, 2009, p. 3.

771. Hausman, Carl, Editor, "The Top 10 Ethics Stories of 2012, *Ethics Newsline*, Jan. 2, 2013. http://www.globalethics.org/newsline/2013/01/02/top-ethics-stories-of-2012/.

772. Diamandis, Peter H., and Kotler, Steven., *Abundance...The Future is Better Than You Think*, Free Press, A Division of Simon & Schuster, Inc., NY, 2012, pp. 55-56.

773. Samuels, Christina A., "Educators Look for Lessons from Cheating Scandals," *Education Week*, March 7, 2012. http://www.edweek.org/ew/articles/2012/03/07/23testsecurity.h31.html.

774. Handelsman, Mitchell M., Example based in part on an item in "5 Ways to Teach Ethics," *Psychology Today*, Feb. 1, 2011. http://www.psychologytoday.com/blog/the-ethical-professor/201102/5-ways-teach-ethics.

775. Murrow, Edward R. quote, *ASCD Smart Brief*, Feb. 8, 2012.

776. Butts, R. Freeman, *The Morality of Democratic Citizenship*, quoted in *Teaching Values and Ethics*, by Kristen J. Amundson, American Association of School Administrators, Arlington, Virginia, 1991, pp. 24-25.

777. Character Counts, Los Angeles, California, "Six Pillars of Character." http://charactercounts.org/sixpillars.html.

778. Josephson, Michael, president of Character Counts, quotes. http://www.goodreads.com/author/quotes/1090178.Michael_Josephson.

779. Kerns, Charles D., "Creating and Sustaining an Ethical Workplace Culture," *Graziadio Business Review*, Pepperdine University, 2003, Vol. 6, Issue 3. http://gbr.pepperdine.edu/2010/08/creating-and-sustaining-an-ethical-workplace-culture/.

780. Colero, Larry, "A Framework for Universal Principles of Ethics," Crossroads Programs, Inc. http://www.ethics.ubc.ca/papers/invited/colero.html.

781. Amundson, Kristen J., "Common Core of Values, Baltimore County Public Schools, Maryland," quoted in *Teaching Values and Ethics*, American Association of School Administrators, Arlington, Virginia, 1991, p. 27.

782. Glaze, Avis, "Character Development: Education at its Best," *ASCD Manitoba Journal Reflections*, The Whole Child, Summer 2011, Volume 11.

783. Uchida, Donna; Cetron, Marvin; McKenzie, Floretta, *Preparing Students for the 21ˢᵗ Century*, American Association of School Administrators, Arlington, Virginia, 1996, p. 20.

784. Prosser, John, and Prosser, Ryan, "Courting Controversy: Why (and How) We Teach Ethics," *Education Week* Teacher Leaders Network (TLN), Sept. 12, 2012. http://www.edweek.org/tm/articles/2012/09/11/tln_johnprosser_ryanprosser.html.

785. "Ethics Bowl," shared online by the Illinois Institute of Technology, explaining the Association for Practical and Professional Ethics (APPE) events. http://ethics.iit.edu/teaching/ethics-bowl.

786. Vitello, Paul, "At Ethics Bowl, L.I. Teenagers Debate Slippery Issues," *New York Times*, Feb 13, 2011. http://www.nytimes.com/2011/02/14/nyregion/14ethics.html?_r=0.

787. Korn, Melissa, "Does An 'A' in Ethics Have Any Value?" *The Wall Street Journal*, Feb. 6, 2013. http://online.wsj.com/article/SB10001424127887324761004578286102004694378.html.

788. Long, Rich, "Countering Corporate Arrogance," *The Strategist*, Public Relations Society of America, Spring 2002, p. 7.

789. Boseley, Sarah, "World Bank's Jim Yong Kim: 'I Want to Eradicate Poverty,'" *The Guardian* global development, July 25, 2012. http://www.guardian.co.uk/global-development/2012/jul/25/world-bank-jim-yong-kim-eradicate-poverty.

790. Tang, Mei-Ying, "Human Rights Education in Taiwan: The Experience of the Workshops for

Schoolteachers." http://www.hurights.or.jp/archives/human_rights_education_in_asian_schools/section2/1999/03/human-rights-education-in-taiwan-the-experience-of-the-workshops-for-schoolteachers.html.

791. State Secretariat for Economic Affairs (SECO), "What is Corruption?" http://www.seco.admin.ch/themen/00645/00657/00659/01387/index.html?lang=en&print_style=yes.

792. Lukaszewski, James E., "The Ethical Practitioner, Dilemmas and Moral Questions: The Heart of Ethical Decision Making," *Tactics* Magazine, February 2002, p. 22.

793. Franklin, Benjamin, "Ben Franklin's Two Daily Questions and 13 Virtues," Ethics Alarms. http://ethicsalarms.com/rule-book/ben-franklins-two-daily-questions-and-13-virtues/.

794. Josephson, Michael, President, Josephson Institute, home of Character Counts interview, KNBC Television, Los Angeles, Dec. 30, 2012. http://josephsoninstitute.org/michael/.

795. Glaze, Avis, "Character Development: Education at its Best," *ASCD Manitoba Journal Reflections, The Whole Child*, Summer 2011, Volume 11, pp. 18-23. http://www.mbascd.ca/documents/Journal2011_Final.pdf.

796. Institute for Global Ethics. http://www.globalethics.org/; Josephson Institute. http://josephsoninstitute.org/ and http://charactercounts.org/sixpillars.html. Santa Clara University Markkula Center for Applied Ethics. http://www.scu.edu/ethics/. Ethics Resource Center, http://www.ethics.org/.

797. Brown, Gordon, "Gordon Brown on Global Ethics vs. National Interest," *TED Global* 2009, Interview with TED Curator Chris Anerson. http://www.ted.com/talks/gordon_brown_on_global_ethic_vs_national_interest.html.

798. *"Plato's Ethics*: An Overview," *Stanford Encyclopedia of Philosophy*, revised May 29, 2009. http://plato.stanford.edu/entries/plato-ethics/.

799. Amundson, Kristen J., *Teaching Values and Ethics*, American Association of School Administrators, Arlington, VA, 1991.

800. Gibson, Rowan, *Rethinking the Future*, Rowan Gibson, "Rethinking Business," Nicholas Brealey Publishing, London, 2002, pp. 10-11.

801. Deming, W. Edwards, *Out of the Crisis*, Massachusetts Institute of Technology, Center for Advanced Engineering Study, Cambridge, Massachusetts, 1991, pp. 23-24.

802. "Continuous Improvement, ASQ Overview, PDCA Cycle." http://asq.org/learn-about-quality/continuous-improvement/overview/overview.html.

803. "The 5 Lean Principles," *LEANING Forward*. http://www.leaningforward.co.uk/principles.htm.

804. "What is Six Sigma?" General Electric: Our Company. http://www.ge.com/en/company/companyinfo/quality/whatis.htm.

805. "Where Does Kaizen Come From?" *Manufacturing Info, Business Knowledge Source*, http://businessknowledgesource.com/manufacturing/where_does_kaizen_come_from_033651.html.

806. Leading Edge Group, "The Ten Basic Kaizen Principles." http://www.leadingedgegroup.com/assets/uploads/The_Ten_Basic_Kaizen_principles.pdf.

807. Ashkenas, Ron, "It's Time to Rethink Continuous Improvement," *Harvard Business Review*, May 8, 2012. http://blogs.hbr.org/ashkenas/2012/05/its-time-to-rethink-continuous.html.

808. Denning, Steve, "The Boeing Debacle: Seven Lessons Every CEO Must Learn," *Forbes*, Jan. 17, 2013. http://www.forbes.com/sites/stevedenning/2013/01/17/the-boeing-debacle-seven-lessons-every-ceo-must-learn/; in the article, Denning quotes from "Restoring American Competitiveness," Pisano, Gary P. and Shih, Willy C. *Harvard Business Review*, July-Aug. 2009, http://hbr.org/2009/07/restoring-american-competitiveness/ar/pr.

809. "Jim Collins: How to Manage Through Chaos," *CNN Money*, Sept. 30, 2011. http://management. fortune.cnn.com/2011/09/30/jim-collins-great-by-choice-exclusive-excerpt/.

810. Ouchi, William, *Theory Z...How American Business Can Meet The Japanese Challenge*, William Ouchi, Addison-Wesley Publishing Company, Inc., Philippines, 1981, pp. 5, 14.

811. Peters, Tom, and Austin, Nancy, *A Passion for Excellence...The Leadership Difference*, Random House, Inc., NY, 1985, p. 98.

812. Newman, Michael E., "Four U.S. Organizations Honored with the 2012 Baldrige National Quality Award," *NIST*. http://www.nist.gov/baldrige/baldrige_recipients2012.cfm

813. "Baldrige by Sector: Education," http://www.nist.gov/baldrige/enter/education.cfm.

814. "Baldrige in Action, Montgomery County Public Schools." http://www.montgomeryschoolsmd. org/info/baldrige/about/overview.aspx.

815. Marx, Gary, *Sixteen Trends...Their Profound Impact on Our Future*, Educational Research Service and Education Week Press, 2006, p. 191.

816. Marx, Gary, *Sixteen Trends...Their Profound Impact on Our Future*, Educational Research Service and Education Week Press, 2006, Gary Rowe quote, p. 195.

817. Withrow, Frank; Long, Harvey; Marx, Gary, *Preparing Schools and School Systems for the 21ˢᵗ Century*, American Association of School Administrators, 1999, p.14.

818. Johnson, Randy, "The Zen of Quality," *Education Week*, November 22, 2000, p. 35.

819. "*LLIS.gov*, Lessons Learned by the Federal Emergency Management Agency." https://www.llis.dhs. gov.

820. Montgomery County Public Schools (MCPS),"Malcolm Baldrige National Quality Award, 2010 Award Recipient, In Pursuit of Excellence." http://www.montgomeryschoolsmd.org/uploadedFiles/info/ baldrige/homepage/Baldrige-Pursuit-of-Excellence.pdf, and "Call to Action," a video program prepared by MCPS to explain its continuous improvement process, http://www.montgomeryschoolsmd.org/info/ baldrige/about/overview.aspx.

821. Holiday, Billie, and Herzog, Arthur Jr., "God Bless The Child," Lyrics to song, written by Billie Holiday and Arthur Herzog, Jr., Carlin America, Inc., Warner/Chappell Music, Inc. EMI Music Publishing. http://www.lyricsfreak.com/b/billie+holiday/god+bless+the+child_20018000.html.

822. National Center for Children in Poverty (NCCP). "Child Poverty." http://www.nccp.org/topics/ childpoverty.html.

823. Berliner, David C., "Poverty and Potential: Out of School Factors and School Success," The Horace Mann League Blog, July 11, 2012. http://horacemannleague.blogspot.com/2012/07/poverty-and-potential-out-of-school.html.

824. Definition of poverty, *Webster's New Collegiate Dictionary*, Merriam-Webster, Inc., 2003, p. 973.

825. Shipler, David K., *The Working Poor...Invisible in America*, Borzoi book published by Alfred A. Knopf, NY, 2004, pp. 3-5.

826. The World Bank, "Poverty Reduction and Equity, At a Glance." http://web.worldbank.org/ WBSITE/EXTERNAL/TOPICS/EXTPOVERTY/EXTPA/0,,contentMDK:20040961~menuP-K:435040~pagePK:148956~piPK:216618~theSitePK:430367~isCURL:Y,00.html.

827. Boseley, Sarah, "World Bank's Jim Yong Kim: 'I Want to Eradicate Poverty," *The Guardian* global development, July 25, 2012. http://www.guardian.co.uk/global-development/2012/jul/25/world-bank-jim-yong-kim-eradicate-poverty.

828. "Remarks as Prepared for Delivery: World Bank Group President Jim Yong Kim at the Annual Meeting Plenary Session," Tokyo, Japan, Oct. 11, 2012. http://www.worldbank.org/en/

news/2012/10/12/remarks-world-bank-group-president-jim-yong-kim-annual-meeting-plenary-session.

829. World Bank, *World Development Indicators 2012*, a progress report including progress in working toward the Millennium Development Goals. http://data.worldbank.org/sites/default/files/wdi-2012-ebook.pdf.

830. U.S. Census Bureau, Statistical Abstract of the U.S., "Table 712, Children Below Poverty Level by Race and Hispanic Origin: 1980 to 2009." http://www.census.gov/compendia/statab/2012/tables/12s0712.pdf, and Table 3, People in Poverty by Selected Characteristics, 2012 and 2011. https://www.census.gov/hhes/www/poverty/data/incpovhlth/2011/table3.pdf.

831. World Bank, "An Update to the World Bank's Estimates of Consumption Poverty in the Developing World." http://siteresources.worldbank.org/INTPOVCALNET/Resources/Global_Poverty_Update_2012_02-29-12.pdf.

832. Ratcliffe, Caroline, and McKernan, Signe-Mary "Childhood Poverty Persistence: Facts and Consequences." Urban Institute *Research of Record.* http://www.urban.org/publications/412126.html.

833. Urban Institute, "Understanding Poverty, Consequences of Poverty." http://www.urban.org/publications/412126.html.

834. *Kids Count Data Book*, 2013, Annie E. Casey Foundation, Baltimore, MD. http://datacenter.kidscount.org/files/2013KIDSCOUNTDataBook.pdf.

835. U.S. Census Bureau, "Child Poverty in the U.S., 2009 and 2010: Selected Race Groups and Hispanic Origin" issued Nov. 2011. http://www.census.gov/prod/2011pubs/acsbr10-05.pdf.

836. U.S Census Bureau, "Table 712. "Children Below Poverty Level by Race and Hispanic Origin: 1980 to 2009." http://www.census.gov/compendia/statab/2012/tables/12s0712.pdf.

837. *Kids Count Data Book*, 2013, Annie E. Casey Foundation, Baltimore, MD. http://datacenter.kidscount.org/files/2013KIDSCOUNTDataBook.pdf.

838. Rothstein, Richard, "Whose Problem is Poverty?" Educational Leadership, ASCD, Volume 65, Number 7, pp. 8-13, April 2008. http://www.ascd.org/publications/educational-leadership/apr08/vol65/num07/Whose-Problem-Is-Poverty%C2%A2.aspx.

839. Gardner, John, *Quotations about Poverty.* http://www.quotegarden.com/poverty.html.

840. National Center for Children in Poverty (NCCP), "Child Poverty." http://www.nccp.org/topics/childpoverty.html.

841. Ansell, Susan, "Achievement Gap," *Education Week* Issue Paper, March 18, 2004. Online: www.edweek.org.

842. Department of Education, "What Do Student Grades Mean? Differences Across Schools," *OERI Research Report*, U.S. Office of Educational Research and Information, January 1994. www.ed.gov/pubs/OR/ResearchRpts/grades.html.

843. Jensen, Eric, *Teaching with Poverty in Mind*, Chapter 2, "How Poverty Affects Behavior and Academic Performance," ASCD, Alexandria, VA, 2009. Available for purchase at http://www.ascd.org/publications/books/109074/chapters/how-poverty-affects-behavior-and-academic-performance.aspx.

844. Sandraluz Lara-Cinisomo; Anne R. Pebley; Mary E. Valana; Elizabeth Maggio; Mark Berends; and Samuel R. Lucas, "A Matter of Class," RAND Corporation, updated Sept. 15, 2010. http://www.rand.org/publications/randreview/issues/fall2004/class.html.

845. Dahl, Gordon B. and Lochner, Lance, "The Impact of Family Income on Child Achievement: Evidence from the Earned Income Tax Credit," National Bureau of Economic Research, NBER Working Paper No. 14599, Dec. 2008. http://www.nber.org/papers/w14599.

846. Rebell, Michael A. and Wolff, Jessica R., "We Can Overcome Poverty's Impact on School Success,"

Education Week Commentary, Jan. 18, 2012. http://www.edweek.org/ew/articles/2012/01/18/17rebell. h31.html.

847. McNeely, Robert, "No Education Reform Without Tackling Poverty, Experts Say" *NEA Today*, April 30, 2012. http://neatoday.org/2012/04/30/no-education-reform-without-tackling-poverty-experts-say/.

848. Office of Head Start, "History of Head Start." http://www.acf.hhs.gov/programs/ohs/about/history-of-head-start.

849. U.S. Department of Agriculture (USDA), "Nation School Lunch Program," Q&A. http://www.fns.usda.gov/cnd/lunch/aboutlunch/nslpfactsheet.pdf.

850. "The Thirteen Most Important Charts of 2013," Economic Policy Institute, chart devoted to "the failed 401(k) revolution," Dec. 23, 2013, http://tinyurl.com/kaez6w.

851. U.S. Department of Health and Human Services (HHS), "2013 HHS Poverty Guidelines." http://aspe.hhs.gov/poverty/13poverty.cfm, published in Federal Register on Jan. 24, 2013.

852. *Kids Count Data Book*, 2013, Annie E. Casey Foundation, Baltimore, MD, p. 24. http://datacenter. kidscount.org/DataBook/2012/OnlineBooks/KIDSCOUNT2012DataBookFullReport.pdf.

853. "Poverty in the United States, 1988 and 1989, Current Population Reports, Consumer Income, Series P-60, No. 171, Poverty Status of Persons by Age, Race, and Hispanic Origin, 1959 to 1989," http://www2.census.gov/prod2/popscan/p 60-166.pdf.

854. U.S. Census Bureau, "Table A People and Families in Poverty by Selected Characteristics, 1999 and 2000." http://www.census.gov/hhes/www/poverty/data/incpovhlth/2000/tablea.pdf.

855. U.S. Census Bureau, "Table 3. People in Poverty by Selected Characteristics: 2010 and 2011." http://www.census.gov/hhes/www/poverty/data/incpovhlth/2011/table3.pdf.

856. Social Security Administration, "Income Sources of Aged Units, 2010," Table 1.A-1, Percentage with Income from Specific Source, by Marital Income and Age, 2010, p. 37. http://www.ssa.gov/policy/docs/statcomps/income_pop55/2010/incpop10.pdf.

857. Social Security Administration, "Income of the Population, 55 or Older, 2010 Table 3, A1, Percentage Distribution; and Table 2,A1, Percentage Income from Specified Source. http://www.ssa.gov/policy/docs/statcomps/income_pop55/2010/incpop10.pdf.

858. Global Finance, "The Richest and Poorest Countries in the World 2012," (Based on IMF data, estimates based on value of the international dollar at the time computed). http://www.gfmag.com/tools/global-database/economic-data/12148-the-richest-countries-in-the-world.html#axzz2JhxHlOnF.

859. World Bank, "Share of World's Private Consumption, 2005," Poverty Facts and Stats, Global Issues. http://www.globalissues.org/article/26/poverty-facts-and-stats, updated Jan. 7, 2013.

860. Wright Edelman, Marian, President, Children's Defense Fund, "Families Struggle: Child Poverty Remains Epidemically High," *Huffington Post*, Sept. 28, 2012. http://www.huffingtonpost.com/marian-wright-edelman/families-struggle-child-p_b_1924179.html.

861. Battistoni, Alussa, "America Spends Less on Food Than Any Other Country," *MotherJones.com*, Feb. 1, 2012. http://www.motherjones.com/blue-marble/2012/01/america-food-spending-less. Based on chart, "The Poor Spend a High Percentage of Their Income on Food," drawn from the World Bank 2009, U.S. Department of Agriculture 2009, and Euromonitor International.

862. World Bank, *"World Development Indicators 2012."* http://data.worldbank.org/sites/default/files/wdi-2012-ebook.pdf.

863. King, Jr., Martin Luther, "Martin Luther King, Jr.'s Speech, *I Have a Dream*, The Full Text," *ABC News*, (Includes video and text of his 1963 "I Have a Dream" speech at the Lincoln Memorial in Washington, DC. (Story aired during 2012 dedication of the MLK Memorial). http://abcnews.go.com/

Politics/martin-luther-kings-speech-dream-full-text/story?id=14358231.

864. *Kids Count Data Book 2013*, Annie E. Casey Foundation, Baltimore, MD. http://datacenter. kidscount.org/files/2013KIDSCOUNTDataBook.pdf for download or purchase.

865. Holiday, Billie and Herzog, Arthur, Jr., "God Bless The Child," sung by Billie Holiday, lyrics to song written by Billie Holiday and Arthur Herzog, Jr., Carlin America, Inc., Warner/Chappell Music, Inc. EMI Music Publishing, written 1939, recorded in 1941. http://www.youtube.com/ watch?v=Z_1LfT1MvzI.

866. Shipler, David K., *The Working Poor . . . Invisible in America,* Alfred A. Knopf, NY, 2004.

867. "History of Head Start," video, (7:08), testimony in both English and Spanish. http://www.acf. hhs.gov/programs/ohs/about/history-of-head-start.

868. *American Experience, PBS,* "LBJ, The War on Poverty," (2:06), aired 2013. http://www.pbs.org/ wgbh/americanexperience/films/lbj/player/.

869. Wilder, Laura Ingalls, *The Long Winter,* Harper and Row, NY, 1940, 1953, p. 214.

870. Gronewold, Nathanial, "One-Quarter of World's Population Lacks Electricity," *Scientific American,* Nov. 24, 2009. http://www.scientificamerican.com/article.cfm?id=electricity-gap-developing-coun-tries-energy-wood-charcoal.

871. Prois, Jessica, and Goldberg, Eleanor, "World Water Day 2013: How Shortages Affect Women, Kids, and Hunger," *The Huffington Post,* March 22, 2013. http://www.huffingtonpost. com/2013/03/22/world-water-day-2013-facts_n_2927389.html?view=print&comm_ref=false.

872. "Water for Ross Bethio, Senegal Project at the World We Want Foundation," video. http://www. youtube.com/watch?v=8qjljBxwJ-E.

873. Rifkin, Jeremy, *The Third Industrial Revolution, How Lateral Power is Transforming Energy, The Economy, and The World,* Palgrave Macmillan, NY, 2011, p. 238.

874. Dickler, Jessica , "Paying for Gas Forces Painful Sacrifices,"*CNN Money,* May 4, 2011. http://mon-ey.cnn.com/2011/05/03/pf/high_gas_prices_hurt/index.htm#.

875. Diamandis, Peter H., and Kotler, Steven, "The Abundance Builders," *The Futurist,* World Future Society, July-Aug. 2012, Vol. 46, No. 4. http://www.wfs.org/futurist/july-august-2012-vol-46-no-4/ abundance-builders.

876. Domhoff, G. William, "Wealth, Income, and Power," *WhoRulesAmerica.net, 2013.* http://www2. ucsc.edu/whorulesamerica/power/wealth.html?print.

877. Nichols, Michelle, "U.S. Charitable Giving Approaches $300 Billion in 2011," *Reuters,* June 19, 2012. http://www.reuters.com/assets/print?aid=USBRE85I05T20120619.

878. "Can We Feed the World and Sustain the Planet (Preview)," *Scientific American,* Oct. 27, 2011, http://www.scientificamerican.com/article.cfm?id=can-we-feed-the-world.

879. Elert, Emily, Writer; De Chant, Tim, Illustrator, "Daily Infographic: If Everyone Lived Like An American, How Many Earths Would We Need?" http://www.popsci.com/environment/arti-cle/2012-10/daily-infographic-if-everyone-lived-american-how-many-earths-would-we-need, based on the work of Global Footprint Network. http://www.footprintnetwork.org/en/index.php/GFN/.

880. U.S, Bureau of Labor Statistics, "Table A-4, Employment Status of the Civilian Population 25 Years and Over by Educational Attainment." http://www.bls.gov/news.release/empsit.t04.htm.

881. Cohn, Scott, CNBC Senior Correspondent, "Student Loan Debt Hits Record High, Study Shows," *NBC News. Com, Business,* 2012. http://www.nbcnews.com/business/student-loan-debt-hits-record-high-study-shows-1C6542975#.

882. *Celebrity Networth,* "Mark Zuckerberg's Net Worth." http://www.celebritynetworth.com/rich-

est-businessmen/ceos/mark-zuckerberg-net-worth/.

883. Jayson, Sharon, "Generation Y's Goal? Wealth and Fame, *USA Today*, Jan. 10, 2007. http://usatoday30.usatoday.com/news/nation/2007-01-09-gen-y-cover_x.htm.

884. Diamandis, Peter H., and Kotler, Steven, *Abundance...The Future is Better Than You Think*, Free Press, A Division of Simon & Schuster, Inc., NY, 2012, pp. 179, 181.

885. Items presented for review as possible lesson plans on teaching about scarcity: "Scarcity, Choice, and Decisions," *The Mint*. http://themint.org/teachers/scarcity-choice-and-decisions.html, "Socialize (scarcity and making choices)," *Council for Economic Education*; and "Books for Teaching Economic Concepts, and Books with Examples of Scarcity," *Scholastic*. http://www.scholastic.com/teachers/lesson-plan/books-teaching-economic-concepts?pImages=n&x=67&y=20, (as well as others).

886. Diamandis, Peter H., and Kotler, Steven., *Abundance...The Future is Better Than You Think*, Free Press, A Division of Simon & Schuster, Inc., NY, 2012.

887. Gore, Al., *The Future: Six Drivers of Global Change*," Random House, NY. 2013.

888. "Water for Ross Bethio, Senegal Project at the World We Want Foundation," video at http://www.youtube.com/watch?v=8qJljBxwJ-E.

889. Baldacci, David, *The Innocent*, Quote by David Baldacci, *GoodReads*. http://www.goodreads.com/quotes/639642-there-s-always-more-to-life-than-money-will-money-is.

890. Creagan, Edward T., "Stress Management, Don't Forget the 'Life' in Work-Life Balance," Stress Blog, *Mayo Clinic.com*, Feb. 19, 2010. http://www.mayoclinic.com/health/work-life-balance/MY01203.

891. Creagan, Edward T., "Stress Management, Don't Forget the 'Life' in Work-Life Balance," Stress Blog, *Mayo Clinic.com*, Feb. 19, 2010. http://www.mayoclinic.com/health/work-life-balance/MY01203.

892. "Turnover Problems? Consider Work-Life Balance," *Premium Incentive Products*, April 2013. http://www.pipmag.com/inspiration-for-motivation-feature-201304.php.

893. Kuntz, Brad, "Be Good to Yourself," *Education Update*, ASCD, Oct. 2012, p. 3.

894. Hemingway, Mollie, "What Should Young People Know about Work-Life Balance?" *Ricochet.com*, March 12, 2013. http://ricochet.com/main-feed/What-Should-Young-People-Know-About-Work-Life-Balance.

895. Whitbourne, Susan Krauss, "The Delicate Work-Family Balance, and How to Have It All, Fulfillment at Any Age," *Psychology Today*, March 2, 2013. http://www.psychologytoday.com/blog/fulfillment-any-age/201303/delicate-work-family-balance-and-how-have-it-all.

896. Uscher, Jen, "Beat Burnout by Making More Time for the Activities and People That Matter Most to You," *WebMD* Feature. http://www.webmd.com/balance/guide/5-strategies-for-life-balance, article also features quotes from psychologist Robert Brooks and productivity expert Laura Stack. The titles of their books are included in the corresponding section of *21 Trends*.

897. "Mindfulness," *Psychology Today (online)*.http://www.psychologytoday.com/basics/mindfulness.

898. Langer, Ellen J., *Mindfulness*, 1989, De Capo Press, Perseus Books Group, Cambridge, MA, p. 71.

899. Goleman, Daniel, *Emotional Intelligence*, Bantam Books, NY, 2005, p. xxii.

900. "How to Teach Children Emotional Intelligence," *GoodTherapy.org*, Aug. 2, 2012, based on sources listed in the article. http://www.goodtherapy.org/blog/teach-children-emotional-intelligence-0802125.

901. Goleman, Daniel, *Emotional Intelligence*, Tenth Anniversary Edition, Book (1995), Introduction (2005), Bantam, NY, p. xxi.

902. "Egovonism," *The Urban Dictionary*, paraphrasing of definition. http://www.urbandictionary.com/define.php?term=Egonovism.

903. Cousins, Norman, Quotes," *Brainy Quote*. http://www.brainyquote.com/quotes/authors/n/norman_cousins.html.

904. "Laughter Remains Good Medicine," *Science Daily*, April 17, 2009. http://www.sciencedaily.com/releases/2009/04/090417084115.htm.

905. Lorenzo, George, "Spirituality is Alive and Well on the Web," *Access Magazine*, December 23, 2000. www.accessmagazine.com, from *GWSAE FastRead*, "Finding Spirituality Online," January 3, 2001.

906. Carlin, George, Comedian, "A Wonderful Message," from email message, "Irony of the New Century," 2001.

907. Easterbrook, Gregg, *The Progress Paradox*, Gregg Easterbrook, Random House Trade Paperbacks Edition, NY, 2004, p. 187.

908. Ford, Daniel, "Employers Value EQ Over IQ," *Associations Now*, ASAE, April 25, 2013. http://associationsnow.com/2013/04/study-emotional-intelligence-affects-worker-engagement/?utm_source=AN%2BDaily%2BNews&utm_medium=email&utm_campaign=20130426%2BFriday.

909. U.S Census Bureau, "Arts, Recreation, and Travel," Section 26, Table 1238, Attendance/Participation Rates for Various Arts Activities: 2008; and Table 1239, Attendance/Participation in Various Leisure Activities: 2008. http://www.census.gov/prod/2011pubs/12statab/arts.pdf.

910. Moore, John, "Arts vs. Sports for Popularity?" *Denver Post*, Oct. 29, 2006. http://www.denverpost.com/ci_4558958.

911. National Institutes of Health, "Science-Based Health & Wellness Resources for Your Community." http://www.nih.gov/health/wellness/.

912. "Wellness in the Workplace, Feb. 4, 2013. http://www.cdc.gov/features/WorkingWellness/index.html, and "The CDC Worksite Health ScoreCard Manual." http://www.cdc.gov/dhdsp/pubs/docs/HSC_Manual.pdf.

913. Santa Barbara Unified School District, "Wellness Program." http://www.sbsdk12.org/plans/wellness/.

914. San Francisco Unified School District, "SFUSD-High School Wellness Program and Public Policy," video. http://www.youtube.com/watch?v=tWtT_vF5a-0.

915. "UC Santa Barbara Living Well, Wellness Programs." http://www.hr.ucsb.edu/employee-services/wellness.

916. Collins, Francis S., "Change Your Lifestyle, Save Your Life?" National Institutes of Health, *A Message from NIH*. http://www.nlm.nih.gov/medlineplus/magazine/issues/spring12/articles/spring12pg2-3.html.

917. Parker, Jennifer Leigh, "Study: Workplace Wellness Programs Help Cut Healthcare Costs," *CNBC*. http://usatoday30.usatoday.com/money/story/2012-01-28/cnbc-at-work-wellness/52824820/1.

918. Gardner, Gary, *Invoking the Spirit*, The Worldwatch Institute, 2002, p. 11.

919. Marx, Gary, *Sixteen Trends…Their Profound Impact on Our Future*, Educational Research Service, Alexandria, Virginia, and Education Week Press, Bethesda, MD, 2006, 2012, Carol Peck Quote, p. 278.

920. "What's EQ?" *EQ At Work*, Online: http://www.eqatwork.com, 1998-2004.

921. Goleman, Daniel, "Daniel Goleman Explains Emotional Intelligence," interview, *You Tube* (26.37). Recorded in Canada. Goleman discusses his book, *Working with Emotional Intelligence*. http://www.youtube.com/watch?v=NeJ3FF1yFyc.

922. Kuntz, Brad, "Be Good to Yourself," *Education Update*, ASCD, Oct. 2012. http://www.ascd.org/publications/newsletters/education-update/oct12/vol54/num10/In-the-Classroom-with-Brad-Kuntz.aspx.

923. Goleman, Daniel, *Emotional Intelligence . . . Why It Can Matter More Than IQ,* Bantam Books, NY, 1997.

924. Easterbrook, Gregg, *The Progress Paradox, How Life Gets Better While People Feel Worse.* (Trade Paperbacks Edition) Random House, NY, 2004.

925. Marsh, Nigel, "How to Make Work-Life Balance Work," *TED Talk,* video, (10:05), taped May 2010, in Sydney, Australia, posted Feb. 2011. http://www.ted.com/talks/nigel_marsh_how_to_make_work_life_balance_work.html.

926. Santa Barbara Independent School District, California, "Wellness Policy Video," (6:42). http://www.youtube.com/watch?v=3VZ2Zq6NrTQ; and San Francisco Unified School District (SFUS-D)-"High School Wellness Program as Public Policy," Video (10:02), http://www.youtube.com/watch?v=tWtT_vF5a-0.

927. Bennis, Warren, and Goldsmith, Joan, *Learning to Lead, B*asic Books, Cambridge, MA, 2008, pp. 8-9.

928. "Famous Last Words." http://www.williamson-labs.com/480_last.htm.

929. Marx, Gary, "Ten Trends: Educating Children for Tomorrow's World," *NCA Journal of School Improvement*, North Central Association on Accreditation and School Improvement, Volume 3, Issue 1, Spring 2002.

930. Rubenstein, Herb, CEO, Growth Strategies, Inc., and co-author with Tony Grundy of *Breakthrough, Inc.-High Growth Strategies for Entrepreneurial Organizations*, "Strategic Planning for Futurists," *Futures Research Quarterly*, Fall 2000, drawn from pp. 7-8.

931. "Scenarios," RAND Europe. www.rand.org/randeurope/fields/scenarios.html, search "scenarios," 2013. Also Marx, Gary, *Future-Focused Leadership*, ASCD, Alexandria, VA, 2006, p. 119.

932. Bezold, Clem; Rhea, Marsha; Rowley, Bill, Institute for Alternative Futures, presentation devoted to "Wiser Futures," World Future 2003 preconference program, World Future Society Annual Conference, San Francisco, July 18, 2003.

933. Marx, Gary, *Future-Focused Leadership: Preparing Schools, Students, and Communities for Tomorrow's Realities*, ASCD, 2005.

934. Gladwell, Malcolm, *The Tipping Point . . . How Little Things Can Make a Big Difference,* First Back Bay paperback ed., Little, Brown, and Co., Boston, MA, 2002, pp. 7-9.

935. Wilson, Edward, .O., *Consilience . . .The Unity of Knowledge,* Borzoi Book, Alfred A. Knopf, NY, 1998, pp. 8-12.

936. Smyre, Rick, paper devoted to "Core Skills for Transformational Learning," 2004, p. 3, quoted in Marx, Gary, *Future-Focused Leadership*, ASCD, Alexandria, VA, 2006, p. 151.

937. Borawski, Paul, and Ward, Arian, "Living Strategy: Guiding Your Association through the Rugged Landscape Ahead,*" Journal of Association Leadership*, Center for Association Leadership, Winter 2004, pp. 6-9, 12-13.

938. Marx, Gary, Three sections on engaging people in planning based in part on Marx, Gary, *Ten Trends...Educating Children for a Profoundly Different Future*, Educational Research Service, Arlington, VA, 2000, pp. 81-82.

939. Withrow, Frank; Long, Harvey; Marx, Gary, *Preparing Schools and School Systems for the 21st Century*, American Association of School Administrators, Arlington, VA, 1999, pp. 100-101.

List of Figures
Page

Acknowledgments

Twenty-One Trends for the 21st Century is based on the reality that all of us need to stay in touch with a fast-changing world. It is also driven by a belief that people can come together to create and constantly re-create an even better future for our schools, education systems, and the whole of society. The world is filled with motivated people who share that hopeful spirit. They are in all walks of life. Many of them were part of the team whose ideas, experiences, and wisdom brought this book to life.

My profound gratitude to members of *Futures Council 21*. This talented and intellectually generous council of 26 people from many parts of the nation and world responded to either one or two modified Delphi questionnaires, helping to identify, sort, and expand on issues and trends. Council member names are listed following these acknowledgments.

I have been truly blessed to work with the seasoned and imaginative professionals who make our publisher, Education Week Press, a division of Editorial Projects in Education (EPE), one of the most respected and far-reaching in its field. Particularly, I extend my thanks to Publisher Michele Givens for her dedication to making our trends publications a part of the *Ed Week* family. Director of Knowledge Services Rachael Delgado, who served as home-base for the project and provided ongoing support and encouragement, gave birth to a son while we were writing and temporarily passed the baton to Vice President for Research & Development Christopher Swanson and Knowledge Services Associate Tim Ebner. Thanks to Jaini Giannovario, who edited and helped shape the manuscript for publication, also to Director of Production Jo Arnone, Creative Director Laura Baker for her cover design, and to all members of the EPE staff for their ongoing contributions to the cause. That, of course, includes EPE President and Editor Virginia Edwards, a longtime and treasured colleague.

Since *Twenty-One Trends* is the third book in this series, I want to express appreciation to staff members at the former Educational Research Service (ERS), including but certainly not limited to: John Forsyth, Nancy Protheroe, Katherine Behrens, Jeanne Chircop, Deborah Perkins-Gough, John Draper, and many others. ERS published the first two installments of *Trends* books in 2000 and 2006.

I am particularly grateful for the intellectual firepower of numerous World Future Society colleagues who are constantly stretching my thinking; people such as Edward Cornish, Tim Mack, Jeff Cornish, the late Susan Echard, Cynthia Wagner, Sarah Warner, Ken Hunter, Joseph Coates, Marvin Cetron, Michel Godet, Mika Mannermaa, Graham T.T. Molitor, David Pearce Snyder, Ken Harris, Edward Gordon, Jerry Glenn, Ted Gordon, Harlan Cleveland, John McDonald, William Halal, Stephen Aguilar-Millan, Jose Cordeiro, James Morrison, the late Jack Sullivan, Edie Weiner, Arnold Brown, Ramez Naam, Herb Rubenstein, Rick Smyre, John Meagher, Steve Steele, John Petersen, Clem Bezold, and Bill Rowley, to name just a few. Some have served as members of advisory councils.

Special thanks to the Center for Civic Education, including Executive Director Chuck Quigley, Associate Director John Hale, and talented CCE staff members. I am grateful to dedicated state Civitas leaders for the opportunity to work with educators and others who provide leadership at all levels of society worldwide. In addition to some who are quoted and a few who served on *Futures Council 21*, there are many others in the international community who have provided ongoing insights and connections to what is happening with the big picture and in the trenches. Among them are: Boubacar Tall in Senegal, Calin Rus in Romania, Silvia Uranga in Argentina, Giedre Kvieskiene in Lithuania, Pablo Zavala in Peru, Rahela Dzidic in Bosnia & Herzegovina, Mona Al Alami in Jordan, Arman Argynov in Kazakhstan; Nimi Walson-Jack in Nigeria, Anash Mangalparsad in South Africa, Jacek Strzemieczny in Poland, Marianne McGill in Ireland, William Ryan in Indonesia; Meera Balachandran in India, the late Pimon Ruetrakul in Thailand; Nanangarel Rinchin in Mongolia; and Elarbi Imad in Morocco. Betsy Lim in Singapore and Judith Harrison in Australia are among the many who arranged for presentations at events with a central focus on trends and their implications for education. I am also grateful to have worked with talented educators during conferences of the Department of Defense Education Activity (DoDEA) in several communities in Europe, in Asia and the Pacific (meeting in Tokyo), and the Mediterranean (meeting in the Padua area of Italy).

Dozens of thoughtful leaders provided information and ideas through their writings and research. I'm especially grateful for the many professionals who work each day with the U.S. Census Bureau, the Social Security Administration, the Bureau of Labor Statistics, and *Education Week.*

As a speaker and workshop leader in many parts of the nation and world, my work has been formed by inspiration and ideas from hundreds of thoughtful people in many walks of life, including education. Your wisdom continues to guide me. If you are among these mentors or those who have come to my presentations, studied my books, and considered or used my ideas, my deepest thanks.

I am especially grateful to my family. They have long supported my quest to seek, develop, explain, and encourage the thoughtful consideration of ideas for creating a more promising future. Their broad interests, keen insights, and frankness long ago earned a standing ovation. My wife, Judy, a seasoned educator, has played a key role in each of these books, asking questions such as, "Have you considered this?" "What are you trying to say here?" and "Why is this important?"

Finally, thanks to you, the reader. You are the one who will lift words and ideas from the page and turn them into actions that will help constantly create the communities and institutions of the future, including the education system. I hope you will read this book, get copies for others, and use it to stimulate thinking about what we need to become. *Twenty-One Trends for the 21st Century* is not an end in itself. It is a place to stand while we move the world. It's your turn. I encourage you to "to run with it."

Gary Marx

Futures Council 21

Beginning in the fall of 2012, members of an international Futures Council 21 agreed to provide ideas and thoughtful counsel for the author in shaping of this book, which became *Twenty-One Trends for the 21ˢᵗ Century*. The 26 members of this distinguished advisory council responded to either one or two rounds of questionnaires in a modified Delphi process. In the first round, these advisors were asked to identify significant trends and issues that might affect education and society during the first three decades of the 21ˢᵗ century. In the second, each was asked to share what they considered implications of a cluster including two more-fully-developed, far-reaching trends. One member of the Council, Matthew Moen, dean of the College of Arts & Sciences at the University of South Dakota and past president of the national Council of Colleges of Arts & Sciences, specifically addressed implications for postsecondary education. As author, I know I speak for all readers, including countless people in education and many other walks of life, in many parts of the world, who will benefit from the thinking of this eloquent group of visionary leaders.

Views expressed in *Twenty-One Trends* do not necessarily reflect the beliefs or opinions of any member of Futures Council 21, the Council as a whole, or their organizations, nor do they reflect the official views of the author, or the publisher, Education Week Press, Editorial Projects in Education. The book is designed to be a marketplace of information and provocative ideas.

Members of Futures Council 21 included: **Stephen Aguilar-Millan**, director of research, The European Futures Observatory, Ipswich, U.K.; **Meera Balachandran**, director, Education Quality Foundation of India, Guragaon, Haryana, India; **Laurie Barron**, 2013 MetLife/NASSP National Middle Level Principal of the Year, who became superintendent of the Evergreen School District in Montana; **Sheldon Berman**, superintendent of the Eugene Public Schools in Eugene, Oregon; **Joseph J. Cirasuolo**, executive director, Connecticut Association of Public School Administrators, West Hartford, Connecticut; **Avis Glaze**, president of Edu-quest International, Inc., and former chief student achievement officer for Ontario, Canada. She is now in British Columbia; **Joseph Hairston**, president and CEO of Vision Unlimited, LLC, and immediate past superintendent of the Baltimore County Public Schools in Maryland; **James Harvey**, executive director, National Superintendents Roundtable, Seattle, Washington; **Debra Hill**, associate professor, Argosy University, Chicago, longtime school administrator and 2012 national ASCD president; **Ryan Hunter**, student leader and future-focused scholar at American University in Washington, D.C.; **Frank Kwan**, director of communications services for the Los Angeles County Office of Education in California; **Damian LaCroix**, superintendent, Howard-Suamico School District, Green Bay, Wisconsin; **Allison LaFave**, a recent Harvard graduate, serving as marketing, communication, and development associate for Legal Outreach in New York City; **Anash Mangalparsad**, director, Centre for Community and Educational Development, South Africa. He called upon Jessica Vinod Kumar, a teacher at M.L. Sultan (Pmb) Secondary School in Pietermaritzburg, South Africa, to respond to questions; **Keith Marty,** superintendent, Parkway School District, Chesterfield, Missouri; **John Meagher**, professional futurist, Manassas, Virginia; **Rebecca Mieliwocki,** 7th grade English teacher at Luther Burbank Middle School in Burbank, California, and 2012 National Teacher of the Year; **Matthew Moen**, dean of the College of Arts & Sciences at the University of South Dakota in Vermillion; **Stephen Murley**, superintendent, Iowa City Community School District, Iowa City, Iowa; **Patrick Newell**, learning activist and "vision navigator" of the Tokyo International School in Japan, also president of the 21 Foundation; **Marcus Newsome**, superintendent, Chesterfield County Public Schools, Chesterfield, Virginia; **Concepción Olavarrieta**, president, Nodo Mexicano, El Proyecto del Milenio, A.C. (Mexico's U.N. Millennium Project) Mexico City; **Gary Rowe**, media producer, communications consultant, and president, Rowe, Inc., Lawrenceville, Georgia; **David Pearce Snyder**, futurist and contributing editor, *The Futurist Magazine*, Bethesda, Maryland; **Michael Usdan**, senior fellow, Institute for Educational Leadership, Washington, D.C.; and **Milde Waterfall**, teacher, retired from Thomas Jefferson High School for Science and Technology, Fairfax County, Virginia.

Index

Notes

Notes

About the Author

Gary Marx, CAE, APR, is a noted author, futurist, executive, consultant, and international speaker. As a futurist, Marx has directed a number of studies and written numerous books and articles. Just prior to *Twenty-One Trends for the 21st Century,* his best-selling books included *Sixteen Trends...Their Profound Impact on Our Future,* which highlighted key forces reshaping our world, and *Future-Focused Leadership,* a motivating guide for creating an even brighter future.

Marx is president of the Center for Public Outreach, in Vienna, Va. an organization he founded that provides counsel on future-oriented leadership, communication, education, community, and democracy. He has been called an "intellectual entrepreneur, who is constantly pursuing ideas" and a "deep generalist." During his career, he has worked with a broad range of educators, business people, and other community and government leaders.

His international speaking and consulting assignments have taken him to six continents. A few examples include the northern Andean region of Peru, desert areas of Jordan, the Pacific coast of Russia, communities across Senegal, the eastern coast of Australia, and many parts of Europe and Asia. Those assignments in North America include all 50 of the United States, most provinces of Canada, and many states of Mexico. He has visited more than 80 countries.

Marx, who acts both locally and globally, has been especially inspired by his work with students and educators who are developing civic education projects to improve life in their communities and countries. He is an on-the-ground communicator who attempts to understand people and issues and offer leadership and counsel on numerous fronts, from advising on the Bicentennial of the U.S. Constitution to building parks and a monument in his hometown. Gary Marx has been recognized with the President's Award from the National School Public Relations Association and the Distinguished Service Award from the American Association of School Administrators. Marx and his wife, Judy, live in Vienna, Va. He can be reached at gmarxcpo@aol.com.